HUMAN CELL CULTURE

Volume I: Cancer Cell Lines Part 1

Human Cell Culture

Volume 1

The titles published in this series are listed at the end of this volume.

Human Cell Culture
Volume I
Cancer Cell Lines Part 1

edited by

John R.W. Masters

University College London, 67 Riding House Street, London, UK

and

Bernhard Palsson

Dept. of Bioengineering, University of California San Diego, USA

SPRINGER-SCIENCE+BUSINESS MEDIA, B.V.

Library of Congress Cataloging-in-Publication data is available.

ISBN 978-94-017-3824-8 ISBN 978-0-306-46872-8 (eBook)
DOI 10.1007/978-0-306-46872-8

Printed on acid-free paper

Contents

Foreword to the Series

This series of volumes is in celebration of Human Cell Culture. Our ability to grow nearly every type of normal and diseased human cell in vitro and reconstruct tissues in 3 dimensions has provided the model systems on which much of our understanding of human cell biology and pathology is based. In future, human cell cultures will provide the tools for tissue engineering, gene therapy and the understanding of protein function. The chapters in these volumes are written by leading experts in each field to provide a resource for everyone who works with human cells in the laboratory.

John Masters and Bernhard Palsson

Introduction

Continuous cell lines derived from human cancers are the most widely used resource in laboratory-based cancer research. The first 3 volumes of this series on Human Cell Culture are devoted to these cancer cell lines.

The chapters in these first 3 volumes have a common aim. Their purpose is to address 3 questions of fundamental importance to the relevance of human cancer cell lines as model systems of each type of cancer:

1. Do the cell lines available accurately represent the clinical presentation?
2. Do the cell lines accurately represent the histopathology of the original tumors?
3. Do the cell lines accurately represent the molecular genetics of this type of cancer?

The cancer cell lines available are derived, in most cases, from the more aggressive and advanced cancers. There are few cell lines derived from low grade organ-confined cancers. This gap can be filled with conditionally immortalized human cancer cell lines. We do not know why the success rate for establishing cell lines is so low for some types of cancer and so high for others.

The histopathology of the tumor of origin and the extent to which the derived cell line retains the differentiated features of that tumor are critical. The concept that a single cell line derived from a tumor at a particular site is representative of tumors at that site is naïve and misleading. It is essential that representative cell lines are selected for study, and it is hoped that the chapters in these volumes will help appropriate selections to be made.

The data on the molecular genetics of cancer cell lines has been difficult to gather as it is widely distributed throughout the literature and in a stage of transition. We do not yet know the identity of many of the altered genes for each type of cancer, or what their individual roles are in the progression of the disease.

Despite being an essential resource for much of cancer research, established cell lines are associated with problems that are often ignored, but which can invalidate the work. The most important problems are cross-contamination between cells of either the same or different species, contamination with microorganisms (usually *Mycoplasma*) and phenotypic and genotypic drift. Both cross-contamination and the presence of *Mycoplasma* are easily checked by PCR-based methods. Phenotypic and genotypic drift can be avoided by good tissue culture practice, ie by growing cells for only short periods before returning to frozen stocks.

Many cell lines are cross-contaminated with other human or animal cell lines. Despite the fact that cell lines called Chang liver, KB and Hep-2 are known to be HeLa, authors often fail to acknowledge the fact. HeLa is just the tip of the iceberg of cell line cross-contamination. For most cell lines there is no proof of origin from a particular individual or tumor by a reliable method (DNA fingerprinting is recommended).

Mycoplasma contamination is a widespread and recurring problem. Laboratories that do not test for *Mycoplasma* contamination often have it, and consequently allow it to spread unchecked. How many putative novel human cancer-associated proteins are derived from *Mycoplasma*?

Many human cancer cell lines are easy to grow and maintain. With simple precautions and good practice they can provide models that are representative of almost every type of clinically advanced human cancer. Many more cell lines are needed to represent low grade, clinically localized cancer.

Chapter 1

Sarcomas

Beverly A. Teicher

Dana-Farber Cancer Institute, 44 Binney Street, Boston, MA 02115, Present address: Lilly Research Laboratories, Drop Code 0546, Indianapolis, IN 46285. Tel: 001-317-276-2739; Fax: 001-317-277-6285. E-mail: teicher_beverly_A@lilly.com

From the late 1960's through the 1970's development of human tumor cell lines representing every type of human malignancy was a major research activity. Innumerable attempts were made to obtain continuous growth of tumor cells derived from surgical specimens in well-defined culture systems by varying the nutrient media, serum, growth factors and hormones and substrata (glass, plastic, suspension, cellular feeder-layer). Most of these attempts failed to produce immortal tumor cell lines; however, those that succeeded resulted in the development of the vast majority of the human tumor cell lines that are in use in cancer research today (Bloom 1972; Giard et al. 1973; Fogh 1975; Fogh et al. 1977a,b). In the 1980's, human tumor lines were established as xenografts in immunodeficient mice. After passage for generations in animals, cell culture lines were developed from some of these (Hazelton et al. 1987). Much of the research on the biochemistry and molecular biology of cancer has been derived from these cell lines, although it may certainly be questioned whether the few cells from which immortal cell culture lines originated are representative of their diseases of origin.

It was recognized early on that some tissue types adapted to growth in cell culture more readily than others. Sarcomas were among the more readily cell culture-adapted cell types, so that most of the widely used cell lines in the sarcoma family have been carried in culture for 20 to 25 years. Tables 1-3 list 54 human sarcoma cell lines including: 4 fibrosarcoma; 13 rhabdomyosarcoma; 26 osteosarcoma; 3 leiomyosarcoma; 1 synovial sarcoma; 3 liposarcoma and 4 chondrosarcoma.

J.R.W. Masters and B. Palsson (eds.), Human Cell Culture Vol. 1, 1–19.

Table 1 Origins and establishment of human sarcoma cell lines

Cell line	Patient Age/Sex	Primary site	Specimen site	Culture method	Availability	Primary reference
			Fibrosarcoma lines			
SW684	68/M	low grade fibrosarcoma		monolayer	ATCC	A. Leibovitz, Scott and White Clinic, Temple, TX 1974
Hs913T	51/M	fibrosarcoma	lung	monolayer	ATCC	
HT-1080	35/M	acetabulum		monolayer	ATCC	(Rasheed et al. 1974)
8387						(Aarsonsen 1970)
			Rhabdomyosarcoma lines			
A673	15/F			soft agar	ATCC	(Giard et al. 1973)
Hs729	74/M	leg	connective tissue	monolayer	ATCC	
A-204	1/F			monolayer	ATCC	(Giard et al. 1973)
RD	7/F	pelvis	biopsy	soft agar	ATCC	(McAllister et al. 1973)
HX170C	5/M	paratesticular	biopsy	monolayer	IN	(Kelland et al. 1989)
Rh10	14/F	posterior mediastinum	peritoneum	from xenograft HXRh10	IN	(Hazelton et al. 1987)
Rh18	2/F	abdomen	lower abdomen	from xenograft HXRh18	IN	(Hazelton et al. 1987)
Rh28	17/M	hand	axillary node	from xenograft HXRh28	ATCC	(Hazelton et al. 1987)
Rh30	16/M	unknown	bone marrow	monolayer	ATCC	(Douglass et al. 1987)
Rh12	12/M	buttock	buttock	from xenograft	IN	(Houghton et al. 1982)
RMS	14/F	right chest wall	malignant effusion	monolayer		(Garvin et al. 1986)
KYM.1D4						(Butler et al. 1994)
TE-671						(McAllister et al. 1977)

Continued on next page

Table 1 (continued)

Cell line	Patient Age/Sex	Primary site	Specimen site	Culture method	Availability	Primary reference
			Osteosarcoma lines			
143B(TK^-)				monolayer	ATCC	
G-292	9/F	bone			ATCC	(Peebles et al. 1978)
MG-63	14/M				ATCC	(Billiau et al. 1977)
SaOS-2	11/F			monolayer	ATCC	(Fogh et al. 1977)
SK-ES-1	18/M		biopsy		ATCC	(Bloom 1972)
U-20S	15/F	tibia		monolayer	ATCC	(Ponten et al. 1967)
HOS/TE85	13/F	femur	femur	soft agar	ATCC	(McAllister et al. 1971; Rhim et al. 1975)
KHOS/NP (R-970-5)		derived from HOS/TE85			ATCC	(McAllister et al. 1971; Rhim et al. 1975)
KHOS-240S		derived from HOS/TE85			ATCC	(Cho 1976)
KHOS-312H		derived from HOS/TE85			ATCC	(Cho 1976)
MNNG/H		HOS/TE85 chemically transformed			ATCC	(Rhim et al. 1975)
OSC15						
OHA	16/M			xenograft		
OST						
OHS	14/M	multiple sites	left femur	monolayer	IN	(Fostad et al. 1986)
KPDX						
791T						
HU09	13/F	femur	femur	xenograft	IN	(Kawai et al. 1990)
MHMX					IN	(Fostad et al. 1986)
OHSX					IN	(Fostad et al. 1986)
KRIB		derived from HOS/TE85			IN	(Samid 1989)
HTLA145		femur	femur			

Continued on next page

Table 1 (continued)

Cell line	Patient Age/Sex	Primary site	Specimen site	Culture method	Availability	Primary reference
HTLA161		femur	femur post chemo			
HTLA195		femur	lung met.			
KT005						
SK-05-10			lung met.	monolayer	IN	James Loveless, Mem. Sloan-Kettering Cancer Laboratory, Rye, NY
RD-ES	19/M	Ewing's sarcoma humerus	humerus	monolayer	ATCC	
			Leiomyosarcoma lines			
SK-LMS-1	43/F	vulva	vulva	monolayer	ATCC	(Fogh 1975)
SK-UT-1	75/F	uterus	uterus	monolayer	ATCC	(Fogh 1975)
SK-UT-1B		derived from SK-UT-1			ATCC	
			Synovial sarcoma line			
SW982	25/F	axilla	axilla	monolayer	ATCC	(Fogh et al. 1977b)
SW 872	36/M	left flank	left flank	monolayer	ATCC	(Fogh et al. 1977a,b)
HS-18						
MFH				monolayer		(Iwasaki et al. 1982)
			Chondrosarcoma lines			
SW1353	72/F	right humerus	right humerus	monolayer	ATCC	
105KC	66/M	myxoma (cardiac)	myxoma	monolayer	IN	(Block et al. 1991)
HCS-2/8	72/M	humerus	humerus	monolayer	IN	(Takigawa et al. 1989)
HCS-2/A	72/M	humerus	humerus	monolayer	IN	(Takigawa et al. 1991)

ATCC = from American Type Culture Collection
IN = from Investigator

Table 2 *In vivo* and *in vitro* pathology of human sarcoma cell lines

Cell line	Tumor pathology	*In vitro* cytopathology	Xenograft pathology
		Fibrosarcoma lines	
SW684	low grade fibrosarcoma	malignant spindle cell consistent with sarcoma	consistent with sarcoma
Hs913T	metastatic fibrosarcoma	fibroblast	–
HT-1080	fibrosarcoma	epithelial-like	–
8387			
		Rhabdomyosarcoma lines	
A673		soft agar	tumorigenic
Hs729	fibrosarcoma	fibroblast-like	–
A-204	epithelial-like	mixed small tumor cells	small cell embryonal rhabdomyosarcoma
RD	malignant embryonal rhabdomyosarcoma	spindle cells and large multinucleated cells	tumorigenic
Hx170c	malignant embryonal rhabdomyosarcoma	pleomorphic mesenchymal	undifferentiated small tumor cells
Rh10	embryonal RMS	poorly differentiated	poorly differentiated embryonal RMS
Rh18	mixed alveolar embryonal	poorly differentiated	embryonal plus small amount of alveolar
Rh28	poorly differentiated alveolar RMS	spontaneously differentiates by passage 60	alveolar RMS, metastatic
Rh30	alveolar RMS	myogenic markers, IGF-1 dependent growth	alveolar RMS
Rh12	well-differentiated embryonal RMS	not studied, senesces by passage 20	embryonal, more spindled, moderately differentiated
RMS	undifferentiated round cell sarcoma	round to polygonal cells	tumor cells with rounded nuclei and clumped chromatin
KYM.1D4			
TE-671	medulloblastoma small uniform cells	polygonal and fusiform cells with short unbranched processes	medulloblastoma
		Osteosarcoma lines	
$143B(TK^-)$			
G-292			

Continued on next page

Table 2 (continued)

Cell line	Tumor pathology	In vitro cytopathology	Xenograft pathology
MG-63		epithelial-like	non-tumorigenic
Saos-2			
Sk-ES-1		epithelial-like	small cell malignant tumor consistent Ewing's sarcoma
U-20S		epithelial-like, polygonal	
HOS/TE85		polygonal/fusiform cells, flat morphology	poorly tumorigenic
KHOS/NP (R-970-5)		HOS virally transformed by Kirsten murine sarcoma virus (KiMSV)	tumorigenic
KHOS-240S		revertant from KHOS/NP	non-tumorigenic
KHOS-312H		revertant from KHOS/NP	non-tumorigenic
MNNG/HOSC15		chemically transformed (MNNG 0.01 μg/ml)	tumorigenic
OHA			
OST	osteosarcoma Takase		
OHS (OHS-4)		polygonal to oval containing cytoplasmic granules	tumorigenic/calcified tumor
KPDX			
791T			
Hu09	osteoblastic osteosarcoma	polygonal and cuboidal cells	tumorigenic and mineralization abundant osteoid formation
MHMX	sarcoma		tumorigenic
OHSX	osteogenic sarcoma		polytumorigenic
KRIB		v-Ki-ras transformed subline of TE-85 (HOS)	highly metastatic histologically similar to human osteosarcoma
HTLA145			
HTLA161			
HTLA195			
KT-005			

Continued on next page

Table 2 (continued)

Cell line	Tumor pathology	In vitro cytopathology	Xenograft pathology
SK-05-10			
RD-ES	osseous Ewing's sarcoma	epithelial-like	
		Leiomyosarcoma lines	
SK-LMS-1	leiomyosarcoma, vulva	fibroblast-like (sarcoma cells)	leiomyosarcoma
SK-UT-1	mesodermal mixed tumor, uterus	epithelial-like (adenocarcinoma)	spindle cell sarcoma
SK-UT-1B		epithelial-like (tumor)	well-differentiated adenocarcinomas consistent with endometrial carcinoma
		Synovial sarcoma line	
SW982		mixed spindle and giant cell tumor cells	
SW872	undifferentiated malignant tumor consistent with liposarcoma	fibroblast-like, spindle and giant cell tumor cells	spindle cell sarcoma consistent with liposarcoma
HS-18			
MFH	pleomorphism and storiform pattern	spindle cells; polygonal cells bizarre giant cells	
		Chondrosarcoma lines	
SW1353	grade V chondrosarcoma	fibroblast-like; sarcoma consistent with chondrosarcoma	
105KC	myxomatous chondrosarcoma	heterogeneous; fibroblast-like to star-shaped myxomatous to polygonal chondrocytic	
HCS-2/8	chondrosarcoma	rounded polygonal cells to fibroblast-like cells	well-differentiated chondrosarcoma
HCS-2/A	chondrosarcoma	elongated polygonal cells	well-differentiated chondrosarcoma

Table 3 Genetic abnormalities observed in human sarcoma cell lines

Cell line	Main genetic changes	Specific genetic features
		Fibrosarcoma lines
SW684	hypertriploid; modal chromosome number = 73	marker chromosomes include: der(2)t(2;6)(p13;q13), der(12)t(8;12)(q11;q24), t(15q21q), 19qt, t(8p21q?); Double minutes were found
Hs913T	aneuploid, female; chromosome counts 100–130	markers and unassignable chromosome comprise 80% of chromosomes
HT-1080	pseudodiploid; modal chromosome number = 46	80% of cells: extra c-group chromosome; missing B-group chromosome; 10% of cells: extra A-2-chromosome; missing c-group
8387		
		Rhabdomyosarcoma lines
A673	Chromosome number = 47	extra F chromosome, 2 abnormal B chromosomes; 4 or more marker chromosomes
Hs729	pseudodiploid; chromosome counts mainly 46–47	
A-204	aneuploid, female; chromosome frequency distribution = 46	most chromosome counts near-diploid; 30% tetraploid; major abnormality in chromosome N22; marker chromosome 22p+
RD	unstable hyperdiploid bimodal chromosome stemline = 49–50	chromosome associations, microchromosomes, chromosome fragments and breaks; achromatic gaps
Hx170C	near diploid; mean chromosome number = 50	
Rh10	hyperdiploid; chromosome number = 87	chromosome markers: t(2;13)(q14), der(12)t(12;?)(q23;?), der (16)t(1,16)(p22;q21); some markers found in HxRh10 xenograft
Rh18	hyperdiploid; chromosome number = 70–80	chromosome markers: der(1)(p22),der(3)t(3;?)(p14;?), del(16)(q24) and der(22)t(1;22)
Rh28	near tetraploid	translocation markers: t(2;13)(q35;q14), der(12)t(3;12)(q21;q24). 25% of metaphases near octaploid
Rh30		
Rh12		
RMS	pseudodiploid modal chromosome number 47 (range 41–49)	60% of cells have double minute chromosomes (mean, 27; range 4–100) Marker chromosomes 2q+, 4p+, 13q-

Continued on next page

Table 3 (continued)

Cell line	Main genetic changes	Specific genetic features
KYM.1D4		
TE-671	polyploid; 25–45% tetraploid; stemline chromosome	chromosome markers: t(1;?)(p36;?), del(1)(:p36→q32), del(2)(q33), der(2)t(2;6)(q23;p23)
	Osteosarcoma lines	
143B(TK$^-$)		
G-292	hypotriploid; modal chromosome number = 57	marker chromosome present, 9% of cells tetraploid or greater
MG-63	hypotriploid; modal chromosome number = 66	marker chromosome common to all cells
Saos-2	hyperdiploid to hypopentaploid; modal chromosome number = 56	numerous chromosome breaks, secondary constrictions and double minutes; identifiable markers: 6p+/q+, 7p+, 11p+ and 12p+
SK-ES-1	aneuploid human male, modal chromosome number = 50	chromosome N9 and N22 absent; chromosomes N1, N16 and N17 are single, N7 is triple
U-20S	hypertriploid; modal chromosome number = 76	chromosomally highly altered; marker chromosomes include: t(9qter→9q21::1p36→1p::?), 7p+, iso(17q), t(15q:?)
HOS/TE85	hypotriploid; chromosome number range 47–183; stemline chromosome number 58–65	30% of chromosomes have breaks
KHOS/NP(R-970-5)		
KHOS-240S		
KHOS-312H		
MNNG/HOSC15		
OHA		expresses the c-ras-Ki 2 gene; this gene appears to be amplified
OST		
OHS	modal chromosome number = 53–56	25% of cells had double minute chromosomes. Marker chromosomes include: M1(1q+), M16(1q-), M2(6p+), M6(11q+)
KPDX		
791T		
Hu09	triploid range; range chromosome number = 69–83	

Continued on next page

Table 3 (continued)

Cell line	Main genetic changes	Specific genetic features
MHMX		
OHSX		
KRIB		HOS/TE-85 cells transformed with v-Ki-ras oncogene
HTLA145		
HTLA161		
HTLA195		
KT-005		
SK-05-10		
RD-ES		
	Leiomyosarcoma lines	
SK-LMS-1	triploid to hypertriploid, chromosome number range 64–72	normal chromosome N1, N4, N13, N17 under-represented; chromosomes N6, N8, N20 over-represented; many abnormalities, fragments, breaks, secondary constructions, minutes
SK-UT-1	hypodiploid, modal chromosome number = 45	except for the monosomic N13, no chromosome aberrations were present
SK-UT-1B	diploid, modal chromosome number = 46	normal karyotype, no subband changes were found
	Synovial sarcoma line	
SW982	hyperdiploid, modal chromosome number = 48	marker chromosomes: t(1q4q), del(5)(q31q33), t(8q21p), der (9)t(4;9)(q11;p24), t(9q13q). Minutes were seen
SW872	chromosome number = 79–80	
HS-18		
MFH		
	Chondrosarcoma lines	
SW1353	hyperdiploid; chromosome number = 47	trisomic N7 chromsome
105KC		
HCS-2/8	mean chromosome number = 48 (range 42–54)	
HCS-2/A	mean chromosome number = 51 (range 44–54)	

1. FIBROSARCOMA CELL LINES

Although four human fibrosarcoma cell lines are listed in the tables, the most frequently studied human fibrosarcoma cell line is HT1080. The HT1080 cell line was initiated from a biopsy of a fibrosarcoma arising adjacent to the acetabulum of a 35 year old male in July 1972 (Rasheed et al. 1974). The patient had never received chemotherapy or radiation therapy. A fine mince of the tumor tissue was seeded into plastic culture flasks and dishes were covered with Eagle's minimum essential medium with 10% fetal bovine serum and antibiotics. "Quick trypsinization" and "picking" procedures were used to eliminate fibroblasts from the cultures.

The primary tumor biopsy was very cellular with little intercellular material and invaded the acetabulum cartilage (Rasheed et al. 1974). Although the major cell type in the tumor was an elongated tumor cell with only a minor population of rounded cells, the HT-1080 cell line that was developed consisted almost entirely of rounded cells. The doubling time of the cell line was 26 hours (Rasheed et al. 1974).

The human fibrosarcoma cell line HT-1080 has been used extensively to study the effect of anti-inflammatory agents such as glucocorticoids on the gene expression of inflammatory mediators (Andreasen et al. 1986, 1987; Medcalf et al. 1986; Neilsen et al. 1986; Walker et al. 1986; Nielsen et al. 1987; Stephens et al. 1989; Cajot et al. 1990; Oikarinen et al. 1990; Barlati et al. 1991). The human fibrosarcoma cell line HT-1080 has also been used extensively in the study of the extracellular matrix proteins involved in attachment, invasion and metastasis (Nurcombe et al. 1989; Palmer et al. 1989; Yamada et al. 1990; Kubota et al. 1991a,b; Mukaida et al. 1991; Watanabe et al. 1991). Human HT-1080 fibrosarcoma cells have been instrumental in examining the role of the ras-oncogenes in the transformed phenotype (Lowe and Goeddel 1987; Paterson et al. 1987) and the role of the expression of the retinoblastoma gene product in cellular response to therapy (Li et al. 1995). It was found that human HT-1080 fibrosarcoma cells readily accept transfected genes carried by a variety of vectors, therefore this cell line has been useful in the study of gene overexpression and potential gene therapy strategies (Nagy and Baker 1987; Tang et al. 1989; Hoglund et al. 1992; Kim et al. 1995; Marini et al. 1995; Hochhauser et al. 1996; Ueda et al. 1996). The human HT-1080 fibrosarcoma cell line has also been useful in the study of therapy related issues in the treatment of fibrosarcoma (Falugi et al. 1986; Hodgkiss and Stratford 1988; Kinsey et al. 1989; Rathod and Kharti 1990; Zwelling et al. 1990; Slovak et al. 1991; Mukaida et al. 1992; Tanimoto et al. 1992; Shaughnessey et al. 1994; Noel et al. 1995).

2. RHABDOMYOSARCOMA CELL LINES

The RD cell line was the first human rhabdomyosarcoma cell line characterized and studied broadly. The RD cell line was established from a malignant embryonal rhabdomyosarcoma of the pelvis of a 7-year-old female in 1968 (McAllister et al. 1969). It has been used to examine the potential of differentiating agents as a cancer therapy (Aguanno et al. 1990; Germani et al. 1994). The RD cell line has also been used extensively in the study of adhesion molecules, cytopathic effects and matrix components (Chan et al. 1991; Espy et al. 1991; Johnston 1991; Lollini et al. 1991; Lindberg et al. 1992; Myohanen et al. 1993).

In 1982, Houghton and colleagues reported the initiation of a continuing effort on the growth and characterization of childhood rhabdomyosarcoma as xenografts in immune-deprived female mice. Eleven surgical specimens of rhabdomyosarcoma, 2 bone marrow samples and 1 sample of cells from ascitic fluid were implanted in immunodeficient animals and seven xenografts were obtained. The xenografts retained the histological characteristics of the tumors. From these and several xenografts established later, human rhabdomyosarcoma cell lines were established from the xenograft tissues by preparing a single cell suspension of the tumor cells and plating them in RPMI 1640 containing 20% fetal calf serum. The serum level was gradually reduced to 10% (Hazelton et al. 1987). Drug resistant variants of the Rh rhabdomyosarcoma xenografts were established to vincristine and melphalan designated HxRh12/VCR-3 and HxRh28/L-PAM13, respectively (Horton et al. 1987; Houghton et al. 1987a,b, 1988; Masurel et al. 1990). The rhabdomyosarcoma xenograft lines established by Houghton have been used extensively in the development of new therapeutic approaches to this disease (Houghton et al. 1987a, 1989, 1991, 1992, 1993, 1995; Horton et al. 1989; Dilling et al. 1994). The human Rh28 rhabdomyosarcoma cell line has been characterized morphologically and molecularly (Shapiro et al. 1990). The Rh28 cell line displayed a constant population doubling time (45-55 hrs) until passage 60 when proliferation gradually ceased. The loss of proliferative capacity was associated with morphological evidence of differentiation into multinucleated myotubes, fusion and the expression of numerous muscle-specific genes. Subsequently, six human Rh rhabdomyosarcoma cell lines were analyzed for myogenic regulatory protein (MyoD1) expression (Dias et al. 1990).

In 1977, McAllister et al. reported the establishment of the human TE671 medulloblastoma cell line from a cerebellar medulloblastoma of a 6-year-old girl. No neural or glial elements were demonstrated in the cultured cells by electron microscopy. However, in 1989, Stratton et al. reported that the TE671 cell line carried an activated N-ras gene. Based on this finding and on the phenotypic characteristics of the TE671 cells such as the presence of muscle-type nicotinic acetylcholine receptors and the intermediate filament protein desmin the identity of the line was further investigated. Cytogenetic analysis and DNA fingerprinting at several loci established that the TE671 cell line and the RD cell line were derivatives of the same cell line. This study established TE671 as a rhabdomyosarcoma. A drug resistant subline of TE671 (TE-671MR) has been developed and its response to several drugs has been compared with that of the parent TE-671 rhabdomyosarcoma (Castellino et al. 1995).

3. OSTEOSARCOMA CELL LINES

Numerous human osteosarcoma cell lines have been developed and these cell lines vary greatly in specific characteristics thus making comparisons between cell lines interesting. The OHA osteosarcoma cell line overexpresses the Ki-ras oncogene (Suarez et al. 1985; Brouty-Boye et al. 1986; Nardeaux et al. 1987). Continuous exposure of OHA cells to interferon-alpha did not alter the expression of Ki-ras (Brouty-Boye et al. 1986). In three cell lines isolated from the same osteosarcoma patient before treatment (HTLA145), after chemotherapy (HTLA161) and from a lung metastasis (HTLA195), it was found that the c-myc protooncogene was amplified and that the overexpression of c-myc in cell culture and in xenografts was not different among the three cell lines (Bogenmann et al. 1987). Asano and Yamashita (1988) assessed the expression of oncogenes in ten human osteosarcoma cell lines grown as xenografts in nude mice. Five of the cell lines overexpressed c-myc; four of the lines overexpressed c-Ha-ras and one line overexpressed c-fos. Although the human TE85 osteosarcoma cell line is not tumorigenic the subline KHOS/NP(R-970-5) transformed with the Kirsten murine sarcoma virus (KiMSV) is tumorigenic while the revertant subline KHOS-240S which has lost the virus is again not tumorigenic (Carloni et al. 1988). Transfection of the human TE85 osteogenic sarcoma cell line with the c-Ha-ras gene also conferred tumorigenicity on this line (Fidler et al. 1991). Transformation of the human osteosarcoma cell line TE-85 with the v-Ki-ras oncogene to form the subline KRIB resulted in a cell line that was tumorigenic and metastatic from an implant into the tibia (Berlin et al. 1993). Mutated p53 was detected in the human osteosarcoma derived cell line HOS-SL (Romano et al. 1989). Transfection of the human Saos-2 osteosarcoma cell line with the retinoblastoma (*Rb*) tumor suppressor gene resulted in a subline which continued to proliferate but more slowly than the parent line and formed tumors but with a longer latency period than the parent line (Zhou et al. 1994). On the other hand, transfection of human Saos-2 osteosarcoma cells with B-myb stimulated DNA synthesis in growth-arrested cultures (Sala et al. 1996).

Osteosarcoma cells are highly responsive to growth factors and cytokines.

Human MG-63 osteosarcoma cells produce interferon-α (Billiau et al. 1977). However, when the human OHA osteosarcoma cell line was continuously exposed to interferon-alpha after several passages there was an inhibition of cell proliferation (Brouty-Boye et al. 1986). Interferon-gamma was a potent inhibitor of DNA synthesis in human Saos-2 osteosarcoma cells. Exposure of osteosarcoma Saos-2 and U-2OS to interferon-alpha, beta, gamma or TNF-alpha produced an increase in class I HLA antigens. Exposure to interferon-gamma and to a much lesser extent exposure to interferon-beta induced an increased expression of class II HLA antigens (Katayama and Hanazawa 1989; Scotlandi et al. 1992). Exposure to TNF-alpha also inhibited proliferation of the human OST osteosarcoma Takase cell line (Sakayama et al. 1996).

The human U-2OS osteosarcoma cell line expresses c-sis mRNA and synthesizes platelet-derived growth factor (PDGF)-like proteins (Graves et al. 1986; Weich et al. 1986). The U-2OS cells also have specific PDGF binding sites; however, PDGF binding to these cell surface receptors is not necessary for the proliferation of these cells (Richter and Graves 1988; Sanchez et al. 1991). Exposure of human MG-63 osteosarcoma cells to transforming growth factor-ß (TGF-ß) inhibited the proliferation of these cells induced by PDGF (Pontbriant et al. 1990). MG-63 cells alone induce platelet aggregation; U-2OS cells induce platelet aggregation only after incubation of platelets with low concentrations of epinephrine (Mehta et al. 1987). Exposure to insulin or insulin-like growth factor enhanced the growth of human OHS-4 and U-2OS osteosarcoma cells (Andress and Birnbaum 1991; Fournier et al. 1993; Raile et al. 1994). Estrogen receptors and progesterone receptors are present on human Saos-2, HOS/TE85 and MNNG/HOS/TE85 osteosarcoma cells. Nuclear androgen receptor and its upregulation by dihydrotestosterone were observed in human MG-63 osteosarcoma cells (Etienne et al. 1990; Zhuang et al. 1992).

Human osteosarcoma cells have been found to express angiogenic factors including vascular endothelial growth factor (VEGF) in MG-63 cells as well as the angiogenesis inhibitor thrombospondin in MG-63 cells and TE-85 cells (Clezardin et al. 1991; Iizuka et al. 1994).

Bone alkaline phosphatase produced in osteoblastic cells plays a key role in the formation and calcification of the skeleton. The expression of alkaline phosphatase (orthophosphoric-monoester phosphohydrolase [alkaline optimum], EC 3.1.3.1) varies widely among the osteosarcoma cell lines. The alkaline phosphatase activity in MG-63 cells is very low (0.02 µmol/min per mg protein), in KT005 cells is moderate (0.2 µmol/min per mg protein) and in Saos-2 and OHS cells is high (2-2.5 µmol/min per mg protein) (Nakamura et al. 1988; Takashi 1988; Farley et al. 1989; Fournier and Price 1991). Exposure of MG-63 cells or Hu09 cells to vitamin D or analogs of vitamin D induced the synthesis of osteocalcin and alkaline phosphatase activity in these cells (Kawai 1990; Valaja et al. 1990; Manonen and Maenpaa 1994).

These cell lines have been used widely to assess therapeutic approaches in osteosarcoma (Tsang et al. 1986; Supino et al. 1987; Tsuchiya and Tomita 1987; Goto et al. 1991; Rani et al. 1993; Kjonniksen et al. 1994; Larsen et al. 1994; Sonoda et al. 1995).

4. LEIOMYOSARCOMA, SYNOVIAL SARCOMA AND LIPOSARCOMA CELL LINES

Leiomyosarcoma is a malignant tumor of the smooth muscle. Leiomyosarcomas of the retroperitoneum and the vena cava occur most commonly in women. Cutaneous and subcutaneous leiomyosarcomas generally affect men. Synovial sarcomas are most often composed of uniform small cells and most frequently occur in the extremities.

Liposarcomas are the most commonly encountered soft-tissue tumors. The pleomorphic or poorly differentiated liposarcomas are characterized by pleomorphism, frequent mitosis, which may contain lipid droplets, and bizarre giant cells, which may appear as lipoblastic tissue. The human HS-18 liposarcoma cell line was established from an untreated patient (Li et al. 1995). The HS-18 cell line has been compared with the human HT-1080 fibrosarcoma cell line and the human Saos-2 osteosarcoma cell line with respect to retinoblastoma tumor suppressor gene status (*Rb*) and response to the antitumor antimetabolite drugs methotrexate and 5-fluorodeoxyuridine. Growth inhibition assays indicated that the IC50 values for methotrexate and 5-fluorodeoxyuridine in HS-18, a liposarcoma cell line lacking retinoblastoma protein (pRb) and Saos-2, an osteosarcoma cell line with a truncated and nonfunctional pRb, were 10- to 12-fold and 4- to 11-fold higher respectively, than for the HT-1080 fibrosarcoma cell line, which has wild-type pRb. These Rb-/- cell lines exhibited a 2- to 4-fold increase in both dihydrofolate reductase and thymidylate synthase enzyme activities as well as a 3- to 4-fold increase in mRNA levels for these enzymes compared to the HT-1080 (Rb+/+) cells. In contrast, there was no significant difference in growth inhibition among these cell lines for the non-antimetabolites VP-16, cisplatin and doxorubicin. These results suggest the Rb gene product pRb may be involved in the mechanisms of interaction between cytotoxic agents and genes involved in cell cycle progression.

5. CHONDROSARCOMA CELL LINES

Chondrosarcoma is a malignant primary tumor of bone characterized by malignant cartilaginous proliferation. The human chondrosarcoma cell line (HCS-2/8) expresses cartilage phenotypes such as production of cartilage-type proteoglycans and collagen type II, but its tumorigenicity is low. A second immortal cell line of human chondrosarcoma, named HCS-2/A, was established from the same tumor. HCS-2/A cells proliferated with a doubling time of 3.5 days in a medium containing 20% fetal bovine serum (FBS). This growth rate was comparable to that of HCS-2/8 cells. However, HCS-2/A cells proliferated more rapidly than HCS-2/8 cells in the presence of 2-10% FBS. Like HCS-2/8 cells, HCS-2/A cells had a polygonal shape in sparse cultures and became spherical as they reached confluence, after which they formed nodules composed of multilayered cells and a large quantity of extracellular matrix showing metachromasia. The nodules formed by HCS-2/A cells were thicker and also larger in diameter than those formed by HCS-2/8 cells. Electron microscopically, the cells in the nodules resembled chondrocytes in vivo, but each cell had an irregular shaped nucleus which is a characteristic of tumor cells. The cells actively synthesized "cartilage-specific" large proteoglycans and their level of proteoglycan synthesis was comparable to that of HCS-2/8 cells. Insulin, which stimulates proteoglycan and DNA synthesis in cultured chondrocytes, markedly increased proteoglycan synthesis in HCS-2/A cells. On the other hand, the hormone only slightly increased proteoglycan synthesis in HCS-2/8 cells. Insulin also stimulated DNA synthesis in cultured HCS-2/A cells, but not in HCS-2/8 cells. Immunostaining revealed that HCS-2/A cells produced type-II collagen but not type-I collagen. However, the level of collagen synthesis in HCS-2/A cells was lower than that of HCS-2/A cells. Inoculation of HCS-2/A cells into athymic mice resulted in the formation of chondrosarcomas that grew faster than those arising from HCS-2/8 cells (Takigawa et al. 1991). The human 105KC chondrosarcoma cell line synthesizes keratan sulfate and produces chondrocytic large-aggregating proteoglycans.

6. CONCLUSIONS

Sarcomas are malignant tumors of mesenchymal origin. These account for about 20% of all malignant tumors and can arise from many tissues including bone, muscle, breast and other gynecologic sites (ovaries, uterus, vulva), head and neck, kidneys and soft tissue. The majority of the cell lines described herein were developed 20 to 30 years ago. The rhabdomyosarcoma lines of Peter Houghton were developed in the early 1980's. More recently, laboratories have developed cell lines of a particular histology or with other specific properties of interest to that laboratory. These cell lines represent an enormous resource to the cancer research community and most are representative of the tumor of origin.

References

Aaronsen (1970). *Exp Cell Res* 61: 1.
Aguanno, A., M. Bouche, et al. (1990). *Cancer Res* 50: 3377.
Andreasen, P., L. Neilsen, et al. (1986). *J Biol Chem* 261: 7644.
Andreasen, P., C. Pyke, et al. (1987). *Mol Cell Biol* 7: 3021.
Andress, D. and R. Birnbaum (1991). *Biochem Biophys Res Commun* 177: 213.
Asano, S. and T. Yamashita (1988). *Nippon Seik Gak Zass* 62: 231.
Barlati, S., F. Paracini, et al. (1991). *FEBS Lett* 281: 137.
Berlin, O., D. Samid, et al. (1993). *Cancer Res* 53: 4890.
Billiau, A., V. Edy, et al. (1977). *Antimicro Agents Chemother* 12: 11.
Block, J., S. Inerot, et al. (1991). *J Bone Joint Surg* 73A: 647.
Bloom, E. (1972). *Cancer Res* 32:960.
Bogenmann, E., H. Moghadam, et al. (1987). *Cancer Res* 47: 3808.
Brouty-Boye, D., J. Wybier-Franqui, et al. (1986). *J Interferon Res* 6: 461.
Butler, D., B. Scallon, et al. (1994). *Cytokine* 6: 616.
Cajot, J., J. Bamat, et al. (1990). *Proc Nat Acad Sci USA* 87: 6939.
Carloni, G., A. Venaut, et al. (1988). *FEBS Lett* 229: 333.
Castellino, S., H. Friedman, et al. (1995). *Br J Cancer* 71:1181.
Chan, B., N. Matsuura, et al. (1991). *Science* 251: 1600.
Cho (1976). *Science* 194: 951.
Clezardin, P., C. Serre, et al. (1991). *Cancer Res* 51: 2621.
Dias, P., D. Parham, et al. (1990). *Am J Pathol* 137: 1283.
Dilling, M., P. Dias, et al. (1994). *Cancer Res* 54: 903.
Douglass, E.C. et al. (1987). Cytogenet. Cell Genet. 45:148.
Espy, M., A. Wold, et al. (1991). *J Clin Microbiol* 29: 2701.
Etienne, M., J. Fischel, et al. (1990). *Eur J Cancer* 1990: 807.
Falugi, C., P. Castellani, et al. (1986). *Basic Appl Histochem* 30: 433.
Farley, J., E. Kyeyune-Nyombi, et al. (1989). *Clin Chem* 35: 223.
Fidler, I., L. Li, et al. (1991). *Anticancer Res* 11: 17.
Fogh, J. (1975). Human tumor cells *in vitro*. J. Fogh. New York, Plenum Press: 115.
Fogh, J., J. Fogh et al. (1977a). *JNCI* 59: 221.
Fogh, J., W. Wright et al. (1977b). *JNCI* 58: 209.
Fostad, O., A. Brogger, et al. (1986). *Int J Cancer* 38: 33.
Fournier, B., J. Ferralli, et al. (1993). *J Endocrin* 136: 173.
Fournier, B. and P. Price (1991). *J Cell Biol* 114: 577.
Garvin, A., W. Stanley, et al. (1986). *Am J Pathology* 125: 208.
Germani, A., C. Fusco, et al. (1994). *Biochem Biophys Res Commun* 202: 17.
Giard, D., S. Aaronson, et al. (1973). *JNCI* 51: 1417.
Goto, H., I. Matusi-Yuasa, et al. (1991). *Arch Biochem Biophys* 286: 316.
Graves, D., A. Owen, et al. (1986). *PNAS USA* 83: 4636.
Hazelton, B., J. Houghton, et al. (1987). *Cancer Res* 47: 4501.
Hochhauser, D., B. Schnieders, et al. (1996). *JNCI* 88: 1269.
Hodgkiss, R. and M. Stratford (1988). *Int J Radiat Biol* 54: 601.
Hoglund, M., T. Siden, et al. (1992). *Gene* 116: 215.
Horton, J., P. Houghton, et al. (1987). *Cancer Res* 47: 6288.
Horton, J., K. Thimmaiah, et al. (1989). *Biochem Pharmacol* 38: 1727.
Houghton, J., P. Houghton, et al. (1982). *JNCI* 68: 437.
Houghton, J., W. Meyer, et al. (1987a). *Cancer Treat Rep* 71: 717.
Houghton, J., L. Williams, et al. (1987b). *Biochem Pharmacol* 36: 81.

Houghton, P., P. Cheshire, et al. (1993). *Cancer Res* 53: 2823.
Houghton, P., P. Cheshire, et al. (1995). *Cancer Chemother Pharmacol* 36: 393.
Houghton, P., P. Cheshire, et al. (1992). *Cancer Chemother Pharmacol* 31: 229.
Houghton, P., J. Houghton, et al. (1987). *Anticancer Drug Des* 2: 165.
Houghton, P., J. Houghton, et al. (1989). *Cancer Chemother Pharmacol* 25: 84.
Houghton, P., D. Shapiro, et al. (1991). *Pediatr Clin North Am* 38: 349.
Houghton, P., R. Tharp, et al. (1988). *Cancer Chemother Pharmacol* 22: 201.
Iizuka, M., M. Yamauchi, et al. (1994). *Biochem Biophys Res Commun* 205: 1474.
Iwasaki, H., M. Kukuchi, et al. (1982). *Cancer* 50: 520.
Johnston, S. (1991). *Diagn Microbiol Infect Dis* 14: 373.
Katayama, H. and S. Hanazawa (1989). *Meika Daigaku Shi Zass* 18: 411.
Kawai, A. (1990). *Clin Orthop* 259: 256.
Kawai, K., N. Kamatani, et al. (1990). *Journal of Biological Chemistry* 265: 13137.
Kelland, L., L. Bingle, et al. (1989). *Br J Cancer* 59: 160.
Kim, N., S. Sekine, et al. (1995). *J Biochem* 117: 359.
Kinsey, B., A. Van den Abbeele, et al. (1989). *Cancer Res* 49: 5986.
Kjonniksen, I., M. Winderen, et al. (1994). *Cancer Res* 54: 1715.
Kubota, S., R. Fridman, et al. (1991a). *Biochem Biophys Res Commun* 176: 129.
Kubota, S., T. Mitsudomi, et al. (1991b). *Biochem Biophys Res Commun* 181: 1539.
Larsen, R., O. Bruland, et al. (1994). *Br J Cancer* 69: 1000.
Li, W., J. Fan, et al. (1995). *Proc Natl Acad Sci USA* 92: 10436.
Lindberg, A., R. Crowell, et al. (1992). *Virus Res* 24: 187.
Lollini, P., C. DeGiovanni, et al. (1991). *Invasion and Metastasis* 11: 116.
Lowe, D. and D. Goeddel (1987). *Mol Cell Biol* 7: 2845.
Manonen, A. and P. Maenpaa (1994). *Biochem Biophys Res Commun* 205: 1179.
Marini, F., J. Cannon, et al. (1995). *Hum Gene Ther* 6: 1215.
Masurel, D., P. Houghton, et al. (1990). *Cancer Res* 50: 252.
McAllister, R., M. Gardner, et al. (1971). *Cancer* 27: 397.
McAllister, R., H. Issacs, et al. (1977). *Int J Cancer* 20: 206.
McAllister, R., J. Melnyk, et al. (1969). *Cancer* 24: 520.
McAllister, R.M. et al. (1973). *Nat New Biol* 242: 75.
Medcalf, R., R. Richards, et al. (1986). *EMBO J* 5: 2217.
Mehta, P., D. Lawson, et al. (1987). *Cancer Res* 47: 3115.
Mukaida, H., N. Hirabayashi, et al. (1991). *International Journal of Cancer* 48: 423.
Mukaida, N., G. Gussella, et al. (1992). *Immunology* 75: 674.
Myohanen, H., R. Stephens, et al. (1993). *J Histochem Cytochem* 41: 1291.
Nagy, A. and R. Baker (1987). *J Cell Sci* 87: 651.
Nakamura, T., K. Nakamura, et al. (1988). *Arch Biochem Biophys* 265: 190.
Nardeaux, P., L. Daya-Grosgean, et al. (1987). *Biochem Biophys Res Commun* 146: 395.
Neilsen, L., P. Andreasen, et al. (1986). *Thromb Haemost* 55: 206.
Nielsen, L., I. Lecander, et al. (1987). *Thromb Res* 46: 411.
Noel, A., V. Borcy, et al. (1995). *Anticancer Res* 15: 1.
Nurcombe, V., M. Aumailley, et al. (1989). *Eur J Biochem* 180: 9.
Oikarinen, A., M. Hoyhtya, et al. (1990). *Arch Dermatol Res* 282: 153.
Palmer, H., V. Maher, et al. (1989). *In Vitro Cell Dev Biol* 25: 1009.
Paterson, H., B. Reeves, et al. (1987). *Cell* 51: 803.
Peebles, P., T. Trisch, et al. (1978). *Pediatric Res* 12: 485.
Pontbriant, C., J. Chen, et al. (1990). *J Cell Physiol* 145: 488.
Ponten, J. and E. Saksela (1967). *Int J Cancer* 2: 434.
Raile, K., A. Holfich, et al. (1994). *J Cell Physiol* 159: 531.
Rani, A., D. Qu, et al. (1993). *Carcinogenesis* 14: 947.

Rasheed, S., W. Nelson-Rees, et al. (1974). *Cancer* 33: 1027.
Rathod, R. and A. Kharti (1990). *J Biol Chem* 265: 14242.
Rhim, J., H. Cho, et al. (1975). *Int J Cancer* 15: 23.
Rhim, J., D. Park, et al. (1975). *Nature* 256: 751.
Richter, M. and D. Graves (1988). *J Cell Physiol* 135: 474.
Romano, J., J. Ehrhart, et al. (1989). *Oncogene* 4: 1483.
Sakayama, K., H. Masuno, et al. (1996). *Biochem J* 316: 813.
Sala, A., I. Cassella, et al. (1996). *J Biol Chem* 271: 9363.
Samid, D. (1989). *Clin Biol* 1: 21.
Sanchez, A., C. Chesterman, et al. (1991). *Gene* 98: 295.
Scotlandi, K., N. Baldini, et al. (1992). *Anticancer Res* 12: 767.
Shapiro, D., P. Houghton, et al. (1990). *Cancer Res* 50: 6002.
Shaughnessey, E., M. Walker, et al. (1994). *Anticancer Res* 14: 513.
Slovak, M., S. Mirski, et al. (1991). *Br J Cancer* 64: 296.
Sonoda, J., H. Hibasami, et al. (1995). *Anticancer Res* 15: 907.
Stephens, R., J. Pollanen, et al. (1989). *J Cell Biol* 108: 1987.
Stratton, M., J. Darling, et al. (1989). *Carcinogenesis* 10: 899.
Suarez, H., L. Grosjeachn, et al. (1985). Characterization of a human osteosarcoma oncogene. Retroviruses and Human Pathology. R. Gallo, D. Stehelin and O. Varnier. Clifton, NJ, Humana Press.
Supino, R., L. Bardella, et al. (1987). *Tumori* 73: 109.
Takashi (1988). *Arch Biochem Biophys* 265: 190.
Takigawa, M., H. Pan, et al. (1991). *Int J Cancer* 48: 717.
Takigawa, M., K. Tajima, et al. (1989). *Cancer Res* 49: 3996.
Tang, F., P. T'so, et al. (1989). *J Histochem Cytochem* 37: 697.
Tanimoto, K., K. Tamura, et al. (1992). *Biochem Biophys Res Commun* 182: 773.
Tsang, K., R. Warren, et al. (1986). *JNCI* 77: 1175.
Tsuchiya, H. and K. Tomita (1987). *Gan To Kagaku Ryoho* 14: 2269.
Ueda, E., S. Ohno, et al. (1996). *J Biol Chem* 271: 9790.
Valaja, T., A. Mahonen, et al. (1990). *Biochem Biophys Res Commun* 169: 629.
Walker, M., C. Lim, et al. (1986). *Cancer Res* 46: 4927.
Watanabe, H., P. Carmi, et al. (1991). *J Biol Chem* 266: 13442.
Weich, H., W. Sebald, et al. (1986). *FEBS Lett* 198: 344.
Yamada, K., D. Kennedy, et al. (1990). *Cancer Res* 50: 4485.
Zhou, Y., J. Li, et al. (1994). *PNAS* 91: 4165.
Zhuang, Y., M. Blauer, et al. (1992). *J Steroid Biochem Mol Biol* 41: 693.
Zwelling, L., M. Slovak, et al. (1990). *J Natl Cancer Inst USA* 82: 1553.

Chapter 2

Neuroblastoma

Carol J. Thiele
Cell & Molecular Biology Section, Pediatric Oncology Branch, National Cancer Institute, National Institutes of Health, Bld. 10 Rm 13N240, 10 Center Dr. MSC1928, Bethesda, MD. 20892-1928; Tel: 001-301-496-1543; Fax: 001-301-402-0575; E-mail: ct47a@nih.gov

Neuroblastoma accounts for approximately 9% of all childhood cancers, occurring once out of 8,000 live births. This results in an annual incidence of approximately 1 in 100,000 children less than 15 years of age world-wide. The median age at diagnosis is approximately 22 months with over one-third diagnosed at less than 1 year of age and over 88% diagnosed by the age of 5. Some studies indicate a bimodal age distribution with one peak at approximately 1 year and the second between 2 and 4 years (1,2).

In children under 5 years of age, neuroblastoma usually presents in the abdominal region involving the sympathetic ganglia of the paraspinal region or the adrenal gland. In infants under a year of age there is a higher incidence of tumors in the thoracic region. In Stage I and II neuroblastoma (NB) where tumor is confined to the originating organ or surrounding tissue, the prognosis is favorable. However in Stage III and IV NB where tumor extends beyond the midline, is metastatic or involves bony lesions the prognosis is poor. It is possible that early stage disease (Stage I,II, IVS) is a distinct entity from late stage (Stage III, IV) disease because their response rate to therapy and their molecular genetic and biologic characteristics are distinct.

Of all human tumors neuroblastomas have one of the highest rates of spontaneous tumor regression. This is primarily due to an unusual type of neuroblastoma called Stage IVS that occurs in infants under a year of age who present with widely disseminated disease that typically resolves with minimal therapy. Stage IVS tumors have intrigued scientists studying neuroblastoma who feel that an understanding of this disease entity will provide clues to the more devastating presentations of neuroblastoma.

J.R.W. Masters and B. Palsson (eds.), Human Cell Culture Vol. 1, 21–53.

A report of familial NB which affected 4 of 5 siblings led Knudson and Meadows to formulate a "Mutation Model for Neuroblastoma" in 1976 (3). This model proposed that all NB derive from a single cell that is transformed from a normal cell by two mutations; one that may arise prezygotically in the germline and the second that arises in a somatic cell of the target tissue. This "two-hit" model of tumor development was hypothesized to describe the genetics of retinoblastoma tumors and subsequently was verified with the identification of *Rb*, the retinoblastoma tumor suppressor gene (4). While the probability of tumor occurrence would be high in familial or germline cases of neuroblastoma, the actual incidence is quite low. This low incidence may be due to the fact that the elaboration of both mutations causes an embryonic lethal condition or that there is a failure to detect affected parents because they have occult tumors that spontaneously regress or benign tumor forms such as ganglioneuromas (5,6). The majority of neuroblastoma arise sporadically and as such germline mutations would not occur and both mutations would arise somatically. Neuroblastomas occur early in life and the age-associated incidence suggests that the target cell may differentiate or does not persist into adult life. For sporadic neuroblastomas to occur, both mutations must occur in a cell before it fully differentiates.

1. NEUROBLASTOMA CELL CULTURE

Short term in vitro culture of neuroblastoma tumors was used as a tool for diagnosis by Murray and Stout (7) who found that explants of tumors grown in plasma-clot cultures readily elaborated axons. The long term culture of neuroblastoma tumors and the differentiated properties these cells express in culture have fascinated investigators who study these cultures in order to understand the propensity of neuroblastoma cells to differentiate in vivo and in vitro (8).

Tumor tissue can be obtained from primary tumor samples obtained from surgical resections, fine needle and bone marrow aspirates and occasionally from peripheral blood. Typically, samples of solid tumor are placed in culture media (Dulbecco's Modified Eagles Medium [DMEM] + 10mM HEPES buffer or RPMI-1640) containing from 3-15% fetal calf serum, 100 IU/ml penicillin and 100 ug/ml streptomycin. Tumor tissue is minced with scissors or a scalpel and may be filtered through a fine wire mesh under sterile conditions (TD; tissue dispersion) and cultured in complete media at 37°C in a humidified atmosphere containing 5% CO_2. Frequently fibroblasts grow out of these preparations. However, if the tumor sample is cultured on extracellular matrix components such as laminin or collagen the growth of fibroblasts may be reduced. Bone marrow samples should be aspirated into a preservative-free heparinized syringe. Techniques for the isolation of the tumor cells from marrow cells have varied from direct culture (C) to density gradient centrifugation (DG). In the direct culture method, bone marrow is diluted into media and red blood cells are washed off during in vitro culture. Red blood cells may be separated from tumor cells by placing the diluted bone marrow sample on a ficoll-hypaque gradient (density 1.077g/l) and centrifuging for 30 min at 500xg at room temperature. The interface layer containing the tumor cells and normal bone marrow cells is collected and washed several times before it is cultured in complete media.

Cultures should be observed several times a week over the first few weeks and the media replenished every few days. Many neuroblastoma tumor cells will attach to the culture dish. However, tumor spheroids may remain in suspension and thus the spent media may contain non-adherent tumor cells and should be centrifuged and the pelleted cells re-cultured in a separate flask. After washing detached cells in media, cells may be resuspended in fresh media in multiple or larger tissue culture flasks. The use of antibiotics minimizes the possibility of microbial contamination during the initial phase of culture, however long-term use in established cultures can lead to occult contamination.

Primary neuroblastoma cells in culture may be small, rounded or tear shaped cells with a relatively immature appearance, although cells frequently elaborate long neurite-like processes. NB cells may grow in cell aggregates that are only loosely substrate-adherent. The morphological appearance of established neuroblastoma cell lines varies (see Biological features) with cells having a neuroblastic, intermediate and substrate-adherent or flat morphology (9,10). Cultures may be homogeneous or heterogeneous for these cell types.

Reports of the adaptation of neuroblastoma cell lines to serum-free conditions in the absence of additional growth factors have been infrequent. The studies of El-Badry et al (11) revealed that the SK-N-AS cell line adapted to serum-free conditions because it constitutively expressed IGF-2, whereas other NB cell lines required IGF-2 to grow in serum-free conditions. Table 1 lists the clinical features from which over 100 neuroblastoma cell lines have been derived.

2. MOLECULAR AND CYTOGENETICS

The most widely characterized cytogenetic alterations in neuroblastoma tumors include the loss or rearrangement of the distal portion of the short arm of chromosome 1 (1p31-term) (12,13), and amplification of the N-myc gene (14). These cytogenetic features are most commonly found in advanced stage tumors and most neuroblastoma cell lines are derived from advanced stage tumors (Table 2 and 2A).

Table 1 Clinical features

Cell line		Patient yr.mo/sex	Primary site	Metastatic site	Origin	Treatment	Culture	Stage	Ref.
1	IMR32	1.1/M	abdom.	unk.	abdom.	none	TD	unk	(72)
2	SK-N-SH	4/F	thorax	BM	BM	+	C	4	(9)
3	SK-N-BE(1)	1.8/M	unk.	BM	BM	none	C	4	(73)
4	SK-N-BE(2)	2.2/M	unk.	BM	BM	+	C	4	(48)
5	SMS-KAN	3/F	pelvic	BM,LN	pelvic	none	TD	4	(49)
6	SMS-KANR	3.8/F	pelvic	BM	BM	+	DG	4	"
7	SMS-KCN	0.11/M	adrenal	LN, bone, BM	adrenal	none	TD	4	"
8	SMS-KCNR	1.2/M	adrenal	BM	BM	+	DG	4	"
9	SMS-MSN	5/M	adrenal	BM	BM	none	DG	4	"
10	SMS-SAN	3/F	adrenal	BM	BM	none	DG	4	"
11	SMS-LHN	2/M	femur mass	BM, bone	femur	+	TD	4	(74)
12	HTLA230	0.11/M	unk	BM	BM	unk.	NM	4	(75)
13	SJNB-1(EB)	2.6/M	adrenal	LN, mediast	adrenal	+	TD	4	(76)
14	SJNB-2	4/F	unk	BM	BM	none	DG	unk.	"
15	SJNB-3	0.10/F	adrenal	BM, bone	BM	unk.	DG	3	"
16	SJNB-4(SD)	1/M	adrenal	BM, bone	BM	+	DG	4	"
17	SJNB-5	2/F	adrenal	BM, bone	BM	unk.	DG	3	"
18	SJNB-6	unk/M	unk.	BM	BM	unk.	DG	4	"
19	SJNB-7	0.07/M	adrenal	testicle	adrenal	unk.	TD	3	(76)
20	SJNB-9	5/F	adrenal	BM	BM	none	TD	3	"
21	SJNB-10	2/M	adrenal	liver	liver	unk.	TD	3	"
22	SJNB-12	0.10/F	kidney	LN	abdomen	unk.	TD	3	"
23	SJNB-13	1/M	adrenal	pleural, LN	LN	unk.	TD	3	"
24	SJNB-14	1/F	adrenal	BM,bone	BM	unk.	DG	3	"
25	SJNB-16	5/F	unk.	BM, optic nerve, bone	BM	unk.	DG	3	"
26	SJNB-17	1/F	adrenal	LN, BM bone	BM	unk.	DG	3	"

Continued on next page

Table 1 (continued)

Cell line		Patient yr.mo/sex	Primary site	Metastatic site	Origin	Treatment	Culture	Stage	Ref.
27	KP-N-RT-BM	1.2/F	adrenal	BM, bone	BM	unk.	DG	4	(53)
28	KP-N-RT-LN	1.2/F	adrenal	BM, LN, bone	LN	unk	TD	4	"
29	KPNRT-BMV	1.2/F	adrenal	BM, bone	BM	unk.	DG	4	(77)
30	KP-N-SI(LA)	5/M	adrenal	LN, bone	LN	unk.	TD	4	(78)
31	KP-N-SI(FA)	5/M	adrenal	LN, bone	bone	unk.	TD	4	(79)
32	KP-N-YN	2/M	adrenal	LN	LN	unk.	TD	3	(78)
33	KP-N-AY	2.6/F	adrenal	LN, BM	BM	none	DG	4	(80)
34	KP-N-AYR	2.6/F	adrenal	LN, BM	BM	+	DG	4	"
35	GI-ME-N	2/F	adrenal	LN, BM	BM	+	DG	4	(81)
36	GI-LI-N	1.11/F	adrenal	BM	P.B.	unk.	DG	4	(82)
37	GI-LA-N	2.3/M	adrenal	LN	LN	+	TD	3	(83)
38	GI-CA-N	0.9/F	adrenal	BM	BM	-	DG	4	(82)
39	IGR-N-835	2/F	adrenal	BM, bone	adrenal	+	TD	4	(84)
40	IGR-N-91	8/M	adrenal	BM, bone	BM	+	DG	4	(85)
41	CA-2-E	1/M	abdomen	BM, bone	BM	+	DG	4	(86)
42	WSN	2.9/F	abdomen	-	abdomen	-	TD	3	(87)
43	MSN	1.2/F	abdomen	BM	BM	-	DG	4	"
44	ACN	3.3/M	abdomen	BM, bone	BM	+	DG	4	(88)
45	POG-1382.2	3.5/F	retroperit.	unk.	retroper.	+	TD	3	P.Houghton
46	POG-1771	2.10/M	adrenal	unk.	adrenal	-	TD	4	"
47	POG-1643	1.7/M	retroperit.	unk.	retroper.	none	TD	4	"
48	POG-1691	1.9/M	retroperit.	unk.	retroper.	+	TD	4	"
49	RN-GA	1.8/F	adrenal	LN	adrenal	unk.	TD	3	(89)
50	NBL-S	3.6/M	adrenal	none	adrenal	none	TD	3	(90)
51	NBL-W	0.6/M	adrenal	liver	adrenal	none	TD	4S/4	(91)
52	NUB-7	0.7/M	adrenal	BM-LN	LN	-	TD	4S/4	(92)

Continued on next page

Table 1 (continued)

Cell line		Patient yr.mo/sex	Primary site	Metastatic site	Origin	Treatment	Culture	Stage	Ref.
53	CLB-Pe	0.6/F	adrenal	BM	BM	+	DG	1	(50)
54	CLB-Ge1	1.6/M	retroperit.	LN, bone	BM	unk.	DG	4	"
55	CLB-Ge2	1.6/M	retroperit.	LN, bone	LN	unk.	TD	4	"
56	CLB-Be	1.6/M	adrenal	LN, BM	LN	+	TD	3	"
57	CLB-Tr	1.3/F	adrenal	liver, bone	BM	+	DG	4	"
58	CLB-Ma1	0.9/F	abdomen	BM	abdomen	none	TD	4	"
59	CLB-Ma2	0.9/F	abdomen	BM	BM	none	DG	4	"
60	CLB-Es	1/M	adrenal	LN, bone	BM	none	DG	4	"
61	CLB-Bac	3.2/M	adrenal	BM, liver, bone	adrenal	+	TD	4	"
62	CLB-BerLud1	5.5/M	adrenal	BM, bone	BM	none	DG	4	"
63	CLB-BerLud2	5.5/M	adrenal	BM, bone	BM	+	DG	4	"
64	CLB-Br	3.7/F	retroperit.	BM, LN, bone	BM	none	DG	4	"
65	CLB-Ca	11.4/F	adrenal	BM, bone	BM	+	DG	4	"
66	CLB-Ba	2.3/M	adrenal	BM, LN, bone	BM	none	DG	4	"
67	CLB-Ga	4/M	retroperit.	BM, LN, bone mediastinum	BM	-	DG	4	
68	CLB-Re	4.6/M	adrenal	BM, bone, LN	BM	+	DG	4	"
69	LA-N-1	2/M	unk.	BM	BM	+	TC	4	(93)
70	LA-N-2	3/F	abdominal	unk.	abdominal	-	TD	4	"
71	LA-N-5	0.4/M	unk.	BM	BM	unk.	DG	unk.	R. Seeger,pc
72	LAN-6	5.8/M	adrenal	BM, bone	BM	+	DG	4	(74)
73	CHP-126	1.2/F	retroperit.	-	retroper.	none	TD	3/4	(94)
74	CHP-126B1	1.9/F	"	-	PB	+	plasma	unk.	(54)
75	CHP-134	1.1/M	adrenal	LN	LN	+	TD	4	(94)
76	NGP	2.6/M	unk.	BM, lung	lung	+	TD	unk.	(95)
77	NMB-7	0.10/F	unk.	BM	BM	+	DG	unk.	"

Continued on next page

Table 1 (continued)

Cell line		Patient yr.mo/sex	Primary site	Metastatic site	Origin	Treatment	Culture	Stage	Ref.
78	NB9	1.10/M	adrenal	-	adrenal	none	TD	4	"
79	NB16	2.11/F	unk.	BM	BM	none	DG	4	"
80	NB19	1/F	unk.	BM	BM	+	DG	4	"
81	NB56	2/M	adrenal	BM pleural effusion	adrenal	+	TD	4	"
82	NB69(CHP270)	1.4/M	adrenal	ascites	adrenal	-	TD	3	"
83	NB69(2)	2.3/M	adrenal	lung liver sternum	lung liver sternum	+	TD	3	"
84	NB76	3/M	adrenal	ascites	ascites	-	TD	3	"
85	NLF	3/M	abdomen	no	abdomen	none	TD	3	(14)
86	CHP-166	2.6/M	unk.	BM	BM	-	DG	unk.	(26)
87	CHP234	3/F	unk.	BM	BM	none	DG	4	(96)
88	CHP212	1.8/M	kidney mass	-	kidney mass	none	TD	unk.	(94)
89	NAP	unk./F	unk.	BM	BM	+	DG	unk.	(26)
90	NJB	2.9/F	unk.	LN	LN	unk.	TD	unk-	(95)
91	NLB	2.3/F	abdomen	-	abdomen	unk.	TD	unk.	"
92	SJ-N-KP	5/?	unk.	BM	BM	unk.	DG	unk.	(97)
93	SJ-N-CG-NCG)	0.6/M	unk.	BM	unk.	unk.	DG	4	"
94	VA-N-BR	6/M	abdomen	serosal, liver, pelvic	serosal	+	TD	unk.	(98)
95	MMH	3/M	unk.	BM	BM	+	DG	unk.	(26)
96	SK-N-AS	8/F	adrenal	BM	BM	+	C	4	L. Helson, pc
97	CHP 901	2/M	unk.	BM	BM	unk.	DG	4	A. Evans, pc
98	NB-19	0.10/M	adrenal	liver, BM	BM	unk.	DG	4	(26)
99	CHP-903	3/F	adrenal	BM	BM	unk.	DG	4	A. Evans, pc
100	PER 106	1.5/M	suprarenal	BM	BM	none	DG	4	(99)
101	PER 107	2.6/M	suprarenal	BM	BM	none	DG	4	"
102	PER 108	2.9/M	suprarenal	BM	BM	+	DG	4	"

Continued on next page

Table 1 (continued)

Cell line		Patient yr.mo/sex	Primary site	Metastatic site	Origin	Treatment	Culture	Stage	Ref.
103	STA-NB-1.1	3.6/M	adrenal	-	adrenal	unk.	TD	3	P.Ambros,pc
104	STA-NB-1.2	3.9/M	adrenal	-	adrenal	none	TD	3	(132)
105	STA-NB-2	1.6/M	adrenal	BM	adrenal	none	TD	4	"
106	STA-NB-3	1.8/F	adrenal	-	adrenal	none	TD	2	"
107	STA-NB-4	1.5/M	adrenal	BM	adrenal	none	TD	4	"
108	STA-NB-5	0.6/M	adrenal	BM, liver, LN	adrenal	none	TD	4S/4	"
109	STA-NB-6	2.1/N	retroperitoneum	-	retroperitoneum	none	TD	3	"
110	STA-NB-7	1.7/M	adrenal	-	adrenal	none	TD	3	"
111	STA-NB-8	2.3/F	adrenal	BM, bone	BM	none	DG	4	"
112	STA-NB-9	0.3/F	adrenal	BM, liver	liver	none	TD	4	"
113	STA-NB-11	-/M	adrenal	BM	adrenal	+	TD	4	"

BM=bone marrow; TD=tissue dispersion; C=direct culture; DG=density gradient; LN=lymph node; PB=peripheral blood; NM=nude mouse; unk=unknown, pc=personal communication

Table 2 Common chromosomal alterations in neuroblastoma cell lines

Cell line	1p alteration		N - myc amplification		N - myc single copy	Chromosome 17	Ref
	1pdel	t(1p)	HSR	DM			
1 IMR32	+	-	+	+			(13, 97)
2 SK-N-SH	-	-	-		+		(26, 97)
3 SK-N-BE(1)	+	-	+	+			(97)
4 SK-N-BE(2)	+	+	+			t(3;17)(p21;q21)	(26)
5 SMS-KAN	+		+	+		del(17),t(17;18)	(49)
6 SMS-KANR	+		+	+			"
7 SMS-KCN	+		+	+		t(17;20)(q21;q13)	"
8 SMS-KCNR	+		+	+		t(17;20)(q21;q13)	"
9 SMS-MSN	-		+	+		i(17q)	"
10 SMS-SAN		+	+	+		t(11;17)(q24;q21)	"
11 SMS-LHN	-		-		+		(74)
12 HTLA-230	unk.		+	+			
13 NB-1(EB)	+	+	-	+			(76)
14 NB-2	?	+	+	+		t(1;17)(p32;q21)	"
15 NB-3	-		-		+		"
16 NB-4(SD)	+	+	+				"
17 NB-5	+	+	+	+			"
18 NB-6	+		+				"
19 NB-7	+		+				"
20 NB-9	-		-		+		"
21 NB-10	+		+	+			"
22 NB-12	+		+	+			"
23 NB-13	+	+	+			t(1;17)(p31;q21)	"
24 NB-14	?		+				"
25 NB-16	+	+	-		+		"

Continued on next page

Table 2 (continued)

Cell line	1p alteration		N - myc amplification		N - myc single copy	Chromosome 17	Ref
	1pdel	t(1p)	HSR	DM			
26 NB-17	+		-		+		"
27 KP-N-RT	+		+	+			(53)
28 KP-N-SI(LA)	-	+	-		+		(78)
29 KP-N-SI(FA)	-	+	-		+		(79)
30 KP-N-YN	+		+	+			(78)
31 KP-N-AY	+		+	+		-17; der(17)	(80)
32 KP-N-AYR	+		+	+		i(17q)	"
33 GI-ME-N	+		-		+		(82)
34 GI-LI-N	+						"
35 GI-CA-N	-		-		+		"
36 IGR-N-835	-	+	-		+	t(11;17)(p11;q11)	(84)
37 IGR-N-91	+		+	+			"
38 CA-2-E	?		+				(86)
39 WSN	+		+	+			(87)
40 MSN	+		+	+			"
41 ACN	-		-		+	c-myc amplification	(88)
42 POG 1382.2	+		+				P. Houghton, pc
43 POG 1771	-	-	+			der(17) t(17;?)(p11;?)	"
44 POG 1643	+		+				"
45 POG 1691	-		+				"
46 RN-GA	+				+		(89)
47 NBL-S	-		-		+		(90)
48 NBL-W	+		+				(91)
49 NUB-7	+		+			$17q^+$	(92)

Continued on next page

Table 2 (continued)

Cell line	1p alteration		N - myc amplification		N - myc single copy	Chromosome 17	Ref
	1pdel	t(1p)	HSR	DM			
50 CLB-Pe	-		+				(50)
51 CLB-Ge1	+		+				"
52 CLB-Ge2	+		+				"
53 CLB-Be	+		+				"
54 CLB-Tr	+		+				"
55 CLB-Ma1	+		+	+			"
56 CLB-Ma2	+		+				"
57 CLB-Es	+		+				"
58 CLB-Bac	+		+				"
59 CLB-Ber Lud1	+		+				"
60 XCLB-Ber Lud2	+		+				"
61 CLB-Br	+		+				"
62 CLB-Ca	+		+				"
63 CLB-Ba	+		+				"
64 CLB-Ga	+		-		+		"
65 CLB-Re	+		+				"
66 LA-N-1	+		+				(97)
67 LA-N-2	+		+	+			"
68 LA-N-5	+		+				
70 LAN-6	-		-		+		(74)
71 CHP-126	-	+	+	+			(13, 97)
72 CHP-134	+	+	+	-			"
73 NGP	+	+	+				(97)
74 NMB	+	+	+				(13)
75 NMB	+		+	+			"

Continued on next page

Table 2 (continued)

Cell line	1p alteration		N - myc amplification		N - myc single copy	Chromosome 17	Ref
	1pdel	t(1p)	HSR	DM			
76 NB16	-	+	+	+		17q⁺	"
77 NB19	-	-	+				"
78 NB56	-	-	+				"
79 NB69	-	+	-	-	+		"
80 NB69 (2)		+	-	-	+		"
81 NB76		+	+				"
82 CHP-166	+		+	+			(26)
83 NAP	-	-	+	+		+17	(26, 97)
84 SJ-N-CG(NCG)	+		+				(97)
85 MMH	+		+	+			(13)
86 SK-N-AS	+	-	-		+		(100)
87 VA-N-BR	+				+?		(98)
88 CHP-903	-	-	+				K. Kisselbach, pc
89 NLF	+		+				(97)
90 STA-NB-1.1	+		+				(132)
91 STA-NB-1.2	+		+				"
92 STA-NB-2	+		-		+	t(1;17)	"
93 STA-NB-3	+		+				"
94 STA-NB-4	+		+				"
95 STA-NB-5	+		+				"
96 STA-NB-6	+		-		+	t(1;17)	"
97 STA-NB-7	+		+				"
98 STA-NB-8	+		+			t(1;17)	"
99 STA-NB-9	+		+			t(1;17)	"
100 STA-NB-11	+		+				"

Table 2A Summary of cell lines with specific genetic alterations

1p alteration			N - myc amplified			
1pdel	(1p)del	t(1p)	HSR	DM	N - myc single copy	Chrom. 17 alteration
IMR32	CLB-Ma(2)	SK-N-BE(2)	SK-N-BE(2)	IMR32	SK-N-SH	SMS-KAN
SK-N-BE(1)	CLB-Es	SMS-SAN	LAN-1	SK-N-BE(1)	NB-17	SMS-KCNR
SMS-KAN	CLB-Bac	NB-16	NB-19	SMS-KAN	NB-69	NB-16
SMS-KCNR	CLB-Ber-Lud1	NB-69	NB-56	SMS-KCNR	NB-1(EB)	SMS-KCN
LAN-1	CLB-Ber-Lud2	NB-76	SMS-KANR	SMS-MSN	NB-9	NB-15
LAN-2	CLB-Br	NB-1(EB)	NB-4 (SD)	NB-9	NB-16	KP-N-AY
NB-9	CLB-Ca	NB-2	NB-6	NB-16	KP-N-SI	KP-N-AYR
SMS-KCN	CLB-Ba	NB-5	NB-7	NB-76	KP-N-SI(FA)	KPNRT-BMVC6
SMS-KANR	CLB-Ga	NB-7	NB-8	SMS-KCN	RN-GA	KP-N-RT
NB-4(SD)	CLB-Re	NB-13	NB-14	SMS-SAN	GI-ME-N	IGR-N-835
NB-5	SMS-LHN	NB-14	NB-19	NB-2	GI-CA-N	POG-1771
NB-6	NB 9	NB-16	KP-N-RT	NB-3	IGR-N-835	NUB-7
NB-7	CHP-166	KP-N-SI	KPNRT-BMV-C6	NB-4	ACN	CLB-Ga
NB-8	MMH	KP-N-SI(FA)	GI-LI-N	NB-5	NBL-S	CLB-Ba
NB-9	NMB	KP-N-SI(LA)	CA-2-E*	NB-6	SMS-LHN	MMH
NB-12	SK-N-AS	KP-N-AYR	POG-1691	NB-9	LAN-6	SMS-MSN
NB-13	CLB-Ge1	NB-69(2)	NBL-W	NB-12	SK-N-AS	SK-N-BE(2)
NB-15	CLB-Ge2	NB-4(SD)	NUB-7	NB-15	CLB-G2	STA-NB-2
NB-16	CLB-Be	NBL-S	CLB-Be	HTLA-230	NB-3	STA-NB-6
NB-17	CLB-Tr	CLB-Ca	CLB-Pe	KP-N-RT	STA-NB-6	STA-NB-8
KP-N-RT	CLB-Ma(1)	CLB-Ga	CHP-126	WSN	STA-NB-2	STA-NB-9
KP-N-YN	NLF	CHP-126	CHP-134	MMH		
KP-N-AY	SJ-N-KP(NKP)	CHP-134	NGP	NMB		
RN-GA	SJ-N-CG(NCG)	NGP	NMB	CHP-126		
KPNAY-BMVC6	VA-N-Br		IGR-N-91	CHP-166		

Continued on next page

Table 2A (continued)

1p alteration			N - myc amplified		N - myc single copy	Chrom. 17 alteration
1pdel	(1p)del	t(1p)	HSR	DM		
GI-ME-N	NB-19		NB-13*	CLB-Ma(1)		
GI-LI-N			POG-1382.2*	CLB-Ma(2)*		
IGRN-91			POG-1771*	CLB-Es		
WSN			POG-1643*	CLB-Bac*		
POG-1382.2			CLB-B2	CLB-Ber-Lud1*		
POG-1643			CLB-Ge2*	CLB-Ber-Lud2*		
NBL-W			CLB-Ge1*	CLB-Br*		
NUB-7			SJ-N-CG(NCG)	CLB-Ca*		
			CLB-Tr*	CLB-Re*		
			CHP-903	KP-N-AY		
			NLF	KP-N-AYR		
			STA-NB-9*	SJ-N-KP(NKP)		
			STA-NB-1.1*	LA-N-2		
			STA-NB-4*	STA-NB-7*		
			STA-NB-5*	STA-NB-11*		
			STA-NB-1.2*	STA-NB-3*		

* unknown if amplification is carried on DMs or HSRs

Alterations and deletions of chromosome 1 are found in many neural crest tumors such as melanoma, neuroepitheliomas, pheochromocytoma, and other tumors including breast cancer, Wilm's tumors and colon carcinoma. This implies that several putative tumor suppressor genes are located on chromosome 1p. In neuroblastoma, chromosome 1p deletions are the most consistent nonrandom genetic alteration and result in chromosome 1p monosomy. The deletion is variable in neuroblastoma but consistently encompassing 1p36.1 to 1pter. At least two putative suppressor genes have been proposed to reside in this region as deletions can be grouped into those whose deletion encompasses 1p36.3 and/or 1p36.1 (15,16). Recently a candidate tumor suppressor gene p73 has been identified that maps to 1p36.33 and is deleted in the cell line (SK-N-AS) that contains the smallest 1p interstitial deletion described to date. Although the p73 gene is an attractive candidate NB tumor suppressor gene as it has structural and functional homologies to p53, it is not typically altered in NB tumors or cell lines. Transfection of p73 into the SK-N-AS cell line suppresses its tumorigenic phenotype. Many NB cell lines such as CHP-212, SMS-KAN and SK-N-BE(2) are monoallelic for p73 and express very low levels of p73 mRNA and no protein for p73. It is not known if the failure to express p73 is due to imprinting of the remaining p73 allele. Some neuroblastomas such as IMR-32 however do express p73 (17).

MycN maps to 2p23-23 and is the common gene contained in the double minute chromosomes (DMs) and homogeneous staining regions (HSRs) that characterize a subset of NB tumors that have a particularly poor prognosis (18). The size of the amplicon varies and other genes found in the amplicon include ornithine decarboxylase, and DDX, a DEAD box protein encoding gene (19). The relationship between DMs and HSRs is not well understood, although there exist in the CHP126 cell line two populations of cells, some of which contain HSRs and some that contain a variable number of DMs. This lead to the hypothesis that DMs were derived from HSRs and were different manifestations of the same underlying genetic alteration (20). The variable number of DMs in tumor cells results from the varying distribution of DMs during cytokinesis. A recent report suggests that only 1 of the possibly 2 tumor suppressor genes localized to 1p is associated with amplification of N-myc (21).

A recent study using comparative genomic hybridization identified that chromosome 17 gain occurred in a high percentage of cases, in Stage 1 and 2 and even 4S suggesting that alterations in this region may be one of the early events involved in NB tumorigenesis (22). Chromosome 17 has tumor suppressor activity in NB cells (23) and mutations and amplification of nm23, a gene on chromosome 17q21 that has a metastasis suppressing ability on some tumor cell types, are detected in poor prognosis patients (24).

Frequently rearrangements on chromosomes 11, 14 and 17 have been noted; 12 of 37 (32%) informative cases contained LOH involving 11q13-

11q23 or 11p and 6 of 27 (22%) informative cases contained loss of heterozygosity (LOH) on 14q (25) while in a sample of 35 NB cell lines 23% had alterations of 17q (26). Typically these genetic alterations have been detected in advanced stage neuroblastomas and derived cell lines and rarely in tumors from low stage disease.

Alterations or mutations in many noted tumor suppressor genes such as p53 (27), Rb, and p16, p18, p27 (CDKN2) (28) and RET (29) have rarely been identified. Occasionally deletions in neurofibromatosis 1 (NF1) have been noted in neuroblastoma cell lines (3/17 analyzed) (30). The prevalence of LOH or mutations in NF1 has not been thoroughly analyzed, although there is a recent report of a neuroblastoma patient with a homozygous deletion in NF1 (31). Translocations and alterations of the region near NF1 on 17q have been described (32).

Although the oncogenic form of N-ras was initially isolated from the SK-N-SH neuroblastoma cell line, mutated N-ras genes are not commonly detected in human neuroblastoma tumors (1/15) (33) and mutations in N-ras, Ki-ras and H-ras were not detected in another series (0/24). Furthermore, it was noted that earlier passages and some sublines of SK-N-SH did not have a mutated N-ras gene. Although a mutation at codon 59 in N-ras was detected in a subline of SK-N-SH, it was different from the codon 61 mutation detected in the original transforming N-ras. This suggests that the mutation in N-ras was not present in the primary tumor and was probably acquired during in vitro culture

3. HISTOPATHOLOGY

The cells from which neuroblastomas are thought to arise are the postganglionic sympathetic neuroblasts of the embryonal neural crest. The neural crest is a transitory structure that arises during the closure of the neural tube. Neural crest cells migrate ventrally and laterally to contribute to a variety of tissues including the peripheral nervous system, medullary cells of the adrenal gland, calcitonin producing cells of thyroid, pigmented cells, and mesectodermal derivatives. Growth and differentiation specific peptides encountered by migratory neural crest cells are thought to influence their development and lineage specific differentiation.

The histopathologic appearance of neuroblastoma represents a spectrum from sheets of monomorphic undifferentiated, small, round blue cells to nests of neuroblasts surrounded by fibrillar bundles to differentiated ganglionic cells. These histologic subtypes correspond to neuroblastoma, ganglioneuroblastoma and ganglioneuroma, respectively. Neuroblastoma is a small, round blue cell tumor characterized by the presence of neuritic processes, neuropil and/or Homer-Wright rosettes which are neuroblasts surrounded by eosinophilic neuropil. Immunohistochemical stains for neurofilaments, synaptophysin and neuron-specific enolase have been utilized to distinguish it from other small round blue cell tumors of childhood. Ganglioneuroblastoma is a heterogeneous group in which tumors contain the spectrum of immature to fully differentiated cells and some stromal component. Ganglioneuroma is composed of mature ganglion cells, neuropil and Schwann cells. Recent evidence indicates that the Schwannian component of these tumors represents an infiltrate of normal cells rather than the differentiation of the tumor cells into Schwannian cells (35). Shimada and colleagues have developed a classification system based on histopathologic features and age which has prognostic utility (36). In this system the presence or absence of Schwann cell stroma; the degree of differentiation and the mitosis-karyorrhexis index (MKI) were considered in addition to age at diagnosis. Favorable histology associated with a good prognosis is: stroma rich without a nodular pattern in any age group; or stroma poor histology in ages 1.5-5 years of age with a MKI < 100; and in ages <1.5 with a MKI of <200. Histology associated with an unfavorable prognosis is: stroma rich with a nodular pattern in any age group; stroma poor over the age of 5; or histology with differentiation (1.5-5 years) or with MKI>100 (1.5-5 years) or MKI>200 (<1.5years). A simplified histopathologic study by Joshi et al (37) found that calcification and low mitotic rate (<10 mitoses/10 high powered fields) predicted a favorable outlook independent of age and stage. Current efforts are underway to develop a unified international neuroblastoma classification system.

Table 3 highlights the pathologic characteristics of a number of neuroblastoma cell lines. The pattern of immunoreactivity was utilized for a time to distinguish neuroblastoma from other small round blue cell tumors. Neuroblastomas and cell lines typically stain poorly for Class I Major Histocompatibility antigens, and are usually positive for neuron-specific enolase and a monoclonal anibody named HSAN1.2 (38). Neuroblastomas form tumors in nude mice although the number of tumor cells required for tumor formation is high (typically 10^8 cells) and the latency for tumor formation is long (several months) (39). A neuroblastoma tumor model in SCID (severe combined immunodeficient disease) mice has been developed (40). Both the SCID and nude mice provide models to study the biologic behavior of tumors in vivo, although the severely impaired immune systems of these animals obviates studies involved with the immune regulation of neuroblastoma cell growth. Other animal models include a neuroblastoma model developed in transgenic animals using the tyrosine hydroxylase promoter to selectively express N-myc in neural cells during development (41) and a transgenic mouse model using the middle T antigen of polyoma virus (42). Although it is too early to determine if the biology of these tumors is similar to that of human neuroblastomas, the utility of these transgenic models is the presence of an intact immune system.

Table 3 Histopathology

Cell line	Original Tumor	Xenograft	Ref
IGR-N-91 – cultured on ECM; both epithelial and neuroblastic cells seen	tumor from BM; catecholamine positive	immature NB with metastases to lymph node, kidney	(85)
IGR-N-835 – floating clusters of small round cells; anchored polygonal cells with few neurites	abdominal tumor tissue was typical immature neuroblastoma after chemotherapy	immature NB	(84)
HTLA-230 – round to bi-polar morphology	unfavorable and stroma-poor histology with a high mitosis-karyorrhexis index	stroma poor histology; high MKI hemorrhagic EM; few processes rare granules	(75)
RN-GA – cells were mainly flat with few neurites, bi-polar rectangular cells	Undifferentiated neuroblastoma; EM showed epithelial-like clusters of tumor cells with neural filaments and dense core granules	typical undifferentiated neuroblastoma, poorly vascularized	(89)
GI-ME-N – extremely substrate adherent with an irregular polygonal shape and few long neuritic processes	undifferentiated, small round cells neuroblastoma tumor	short latency in nude mice; grew as undifferentiated, small, round cell, neuroblastoma	(81)
VA-N-BR – spindle shape with uni- or bi-polar cytoplasmic extensions	monomorphic sheets of polyhedral nucleated cells with occasional fibrovascular stroma-primitive neuroblastoma, positive immunostain for neurofilament and chromogranin, vimentin and desmin	dense pattern of irregularly nucleated cells with moderate neovascularization; stains for desmin and vimentin but not neurofilament or chromogranin	(101)
SK-N-SH – round, tear shaped cells with 'epithelial-like' or flat cells, dense core granules by EM	not available	tumors in cheek pouch of hamsters are small cells, compactly arranged with many mitotic figures	(9)
SMS-KAN – loosely adherent, teardrop shaped cells that grow in clumps with neuritic processes; epithelial-like cell component present	undifferentiated neuroblastoma with poor pseudorosette formation; EM showed neurosecretory granules	NA	(49)

Continued on next page

Table 3 (continued)

Cell line	Original Tumor	Xenograft	Ref
SMS-KANR – smaller cells with few neurites; no epithelial cell component	tumor cells obtained from BM after chemotherapy	NA	"
SMS-MSN – loosely adherent polygonal cells which after crisis had few neurites	tumor from BM was positive for cathecholamine fluorescence; EM revealed neurosecretory granules	NA	"
SMS-KCN – teardrop shaped cells with few neurites; epithelial cell component present	undifferentiated neuroblastoma with no rosettes or ganglion cells; EM showed neurosecretory granules	NA	"
SMS-KCNR – teardrop to polygonal shaped cells with few neurites; rare epithelial cells	tumor cells from BM after chemotherapy	NA	"
SMS-SAN – small round to teardrop shaped cells with some neurites	undifferentiated small-round blue cell tumor originally diagnosed as embryonal rhabdomyosarcoma; tumor from BM had catecholamine fluorescence and neurosecretory granules by EM	NA	"
NBL-S – small teardrop cells, some with short delicate neurite-like processes, weakly substrate adherent, aggregate growth	undifferentiated neuroblastoma with little stroma	NA	(90)
NBL-W – small teardrop neuroblastic cells and substrate-adherent flat cells	undifferentiated neuroblastoma, favorable histology: stroma poor with a MK1 = 120	NA	(91)
PER-106 – suspension culture with large aggregates and few substrates adherent cells	unfavorable and stroma poor histology with a high mitosis-karyorrhexis index	NA	(99)
PER-107, PER-108 – derived after chemotherapy adherent PER-107 has small round cells with few neuritic processes PER-108 have epithelial shape	EM showed occasional neurosecretory granules and neuritic processes with microtubules		

Continued on next page

Table 3 (continued)

Cell line	Original Tumor	Xenograft	Ref
KP-N-RT – small round cells loosely substrate adherent with few neuritic processes	undifferentiated neuroblastoma; EM on cell line showed few dense core granules	NA	(102)
NUB-7 – predominantly growth in tightly adherent colonies; heterogeneous morphology of isolated cells predominantly I-type	nests of neuroblasts surrounded by fibrovascular stroma, numerous Homer-Wright rosettes, isolated areas with ganglioneuroblastoma	NA	(103)

ECM = extracellular matrix; EM = electron microscope; MKI = mitosis-karyorrhexis index; BM = bone marrow

4. BIOLOGICAL FEATURES

Several lines of evidence suggest that neuroblastoma cell lines, like their neural crest cell antecedents, express neuronal and/or neuroendocrine properties. One of the hallmarks of NB cells in culture is their spontaneous or induced elaboration of neuritic processes. Other neuronal properties include; the synthesis of neurotransmitter biosynthetic enzymes; expression of neurofilaments; opioid, muscarinic and neurotrophin receptor expression; dense core granules, which are the presumed sites of catecholamine storage; immunoreactivity to neuron specific enolase. Neuroblastoma cell lines tend to have an adrenergic phenotype producing relatively high levels of tyrosine hydroxylase and dopamine-ß-hydroxylase while peripheral neuroepitheliomas or Ewing's cell lines tend to produce choline aceltyltransferase (43). There are few lineage specific markers discriminating neuronal from neuroendocrine cell types.

Most neuroblastoma cell lines are derived from tumor tissue of patients with advanced stage disease. Exceptions to this are NBL-W (91) and NUB-7 (92) which were derived from stage IVS disease. However, these lines contain 1pdel and N-*myc* amplification and the patients from whom these cell lines were derived eventually died. Thus, they are not typical of stage IVS and may have been derived fom occult stage IV disease. While CLB-Pe is derived from a stage I neuroblastoma, the patient relapsed rapidly with metastasis to the bone marrow and consequently may not be representative of tissue derived from a stage I tumor (50).

Many established neuroblastoma cell lines contain at least 3 morphological variants that contribute to the heterogeneity in these cell lines; neuroblastic (N), flat or substrate adherent (S) and intermediate (I) cell types (10,44,45). Morphologic subtypes from a number of neuroblastoma cell lines have been cloned (Table 4) and also can be distinguished by characteristic biochemical markers. Some clonal populations have the capacity to spontaneously interconvert or transdifferentiate from one morphologic type to another. It is thought that this heterogeneity may reflect their derivation from multi-potent neural crest precursors. N-type and I-type cells express neurofilament proteins while I-type and S-type cells are more strongly positive for vimentin than N-type cells. S-type cells synthesize collagen and fibronectin similar to Schwannian cells (46,47). Most S-type cells do not synthesize readily detected levels of tyrosine hydroxlyase or dopamine-ß-hydroxylase although some produce tyrosinase. A study of cell surface antigen expression indicated that the S-type cells shared antigenic characteristics more in common with a fibroblast-like meningeal cell rather than a Schwannian cell. Neural crest cells can give rise to ectomesenchyme, including skeletal and connective tissues of the head and face which also includes meninges. These features have led to a model in which N-type cells are proposed to resemble embryonic sympatho-

Table 4 NB cell variants – biologic characteristics

Parental cell line	Subline	Type	Enzymes	Colony formation	Ref
SKNSH	SH-SY5Y	N	TH, DβH	+	(10)
	SH-EP	S	-	-	"
	SH-EPI	S	Tyr	NT	"
	SH-EPIE	S	-	NT	"
	SH-IN	I	TH, DβH	+	(10)
LA-N-I	LAI-15n	N	TH	+	(10)
	LAI-19n	N	TH	+	"
	LAI-21n	N	TH	+	"
	LAI-5s	S	Tyr	-	"
	LAI-6s	S	ND	-	"
	LAI-22n/i	N/I	DβH	NT	"
SK-N-BE(2)	BE(2)-M17	N	TH	NT	(10)
	BE(2)-M17V	N	-	NT	"
	BE(2)-M17F	S	-	NT	"
	BE(2)-M17M	I	-	NT	"
	BE(2)-7S	S	Tyr	NT	(10)
	BE(2)-C	I	DβH, TH	+	"
SK-N-BE(1)	SK-N-BE(1)n	N	-	NT	(10)
	SK-N-BE(1)s	S	-	NT	
SMS-KCN	KCN-62n	N	TH	NT	(10)
	KCN-65n	N	TH	NT	"
	KCN-71n	N	TH	NT	"
	KCN-83n	N	TH	NT	"
	KCN-9s	S	Tyr	NT	"
NAP	NAP(H)n	N	TH	NT	(10)
	NAP(H)s	S	ND	NT	"
NBL-W	NBL-W-N	N	TH, DβH	NT	(91)
	NBL-w-S	S	ND	NT	"
RT-BMV	RT-BMV-1	N	NT	NT	(77)
	RT-BMV-C6	I	NT	NT	"

N = neuronal subtype; S = substrate adherent; flat subtype; I = intermediate subtype; TH = tyrosine hydroxylase; DBH = dopamine-β-hydroxylase; tyr = tyrosinase; ND = not detected; NT = not tested; - = negative

blasts, S-type cells resemble Schwannian, glial or melanocytic progenitor cells or ectomesenchymal derivatives and the I-type cells have an intermediate phenotype and the potential to differentiate to N- or S-type cells (45).

An interesting characteristic of neuroblastoma tumors is that even in advanced stage disease the primary tumors are sensitive to chemotherapy, although clinical progression inevitably occurs. A number of investigators have isolated paired neuroblastoma cell lines from primary tumor tissues prior to treatment and from samples taken at relapse after intensive chemotherapy or radiotherapy. These include pairs such as SK-N-BE(1) and SK-N-

BE(2) (48); SMS-KCN and SMS-KCNR (49); SMS-KAN and SMS-KANR (49); CLB-BerLud1 and CLB-BerLud2 (50); and NB69 and NB69(2) (55). Many of these cell lines have been studied to examine tumor cell genetics and biology in the setting of progressive disease. For example, increased expression of N-myc was detected in cell lines from patients with progressive disease. However, it has been difficult to distinguish potential treatment related changes from innate tumor cell heterogeneity. A number of NB cell lines have been cultured in cytotoxic drugs to generate drug-resistant cell lines in vitro (52). These drug-resistant cell lines have been valuable in studying mechanisms of drug resistance yet it is unclear how these in vitro generated cell lines reflect the pathologic processes that occur in patients with progressive disease.

A number of cell lines have also been developed from tumors located at different sites in the same patients; KP-N-SI(LA) from lymph node and KP-N-SI(FA) from bone (53); CHP126 from a primary retroperitoneal tumor and CHP126(BL) from the peripheral blood post therapy (54); CLB-Ge1 from bone marrow and CLB-Ge2 from lymph node; CLB-Ma1 from primary tumor tissue and CLB-Ma2 from bone marrow (50) and NB69(2) lung, NB69(2) liver and NB69(2) sternum (55). In the case of NB69(2) the cell lines from the metastatic sites after therapy essentially contained identical karyotypes and cytogenetic alterations, and these were both common and distinct from the cytogenetic alterations observed in the cell line derived from the primary tumor. In another study, using a panel of nine monoclonal antibodies, the immunophenotypes of the parental and 2 cell lines from different metastatic sites were stable and relatively uniform although growth rates differed (53).

The ability of many neuroblastoma cell lines to differentiate in response to a variety of biologic response modifiers has led to the use of neuroblastoma cell lines as model systems to study neuronal and neuroendocrine cell development as well as regulation of catecholamine biosynthesis. Oncologists have searched for agents that induce terminal differentiation, which may have less or different toxicities from conventional cytotoxic agents. Table 5 lists a number of agents that reduce cell growth and induce differentiation or even apoptosis in selected neuroblastoma cell lines. Phenylacetate and retinoids (all-trans-retinoic acid, 13-cis retinoic acid and 9-cis retinoic acid) are or have been in Phase I trials as single agents or in combination with Interferonα.

Retinoids induce marked neuronal differentiation and arrest of cell growth in some neuroblastoma cell lines (i.e. LA-N-5, LA-N-1, SMS-KCNR) but have little effect on others (i.e. SK-N-AS, SH-EP) (56,57,58). Neuroblastoma cells constitutively express the retinoic acid receptors RARα, RARγ and RARβ and RXRs, although the levels of RARβ are reduced compared to RARα and RARγ (59). Treatment with nanomolar concentrations of retinoids is sufficient to increase RARβ expression and induce neurite extension (60). However micromolar concentrations are required to significantly inhibit cell

Table 5 Effect of differentiation agents on selected neuroblastoma cell lines

Cell line	Treatment	Cell types	Ref
SMS-KCNR	INFγ	N, A	(104)
	RA	N, S	(57)
	BrdU	S	(105)
	cAMP	N, A	(106)
LA-N-5	RA	N	(56)
	cAMP	N	(107)
	VIP	N	(108)
	Vit D	N	(109)
	Phenylacetate	N	(110)
	INFγ	N, A	(111, 112)
	Ara-C	N	(112, 113)
	TPA	Nb	(112)
	TNF	N	(112)
SK-N-SH/SH-SY5Y	RA, +BDNF	N, NE	(60)
	TPA	N	(114)
	NGF	N	(62, 115)
	cAMP	N	(116)
	Herbimycin A	N	(117)
GI-ME-N	INFγ	N, A	(118)
	RA	N	(66)
	Ara-C	N	(113)
	HPR	A	(66)
LA-N-1	RA	N	(66)
	INFγ	N	(118)
	Ara-C	Nb	(113)
	HPR	A	(66)

INFγ = interferon gamma; RA = retinoic acid; Ara-C = cytosine arabinoside; BDNF = brain derived neurotrophic factor; HPR = hydroxyphenylretinamide; TPA = phorbol esters; VIP = vasoactive intestinal peptide; BrdU = bromodeoxyuridine; cAMP = 5′-cyclic adenosine monophosphate; Vit D = vitamin D; TNF = tumor necrosis factor; N = neuronal; A = apoptotic; Nb = neuroblastic; NE = neuroendocrine; S = Schwannian

growth under standard culture conditions (60). The use of receptor selective agonists provides evidence that both RAR and RXR are required for maximal effects on growth and differentiation. Differentiation includes elaboration of extensive neuritic processes that are ultrastructurally and electrophysiologically similar to normal neurons, decreases in a number of proto-oncogenes including N-myc, c-myb (61), and increases in TrkB (62), and RET receptors (63) and their protein kinase activity. The increases in TrkB and RET kinase activity have been shown to induce neurites in NB cells.

The ability of many of the biologic response modifiers to induce apoptosis or cell death in NB cell lines has only recently been appreciated. While a prominent effect of retinoids is to induce neuronal differentiation, RA has been shown to induce apoptosis in a subset of cells by inducing tissue transglutaminase in SK-N-BE2(C) cells (64,65). N-(4-Hydroxyphenyl)retinamide (HPR) has been shown to induce apoptosis in neuroblastoma cells (66). Interferonγ has been shown in a number of cell lines to induce cell death in a subpopulation of cells within 24-36 hours after treatment, although the mechanism is unknown (67).

A number of genetically modified neuroblastoma cell lines have been generated (partially listed in Table 6) that are valuable tools with which investigators can probe the role of particular genes that may be important in the biology of neuroblastoma. The amplification of N-myc has been proposed to be a progression related event as it occurs in only a subset of NB tumors. By constitutively expressing a vector containing anti-sense N-myc and thus decreasing cellular N-myc levels, the growth (68) as well as the invasiveness (69) of NB cells decreased. A limitation of genetically modified cell lines constitutively expressing the gene of interest is clonal heterogeneity, which may be problematic depending on the biologic or biochemical trait being assessed. The use of regulated expression vector systems or retroviral systems with a high transfection efficiency enables assessment without the potentially confounding issue of clonal variability. Using a tet-regulated N-myc expression vector, it was recently shown that increased N-myc expression increases cell growth by decreasing the time cells transit through the cell cycle (70), thus confirming studies that indicated a role for N-myc in cell growth. The use of genetically modified cells has also shown that activation of TrkA (71), TrkB (62) or RET (63) signal transduction paths are capable of inducing neurite extension in neuroblastoma cells.

Although a systematic analysis of neuroblastomas has not been undertaken, a number of cytokines are differentially expressed and active in neuroblastoma cell lines and tumors. For example, using RT-PCR, Stem cell factor (SCF) and its receptor c-kit were shown to be expressed in most NB cells and blockade of the SCF signal transduction path inhibited cell growth (128). However, another study could not reproduce these data using less sensitive assays (129). Some neuroblastoma cell lines, such as SK-N-SY5Y and IMR-32 respond to cytokines such as ciliary-neurotrophic factor (CNTF) and leukemia inhibitory factor (LIF) and increase production of peptides such as vasointestinal peptide or neurotransmitters (130). Members of a heparin binding family of neurotrophic factors, pleiotrophin and midkine, are differentially expressed in neuroblastomas. Midkine is expressed by all primary neuroblastomas and cell lines, yet shows no correlation with disease stage. Pleiotrophin shows a pattern of expression similar to that of TrkA, being expressed in tumors with a good prognosis and consequently not in most cell lines (131).

Table 6 Examples of genetically modified neuroblastoma cell lines

Cell line	Parental line	Genetic modification	Ref
IMR5-BCL2	IMR5 (subclone IMR32)	Bcl2; neo	(119)
NBASI-e	NBL-S	antisense N-myc; neo	(68)
NBS-1	"	sense N-myc; neo	"
NBS-2	"	"	"
AS-14.2N+	SK-N-AS	N-myc; neo	(120)
AS-11.4N+	"	N-myc; neo	"
15N-TrkB	"	TrkB; neo	(62)
411	SMS-KCNR	N-myc; neo	(61)
512	"	N-myc; neo	"
422	"	N-myc; neo	"
1810	HTLA230	TrkA; neo	(121)
c1.4M	SK-N-BE	MAX; neo	(122)
c1.6M	"	MAX; neo	"
c1.3N	"	N-myc; neo	"
c1.9N	"	N-myc; neo	"
SK-N-AS:γIGN	SK-N-AS	IFNg; neo	(123)
LA-N-6:γIGN	LA-N-6	IFNγ; neo	"
LA-N-5:γIFN	LA-N-5	IFNγ; neo	"
LA-N-1:γIFN	LA-N-1	IFNγ; neo	"
SK-N-FI:γIFN	SK-N-FI	IFNγ; neo	"
TET-2N	SH-E-P	tet regulated N-myc; hygro.	(70)
TET-2	"	tet transactivator; neo	"
TET-21N	"	tet regulated N-myc; hygro.	"
TET-21	"	tet transactivator; neo	"
SH-400;310;907 SH-802;803	SK-N-SH	N-myc; neo	(124)
15NA-P4;E1;D1	LAI-15N	TrkA; neo	(125)
tTA sense tTA-antisense	SK-N-BE(2)	tissue transglutaminase; neo	(65)
RET/PTC1 RET/PTC3 RET/MEN2A RET/MEN2B	SK-N-BE(2)	activated RET genes; neo	(63)
B/myb	LA-N-5	B-myb; neo	(126)
B-myb antisense	"	antisense B-myb	"
SY5Y-TrkA	SH-SY5Y	TrkA; neo	(71)
LAN5-TrkA	LA-N-5	TrkA; neo	"
SKMYC2 SKMYC6	SK-N-SH	N-myc	(127)
SKMYAS	SK-N-SH	antisense N-myc	"
SH400;310;907 SH802;803	SK-N-SH	N-myc; neo	(124)

neo = neomycin resistant; tet = tetracycline regulated; hygro = hygromycin resistant

ACKNOWLEDGMENTS

I would like to thank my colleagues for taking the time to send me reprints and additional data on their cell lines;

Dr. Jean Benard
Unit of Genetic Markers in Cancer
Dept. Of Clinical Biology
Insitute Gustave Roussy
Rue Camille Desmoulins
F-94805 Villjuif Cedex
France
ph: 33-42-11-48-18
FAX:33-42-11-52-80

Dr. Nicole Gross
Onc.-Hema. Unit Pediatrics
University Hospital (CHUV)
Lausanne, Switerland 1011
ph: (21) 314-3622
FAX (21) 314-3558
email Nicole.Gross@chuv.hospvd.ch

Dr. Marie C. Favrot
Lab of Tumor Biology
Centre Leon Berard
28 rue Laennec
Lyon, Cedex 08, France 69373
ph: (7) 878-2669
FAX: (7) 878-2717
email: FAVROT@LYON.FNCLCC.FR

Dr. Garrett M.Brodeur
Div. of Oncology
Children's Hosp. of Philadelphia
ARC, Room 902-D
324 S. 34th St.
Philadelphia, PA 19104-4318
ph: (215) 590-2817
FAX: (215) 590-3770

Dr. Audrey E. Evans and Ms. Kirsten Kisselbach
Children's Hosp. of Philadelphia
Abramson Ped. Research Ctr.
324 S. 34th St.
Philadelphia, PA 19104-4318
ph: (215) 590-2250
FAX: (215) 590-3770
email: EVANS@VIERMIT.ONCOL.CHOP.Edu

Dr. Robert Seeger
Div. of Hema./Onc.
Children's Hosp. of Los Angeles
MS-57
4650 Sunset Blvd.
Los Angeles, CA 90027
ph: (213) 669-5618
FAX (213) 664-9455
email: seeger%smtpgate@CHLAIS.USC.EDU

Dr. Herman Yeger
Dept. of Ped. Lab. Medicine
Hospital for Sick Children
Div. of Pathology
555 University Ave.
Toronota, Ontario, M5G 1X8 Canada
ph: (416) 813-5958
FAX: (416) 813-5974

Dr. Paolo Cornalglia-Ferraris
Ospedale Pediatrico Bambino Gesu
Istituto Di Richerca Scientifica
Rome Italy
FAX: 6-6830832

Drs. Thomas Look, David Reardon and Susan Ragsdale
St. Jude Children's Res. Hosp.
332 N. Lauderdale
Memphis, TN 38105
ph: (901) 495-3514
FAX: (901) 495-2032
email: thomas.look@stjude.org

Drs. Peter J. Houghton and Joyce Thompson
Dept. of Molecular Pharm
St. Jude Children's Res. Hosp.
332 N. Lauderdale
P. O. Box 318
Memphis, TN 38101
ph: (901) 495-3440
FAX: (901) 521-1668
email: Peter.Houghton@Stjude.ORG

Dr. Susan Cohn
Division of Hematology
Children's Memorial Hospital
2300 Children's Plaza
Chicago, IL 60614
ph: (312) 880-4586
FAX: (312) 880-3053
email: scohn@merle.acns.nwu.edu

Dr. Tohru Sugimoto
Dept. of Ped
Miyazaki Medical College
Kihara, Kiyotake
Miyazaki, Japan 889-16
ph: (985)850989
FAX: (718) 817-3645

Dr. Robert Ross
Dept. of Biological Sciences
Fordham University
East Fordham Rd.
Bronx, NY 10458
ph: (718) 817-3654
FAX: (718) 817-3645

Dr. Carlo Dominici
Dept. Pediatrics
Univ. La Sapienza
Viale Regina Elena 324
Rome, Italy 00161
ph: (6) 497891
FAX: (6) 446-2767

Dr. Mirco Ponzoni
Pediatric Oncology Res. Lab
G. Gaslini Children's Hosp.
Largo Gerolamo Gaslini, 5
Genoa, Italy 16148
ph: (10) 377-6590
FAX: (10) 377-4530

Dr Peter Ambros
Children's Cancer Research Institute
St Anna Kinderspital
Kinderspitalgasse 6
A-1090 Vienna
Austria
ph: 43-1-40470/411
FAX: 43-1-408-7230
E-mail: ambros@ccri.univie.ac.at

REFERENCES

1. Young, J. *Cancer*. 58:598, 1986.
2. Voute, P. In: W. Sutow, D. Fernbach, and T. Vietti (eds.), *Clinical Pediatric Oncology*, pp. 599. St. Louis: Mosby, CV, 1984.
3. Knudson, A. and Meadows, A. J. *Natl Cancer Inst*. 57:675, 1976.
4. Knudson, A. *Proc Natl Acad Sci*. 68:8820, 1971.
5. Everson, T. and Cole, W. In: T. Everson and W. Cole (eds.), *Spontaneous regression of cancer*, Vol. 88. Philadelphia: WB Saunders, 1966.
6. Schwartz, A et al. *J Pediatr*. 85:760–770, 1974.
7. Murray, M. and Stout, A. *Amer J Path*. 23:429–441, 1947.
8. Goldstein, M. N. J. *Pediatric Surg*. 3:166–169, 1968.
9. Biedler, J. L. et al. *Cancer Res*. 33:2643–2652, 1973.
10. Ciccarone, V. et al. *Cancer Res*. 49:219–225, 1989.
11. El-Badry, O. et al. J. *Clin Invest*. 87:648–657, 1991.
12. Brodeur, G. M. et al. *Cancer*. 40:2256–2263, 1977.
13. Gilbert, F. et al. *Cancer Genetics and Cytogenetics*. 7:33–42, 1982.
14. Schwab, M. et al. *Nature*. 305:245–248, 1983.
15. Caron, H. et al. *Nature Genetics*. 4:187–190, 1993.
16. Cheng, J. et al. *Nature Genetics*. 4:191–194, 1993.
17. Kaghad, M. et al. *Cell*. 90:809–819, 1997.
18. Brodeur, G. et al. *Science*. 224:1121–1124, 1984.
19. Brodeur, G. M. and Castleberry, R. P. In: P. A. Pizzo and D. G. Poplack (eds.), *Principles and Practice of Pediatric Oncology*, pp.761–797. Philadelphia: Lippincott-Raven, 1997.
20. Balaban-Malenbaum, G. and Gilbert, F. *Science*. 198:739–741, 1977.
21. Cheng, N. et al. *Oncogene*, 10:291–297, 1995.
22. Plantaz, D. et al. *Am J Pathol*. 150:81–89, 1997.
23. Bader, S. et al. *Cell Growth Differ*. 2:245, 1991.

24. Leone, A. et al. *Oncogene*. 8:855–865, 1993.
25. Srivatsan, E. et al. *Clin Biolog Res*. 366:91–98, 1991.
26. Gilbert, F. et al. *Cancer Res*. 44:5444–5449, 1984.
27. Imamura, J. et al. *Cancer Res*. 53:4053–8, 1993.
28. Kawamata, N. et al. *Cancer*. 77:570–575, 1996.
29. Hofstra, R. et al. *Hum Genet*. 97:362–364, 1996.
30. Johnson, M. et al. *Proc Natl Acad. Sci*. 90:5539–5543, 1993.
31. Martinsson, T. et al. *Cancer Genet Cytogenet*. 95:183–189, 1997.
32. Laureys, F. et al. *Oncogene*. 10:1087–1093, 1995.
33. Ireland, C. *Cancer Res*. 49:5530–3, 1989.
34. Moley, J. et al. *Cancer Res*. 51:1596–9, 1991.
35. Ambros, I. et al. *N. Eng J Med*. 334:1505–1511, 1996.
36. Shimada, H. et al. *J Natl Cancer Inst*. 73:405, 1984.
37. Joshi, V. et al. *Proc Am Soc Clin Oncol*. 10:311, 1991.
38. Donner, L. et al. *Proc Clin Biolog Res*. 175:347–366, 1985.
39. Reynolds, C. et al. *Prog Clin Biolog Res*. 271:291–306, 1988.
40. Bogenmann, E. *Int J Cancer*. 67:379–385, 1996.
41. Weiss, W. et al. *EMBO J*. 16:2985–95, 1997.
42. Aguzzi, A. et al. *Brain Pathol*. 2:195–208, 1992.
43. Thiele, C. et al. *J Clin Invest*. 80:804–11, 1987.
44. Ross, R. et al. *Cell Growth & Differentiation*. 6:449–456, 1995.
45. Ross, R. A. et al. *Adv Neuroblastoma Res*. 4:253–259, 1994.
46. Rettig, W. J. et al. *Cancer Res*. 47:1383–1389, 1987.
47. Rettig, W. J. et al. *Cancer Research*. 53:3327–3335, 1993.
48. Barnes, E. N. et al. *In Vitro*. 17:619–630, 1981.
49. Reynolds, C. P. et al. *J Natl Cancer Inst*. 76:375–387, 1986.
50. Combaret, V. et al. *Int J Cancer*. 61:185–191, 1995.
51. Rosen, J. et al. *Cancer Res*. 46:4139–4142, 1986.
52. Meyers, M. and Biedler, J. *Prog Clin Biolog Res*. 271:449–461, 1988.
53. Sugimoto, T. et al. *Cancer Research*. 46:4765–4769, 1986.
54. Gerson, J. M. et al. *Cancer*. 39:2508–2512, 1977.
55. Feder, M. K. and Gilbert, F. *J Natl Cancer Inst*. 70:1051–1056, 1983.
56. Sidell, N. et al. *Exp Cell Res*. 148:21, 1983.
57. Thiele, C. et al. *Nature*. 313:404, 1985.
58. Gaetano, C. et al. *Cell Growth Differn*. 2:487–493, 1991.
59. Li, C. et al. *Prog. Clin Biolog Res*. 385:221–227, 1994.
60. Matsumoto, K. et al. *Cancer Res*. 55:1798–1806, 1995.
61. Thiele, C. J. and Israel, M. A. *Expl Cell Biol*. 56, 321–333, 1988.
62. Kaplan, D. et al. *Neuron*. 11:321–331, 1993.
63. D'Alessio, A. et al. *Cell Growth Differn*. 6:1387–1394, 1995.
64. Melino, G. et al. *Exp Cell Res*. 179:429–436, 1988.
65. Melino, G. et al. *Mol Cell Biol*. 14:6584–6596, 1994.
66. Ponzoni M. et al. *Cancer Res*. 55:853–861, 1995.
67. Montaldo, P. G. et al. *Cell Death and Differentiation*. 4:150–158, 1997.
68. Schmidt, M. L. et al. *Cell Growth & Differentiation*. 5:171–178, 1994.
69. Goodman, L. et al. *Clin Exp Metastasis*. 15:130–139, 1997.
70. Lutz, W. et al. *Oncogene*. 13:803–812, 1996.
71. Poluha, W. et al. *Oncogene*. 10:185–189, 1995.
72. Tumilowicz, J. J. et al. *Cancer Res*. 30:2110–2118, 1970.
73. Biedler, J. L. et al. *Cancer Res*. 38:3751–3757, 1978.
74. Wada, R. K. et al. *Cancer*. 72:3346–3354, 1993.

75. Matsushima, H. and Bogenmann, E. *Int J Cancer*. 51:250–258, 1992.
76. Shapiro, D. et al. *Am J Path*. 142:1339–1346, 1993.
77. Hino, T. et al. *Int J Cancer*. 44:286–291, 1989.
78. Sugimoto, T. et al. *Int J Cancer*. 48:277–283, 1991.
79. Sugimoto, T. et al. *Cancer Res*. 49:1824–1828, 1989.
80. Kuroda, H. et al. *Int J Cancer*. 47:732–737, 1991.
81. Cornaglia-Ferraris, P. et al. *Pediatric Res*. 27:1–6, 1990.
82. Longo, L. et al. *J Cancer Res Clin Oncol*. 114:636–640, 1988.
83. Donti, E. et al. *Cancer Genet Cytogenet*. 30:225–231, 1988.
84. Bettan-Renaud, L. et al. *Int J Cancer*. 44:460–466, 1989.
85. Ferrandis, E. and Bénard, J. *Int J Cancer*. 54:987–991, 1993.
86. Suardet, L. et al. *Int J Cancer*. 44:661–668, 1989.
87. Gross, N. et al. *Cancer Res*. 57:1387–1393, 1997.
88. Gross, N. et al. *Int J Cancer*. 52:85–91, 1992.
89. Scarpa, S. et al. *Int J Cancer*. 43:645–651, 1989.
90. Cohn, S. et al. *Oncogene*. 5:1821–1827, 1990.
91. Foley, J. et al. *Cancer Res*. 51:6338–6345, 1991.
92. Dimitroulakos, J. et al. *Cell Growth & Differentiation*. 5:373–384, 1994.
93. Seeger, R. C. et al. *Cancer Res*. 37:1364–1371, 1977.
94. Schlesinger, H. R. et al. *Cancer Res*. 36:3094–3100, 1976.
95. Brodeur, G. M. and Goldstein, M. N. *Cytobios*. 16:133–138, 1976.
96. Schlesinger, H. R. et al. *Cancer Res*. 41:2573–2575, 1981.
97. Brodeur, G. M. et al. *Cancer Res*. 41:4678–4686, 1981.
98. Helson, L. and Helson, C. *Anticancer Res*. 12:467–472, 1992.
99. Kees, U. R. et al. *Cancer Genet Cytogenet*. 59:119–127, 1992.
100. White, P. S. et al. *Proc Natl Acad Sci USA*. 92:5520–5524, 1995.
101. Helson, L. and Helson, C. J. *Neuro-Oncology*. 3:39–41, 1985.
102. Sugimoto, T. et al. *Cancer Res*. 47:5433–5438, 1987.
103. Yeger, H. et al. *Differentiation*. 39:216–227, 1988.
104. Lucarelli, E. et al. *J Biol Chem*. 270:24725–24731, 1995.
105. Reynolds, C. and Maples, J. *Prog Clin Biol Res*. 175:13, 1985.
106. Gaetano, C. et al. *Cancer Res*. 52:4402–4407, 1992.
107. Lando, M. et al. *Cancer Res*. 53:722, 1990.
108. Pence, J. and Shorter, N. *Cancer Res*. 50:5177, 1990.
109. Moore, T. et al. *Clin Exp Metastasis*. 14:239–245, 1996.
110. Sidell, N. et al. *Int. J Cancer*. 60:507–514, 1995.
111. Wuarin, L. et al. *Int J Cancer*. 48:136–141, 1991.
112. Lanciotti, M. et al. *Cell Mol Neurobiol*. 12:225–240, 1992.
113. Ponzoni, M. et al. *Exp Cell Res*. 50:5177, 1989.
114. Pahlman, S. et al. *Int J Cancer*. 28:583, 1981.
115. Sonnenfeld, K. and Ishii, D. *J Neurosci Res*. 8:375, 1982.
116. Schulze, I. and Perez-Polo, J. *J Neurosci Res*. 8:393, 1982.
117. Preis, P. et al. *Cancer Res*. 48:6530, 1988.
118. Ponzoni, M. et al. *Cancer Res*. 52:931–939, 1992.
119. Hanada, M. et al. *Cancer Res*. 53:4978–4986, 1993.
120. Feltner, D. E. et al. *J Immunol*. 143:4292–4299, 1989.
121. Matsushima, H. and Boegenmann, E. *Mol Cell Biol*. 13:7447–56, 1993.
122. Peverali, F. et al. *Oncogene*. 12:457–462, 1996.
123. Ucar, K. et al. *Cancer Gene Therapy*. 2:171–181, 1995.
124. Gross, N. et al. *Int J Cancer*. 59:141–148, 1994.
125. Lucarelli, E. et al. *Submitted*, 1998.

126. Raschella, G. et al. *J Biol Chem.* 270:8540–8545, 1995.
127. Judware, R. and Culp, L. *Oncogene.* 14:1341–1350, 1997.
128. Cohen, P. S. et al. *Blood* 84:3465, 1994.
129. Beck, D. et al. *Blood* 86:3132–3138, 1995.
130. Johnson, J. A. and Nathanson, N.M. *J Biol Chem* 269:18856–63, 1994.
131. Nakagawara, A. et al. *Cancer Res.* 55:1792–1797, 1995.
132. Ambros, I. M. et al. *Eur J Cancer* 33:2043–2049, 1997.

Chapter 3

Ewing's Sarcoma Family of Tumors

Frans van Valen
Head, Laboratory for Experimental Orthopedic Research, Department of Orthopedic Surgery, Westfälische-Wilhelms-University of Münster, Domagkstrasse 3, 48149 Münster, Germany. Tel: 0049-251 835-6754. Fax: 0049-251 835-6758. E-mail: valen@uni-muenster.de

In 1921 James Ewing described a non-osteogenic small-round-cell tumor of bone and considered the tumor to be a diffuse endothelioma arising from vascular endothelium of the bone marrow [1]. Ever since Ewing's original postulation, the histogenesis of Ewing's sarcoma has been a matter of dispute. A variety of cells have been proposed as stem cells, including the reticular cell [2], the myeloid cell [3], a transitional cell developing from pericytes to vascular smooth muscle [4], as well as a pluripotential uncommitted mesenchymal cell [5,6]. Neural features were observed in some putative Ewing's sarcomas and it was concluded that the category of Ewing's sarcoma may be broader than had previously been recognized [7]. Though lacking any differentiation markers *in vivo*, Ewing's sarcoma cells *in vitro* either spontaneously express or can be induced to express morphological features characteristic of neural differentiation, such as neural cell adhesion molecule [NCAM], neuron-specific enolase [NSE], Leu-7, S100, neurofilament protein [NFP], and primitive neurites with neurosecretory granules [8-13]. These data suggested that Ewing's sarcoma might be a highly undifferentiated neural tumor and related to osseous and extraosseous peripheral primitive neuroectodermal tumors. The latter tumor, originally described in 1918 by Stout [14], has been frequently referred to as peripheral neuroepithelioma and peripheral neuroblastoma. Here the term peripheral primitive neuroectodermal tumor (pPNET) is used according to Dehner's classification [15]. pPNETs are believed to originate from the neural crest, which is a transitory embryonal structure arising during the formation of the neural tube. Besides neural differentiation markers, Ewing's sarcoma and pPNET cells may also develop epithelial features (e.g. expression of vimentin, cytokeratin,

J.R.W. Masters and B. Palsson (eds.), Human Cell Culture Vol. 1, 55–85.

desmoplakin) [16,17] as well as mesenchymal features (e.g. osteoid and chondrocytic formation; extracellular matrix synthesis) [18-22]. The mesenchymal propensity of Ewing's sarcoma and pPNET cells is in keeping with the unique ability of cells of the neural crest or 'ectomesenchyme' to differentiate into both neuroectodermal and mesenchymal phenotypes [23]. Biochemical studies have shown that Ewing's sarcoma and pPNET display an identical pattern of protooncogenes (c-myc; N-myc; c-myb; c-mil/raf-1) and exhibit high levels of choline acetyltransferase, an enzyme important in the biosynthesis of cholinergic neurotransmitters [24,25]. Immunocytochemical studies on the presence of cell-surface glycoprotein gp30/32 (MIC2; CD99) using monoclonal antibody HBA-71 have indicated that this antigen is highly expressed in 98% of Ewing's sarcoma and pPNET and is rarely present in other small-round-cell tumors of childhood, e.g. retinoblastoma, neuroblastoma, small cell osteosarcoma and lymphoma [26-29]. Cytogenetically, Ewing's sarcoma and pPNET share a unique feature: a reciprocal chromosomal translocation t(11;22)(q24;q12), which was identified in 85% of cases [30-32]. This translocation fuses the FLI-1 gene on chromosome 11q24 to the EWS gene on chromosome 22q12, resulting in EWS/FLI-1 transcripts identifiable in 90% of Ewing's sarcoma and pPNET [33-35]. The remaining percentage of cases have other aberrations such as chromosome deletion del(22)(q12) and the translocation t(21;22)(q22;q12) associated with the EWS/ERG hybrid transcript [34,36,37]. Together, these observations imply that Ewing's sarcoma and pPNET represent different stages of differentiation in a malignancy of neuroectodermal origin with pPNET as the most differentiated tumor. Ewing's sarcoma and pPNET are now considered members of the same family: the Ewing group of childhood tumors [38].

Askin tumor, a primitive small-cell tumor of the thoracopulmonary region typically involving chest wall and peripheral lung in children and young adults [39], and olfactory neuroblastoma or esthesioneuroblastoma, an exeedingly rare malignancy of the nasal cavity, may also be categorized within the Ewing tumor family. Askin tumor and esthesioneuroblastoma exhibit neuroectodermic properties, carry the translocation t(11;22)(q24;q12) and express the EWS/FLI-1 hybrid transcript identical with Ewing tumors [39–42]. Recently, the translocation t(11;22)(q24;q12) and the concomitant EWS/FLI-1 transcript have been described in a set of childhood tumors diagnosed as primitive rhabdomyosarcomas and malignant ectomesenchymomas showing myogenic and neural differentiation [43–45]. These so-called biphenotypic tumors might, therefore, also be members of the Ewing family of tumors. The lack of staining for CD99 in esthesioneuroblastoma and biphenotypic tumor provides a diagnostic tool for separating these neoplasms from Ewing's sarcoma, pPNET and Askin tumor [45–47]. Accordingly, small-round-cell childhood malignancies, in particular those belonging to the Ewing tumor family, are most accurately classified when clinical, morphological and biological characteristics can be integrated and used to decide optimal diagnosis and treatment.

1. CELL CULTURE

Solid tumor tissue samples obtained during surgery are minced with scalpels in pieces of approximately 2 mm in diameter, and incubated at 37°C for 1 hr in 0.05% trypsin/0.02% EDTA in PBS. Every 15 min the solution is vortexed briefly to facilitate the detachment of tumor cells from the stroma. The supernatant is then collected by filtration through sterile gauze and centrifuged at 1200 rpm for 5 min. Tumor samples obtained from pleural effusions can be centrifuged immediately without the trypsin treatment. Cells are washed twice in RPMI-1640 medium supplemented with 10% fetal calf serum (FCS), 2 mmol/l L-glutamine, 100 U/ml penicillin, and 100 μg/ml streptomycin (i.e. growth medium). Since the majority of Ewing tumor cell lines have the tendency to grow as loosely adherent cells or as tightly packed floating aggregates, cells are usually seeded in collagen-coated (5 $\mu g/cm^2$) or fibronectin-coated (2 $\mu g/cm^2$) tissue culture plastic ware. Cultures are incubated at 37°C in a humidified atmosphere of 5% CO_2 in air. At confluence, cultures are overlayed with a film of 0.05% trypsin/0.02% EDTA solution. After 5-10 min at 37°C the cells are dislodged by tapping the culture vessel, and subcultured in growth medium at a 1:4 (e.g. cell line WE-68) to 1:20 (e.g. cell line VH-64) split ratio. For culturing tumor specimens by the explant method, pieces of minced specimens are evenly dispersed over the surface of culture vessels and placed at room temperature for 15-30 min to allow attachment. The tissue is then covered with growth medium. For culturing by the xenograft method, minced tumor specimens diluted with medium (1:1; final volume of 500 μl) or single-cell suspensions of cultured tumor cells (approx. 1×10^7 cells/250 μl) are inoculated subcutaneously into the flanks of BALB/c athymic nude mice. Tumor growths of 1 cm in diameter are aseptically removed, minced into 2 mm fragments, and transplanted into the flanks of other nude mice using a 16-gauge trocar needle. The proliferation of Ewing tumor cells *in vitro* is dependent on the presence of serum (minimally 5% FCS) in the culture medium and ceases under serum deprivation. Cells can survive under serum-free conditions in short-term (48 h) cultures using RPMI-1640 medium supplemented with 2 mmol/l L-glutamine, 100 U/ml penicillin, 100 μg/ml streptomycin, and 50–100 ng/ml insulin-like growth factor-I. *In vitro* induction of Ewing tumor cell differentiation (e.g. formation of neurite-like cytoplasmic extensions; immunoreactivity for NSE, NFP and choline esterase; production of cell-surface-associated fibronectin) is accomplished within 1–12 days in RPMI-1640 medium containing 25 nM – 2.5 mM dibutyryl cyclic adenosine monophosphate (dbcAMP) and 0-10% FCS [9,11-13,21,48]. Cells to be kept as stock under liquid nitrogen are suspended in growth medium supplemented with 10% dimethyl sulfoxide (DMSO). After thawing, cells are initially cultured in growth medium containing 20% FCS.

Cell lines established from Ewing tumors (Ewing's sarcoma; pPNET; Askin tumor; esthesioneuroblastoma; biphenotypic tumor) are described in Tables 1a and 1b.

2. PATHOLOGICAL FEATURES

By definition, Ewing's sarcoma is a morphologically undifferentiated lesion of bone and soft-tissues arising in children between usually 10 and 15 years of age with a slight predisposition to males. Approximately 20% of patients present with visible metastases at diagnosis. Of these about 50% have lung metastasis, 40% have multiple bone involvement, and about 10% show lymphatic spread [49]. Histologically, the principal cells of typical Ewing's sarcoma are homogeneously rounded (about $8-12\mu$ in size) with clear cytoplasm, indistinct cell boundaries and dented nuclei containing faint chromatin and inconspicuous nucleoli, and moderate mitotic activity. In addition to this conventional type of Ewing's sarcoma a small group of lesions, on the bases of morphology and ultrastructure, have been classified as large-cell or atypical Ewing's sarcoma, consisting mainly of large principal cells ($15-20\mu$ in size) exhibiting prominent nuclei with condensed chromatin and one or more nucleoli, and high mitotic activity [50,51]. A frequent histological finding in both typical and atypical Ewing's sarcoma and in pPNET is the coexistence of secondary, smaller dark cells. The secondary cells have been interpreted as degenerative or viable tumor cells, local reactive stromal cells or more differentiated forms of principal cells [3,52–54]. Immunohistochemistry is significant for distinguishing Ewing's sarcoma from its neural variant pPNET. According to most authors, the current diagnosis of pPNET demands at least two neural differentiation antigens or the presence of Homer-Wright rosettes in which the tumor cells are coherently arranged in a lobular, alveolar-like pattern [55]. The center of the rosettes may be composed of connective tissue fibrils or unipolar-positioned glycogen-containing tumor cell cytoplasms [56]. Occasionally, Ewing's sarcoma cells may be disposed in a circular fashion surrounding a collagenous center ('pseudorosettes') [52,56].

The pathological features of some Ewing tumors are recorded in Table 2. From the data it is inferred that distinct morphological, cytogenetic and molecular characteristics of Ewing tumor cell lines either cultured *in vitro* or as xenograft are identical to those of the native tumor specimen. On the other hand, differentiated characteristics towards neural and mesenchymal

Table 1a Cell lines derived from Ewing tumors

Cell line	Patient age(yrs)/sex	Primary site	Metastatic site	Tumor specimen site	Culture method	Tumor class	Reference
A4573	17/F	clavicula	bone; lung	PE	D	ES	[5]
5838	27/M	radius	lung	PE	D	ES	[5]
6647	14/F	tibia	lung	PE	D	ES	[5]
TC-71	22/M	humerus		humerus (R)	D	ES	[9]
TC-106	19/M	pelvis	scalp; lungs	scalp	D	ES	[9]
NCR-EW1	10/M	neck		neck	D	ES	[12]
NCR-EW2	13/M	pelvis		pelvis	D	ES	[12]
NCR-EW3	10/M	rib		rib	D	ES	[12]
NCR-EW4	14/F	rib	lung	PE	X	ES	[12]
H-15 [a]	22/F	thigh	BM	BM	X	pPNET	[13]
H-22 [a]	16/M	ankle	knee	poples	X	pPNET	[13]
H-825	14/M	thigh	lung	lung	D	pPNET	[13]
KU-SN	4/F	scapula	bone; lung	PE	E	pPNET	[20]
CADO-ES-1 [b]	19/F	n.d.	lung	PE	D	ES	[21]
SIM-1	16/M	metatarsus	skin	skin	D	pPNET	[26]
SAL-1	13/F	humerus		humerus (R)	D	pPNET	[26]
EMM [c]	12/F	femur		femur	D	ES	[26]
STA-ET-1 [d]	13/F	humerus		humerus (R)	D	pPNET	[28]
STA-ET-2.1 }	15/M	fibula		fibula	D	pPNET	[28]
STA-ET-2.2 }			BM	BM	D	pPNET	[28]
STA-ET-3	9/F	femur	scalp	scalp	D	pPNET	[28]
IARC-EW12	16/F	iliac crest		iliac crest (R)	D	ES	[31]
DES1	23/F	iliac crest		iliac crest	D	ES	[31]
TC-32	17/F	pelvis		pelvis	D	pPNET	[32]

Continued on next page

Table 1a (continued)

Cell line	Patient age(yrs)/sex	Primary site	Metastatic site	Tumor specimen site	Culture method	Tumor class	Reference
A9423	13/F	pubic bone	bone; lung	lymph node	D	ES	[32]
N1000	14/M	femur		femur	D	pPNET	[32]
N1001	11/F	ilium	bone	BM	D	ES	[32]
N1002	25/M	rib	lung	BM	D	ES	[32]
N1003	18/M	tibia	lung	BM	D	ES	[32]
N1007	19/M	humerus	BM	BM	D	ES	[32]
N1008	33/F	chest wall	lung	PE	D	AT	[32]
N1016	16/F	rib	bone	BM	D	pPNET	[32]
SK-N-MC [bef]	14/F	rib	supraorbital	supraorbital	D	pPNET	[32]
CHP100 [g]	12/F	mediastinum	bone	mediastinum	E	pPNET	[32]
SK-PN-DW	17/M	presacrum	lung	presacrum	D	pPNET	[39]
TC-131 [h]	19/M	thorax		thorax	D	BPT	[44]
TC-174 [h]	9/F	thigh		thigh	D	BPT	[44]
TC-206 [i]	5 mos/M	upper arm		upper arm	D	BPT	[44]
TC-253 [h]	25/M	peritoneum		peritoneum	D	BPT	[44]
TTC-547 [h]	13/F	peritoneum		peritoneum	D	BPT	[44]
WE-68	19/F	fibula	lung	fibula	D	ES	[82]
RM-82	8/M	femur	tibia	femur	D	ES	[82]
NT-68	20/M	rib	lung	PE	D	ES	[88]
VH-64	24/M	metatarsus	lung	PE	D	ES	[93]
GG-62	25/F	tibia	femur	tibia	D	aES	[93]
JS-73	19/F	humerus	lung	humerus	D	pPNET	[93]
MC-77	13/F	pelvis	BM	pelvis	D	pPNET	[93]

Continued on next page

Table 1a (continued)

Cell line	Patient age(yrs)/sex	Primary site	Metastatic site	Tumor specimen site	Culture method	Tumor class	Reference
KAL	36/M	heel	skin	skin	D	pPNET	[98]
IARC-EW1 }	19/M	rib	lung	PE	D	ES	[119]
IARC-EW2 }			blood	blood	D	ES	[119]
IARC-EW3	14/M	epidural	lung	PE	D	ES	[119]
IARC-EW7	20/F	scapula	lung	PE	D	ES	[119]
IARC-EW11	11/F	rib	lymph node	lymph node	D	ES	[119]
W-ES	14/F	chest wall	lungs	PE	D	AT	[120]
SCCH-196	16/F	arm biceps	lymph node	arm biceps	D	pPNET	[121]
LAP-35	12/F	tibia		tibia	D	pPNET	[122]
NUB-20	11/M	vertebra	occipital lobe	occipital lobe	D	pPNET	[123]
STA-ET-6	8/M	fibula	BM; lung	PE	D	pPNET	[124]
STA-ET-8.1 }	18/M	pelvis	bone	pelvis	D	pPNET	[124]
STA-ET-8.2 }			lung	PE	D	pPNET	[124]
STA-ET-12	18/M	pelvis	bone; BM	pelvis	D	pPNET	[124]
KU-9	27/M	peritoneum		peritoneum	D	pPNET	[125]
TTC-466	5/F	neck	lung	lung	D	pPNET	[126]
TTC-633	8/F	ischium		ischium	D	pPNET	[126]
RD-ES bf	19/M	humerus		humerus	D	ES	[127]
JFEN	22/M	nasal cavity	chest wall	chest wall	D	ENB	[128]
TC-268	22/F	nasal cavity	paraspinum	paraspinum	D	ENB	[128]
ICB-112	24/M	knee		knee (R)	D	pPNET	[129]
ES-1	45/F	posterior thigh	lung	posterior thigh	D	ES	Look/Ragsdale (pc)

Continued on next page

Table 1a (continued)

Cell line	Patient age(yrs)/sex	Primary site	Metastatic site	Tumor specimen site	Culture method	Tumor class	Reference
ES-2	14/F	ilium	BM	BM	D	ES	Look/Ragsdale (pc)
ES-3	12/M	pubic bone		pubic bone	D	ES	Look/Ragsdale (pc)
ES-4	18/M	rib	pleural cavity	pleural cavity	D	ES	Look/Ragsdale (pc)
ES-5	16/M	rib	lung	pleura	D	ES	Look/Ragsdale (pc)
ES-6	17/M	tibia	lung, humerus	tibia	D	ES	Look/Ragsdale (pc)
ES-7	15/M	fibula	humerus, femur	BM	D	ES	Look/Ragsdale (pc)
ES-8	10/M	humerus	femur	BM	D	ES	Look/Ragsdale (pc)
PNET-1	19/M	pelvis	peritoneum	ilium	D	pPNET	Look/Ragsdale (pc)
PNET-2	15/F	chest wall		chest wall	D	AT	Look/Ragsdale (pc)
A17/95	23/F	tibia	lung	lung	D	ES	Anderer (pc)
TC-83	13/F	chest wall		chest wall	D	AT	(this paper)

ES, Ewing's sarcoma; aES, atypical Ewing's sarcoma; pPNET, peripheral primitive neuroectodermal tumor; AT, Askin tumor; BPT, biphenotypic tumor; ENB, esthesioneuroblastoma; }, cell lines derived from same patient; R, local recurrence; BM, bone marrow; PE, pleural effusion; D, dissociated tissue; E, explant; X, xenograft; [a]derived from xenografts NU15, respectively, NU22 [130]; [b]available from DSMZ [131]; [c]no longer available (Hamilton, p.c.); [d]originates from the same tumor as SAL-1 (Ambros, Hamilton, pc); [e]originally diagnosed as neuroblastoma [132]; [f]available from ATCC [127]; [g]originally diagnosed as neuroblastoma [133]; [h]originally diagnosed as alveolar rhabdomyosarcoma [44]; [i]originally diagnosed as primitive rhabdomyosarcoma [44]; n.d., not determined

Table 1b Other Ewing tumor cell lines

Cell line	Reference
IARC-EW13	[8]
IARC-EW16	[8]
IARC-EW17	[8]
N1043	[25]
N1046	[25]
N1050	[25]
T50	[61]
KP-EW-Y1	[61]
KP-EW-MS	[61]
IARC-EW24	[62]
TC-135	[95]
CB-AGPN	[95]
IOR-EW4	[96]
STSAR-33	[104]
PD-E02	[108]
PD-N12	[108]
SK-ES-1 [a]	[134]
A673 [a]	[135]
SK-N-LO	[136]
ES-12	[137]
ES-14	[137]
ES-15	[137]
STA-ET-7.1	[138]
STA-ET-7.2	[138]
SMB [b]	[138]
ES-1-OT	[139]
SCMC-ES1	[139]
SCMC-ES2	[139]
SK-PN-LI	[140]
SK-PN-DW	[140]
SK-PN-AG	[140]
SK-PN-WA	[140]

[a]available from ATCC [127]; [b]identical with cell line SIM-1 ([26]; Ambros, Hamilton, pc)

phenotype may be increased in pPNET cells and induced in Ewing's sarcoma cells once grown *in vitro* or as xenograft. Neural differentiation of tumor cells becomes even more evident with dbcAMP present in the culture medium.

3. MOLECULAR AND CYTOGENETIC FEATURES

The translocations t(11;22)(q24;q12) and t(21;22)(q22;q12) result in several different EWS/FLI-1, respectively, EWS/ERG fusion transcripts which vary in the size of the resulting polymerase chain reaction (PCR) products [36,37,57-62]. The molecular-genetic approach can be used as an

Table 2 Pathological features of Ewing tumors

Cell line	Original tumor	Xenograft	Ref
NCR-EW1: round, spindle-shaped cells; weak adhesion to substrate; negative for NFP and 5C11; cell line derived from xenograft	Small-round-cells with clear cytoplasm, round to oval nuclei; glycogen-containing cytoplasmic granules; no neurosecretory granules; cells positive for 5C11; negative for NFP and desmin	Histology similar to that of the original tumor; no neural differentiation; positive for 5C11	[12]
NCR-EW2: polygonal cells firm adhesion to substrate; positive for 5C11; negative for NFP; after dbcAMP treatment: elongated cytoplasmic processes and positive for NFP	Small-round-cells without neural features; cells positive for 5C11; negative for NFP and desmin	Histology similar to that of the original tumor; no neural differentiation; positive for 5C11	[12]
NCR-EW3: round, spindle-shaped cells; firm adhesion to substrate; positive for 5C11; negative for NFP	Small-round-cells without neural features; cells positive for 5C11; negative for NFP and desmin	Histology similar to that of the original tumor; no neural differentiation; positive for 5C11	[12]
NCR-EW4: attached growth; cells positive for 5C11 and NFP; negative for desmin	Small-round-cells without neural features; cells positive for 5C11; negative for NFP and desmin	Histology similar to that of the original tumor; no neural differentiation; positive for 5C11	[12]

Continued on next page

Table 2 (continued)

Cell line	Original tumor	Xenograft	Ref
H-15: homogeneous small, round cells with short cytoplasmic processes; population consisted of small portion of binuclear cells; positive for NSE, S100, NFP, vimentin, HBA-71; after dbcAMP treatment: neural differentiation; cell line derived from xenograft (NU-15)	Homogeneous fields with small-round-cells distributed in diffuse pattern; cells with dense nuclear chromatin and sparse cytoplasm; abundant glycogen; no Homer-Wright rosettes; few neural figures; cells positive for NSE, S100, vimentin, cytokeratin	Grafted tumor (NU-15) showed morphologic features similar to those seen in the original tumor, with an improved neural phenotype: cells strongly positive for NSE, S100, NFP	[13]
H-22: small epithelial-like cells with large nuclei and prominent nucleoli; cells positive for NSE, S100, NFP, cytokeratin; after dbcAMP treatment cells gained neurites and ramifying neuritic processes; cell line derived from xenograft (NU-22)	Undifferentiated small-round-cells; no Homer-Wright rosettes; no cytoplasmic projections; nuclei round to elongated; condensed chromatin; single nucleolus; cytoplasm with scarce organelles; poorly developed Golgi complex; no neurotubules; no secretory granules; intermediate filaments occasionally seen; cells positive for NSE and NFP; negative for S100 and cytokeratin	Grafted tumor cells (NU-22) showed neural differentiation (neurosecretory granules, cytoplasmic extensions); cells positive for NSE and NFP; negative for cytokeratin	[13]
H-825: cells grow as suspension, on FN flat adherence with short	Small-round-cells with neural differentiation; cytoplasmic extensions; neurosecretory granules; Homer-Wright rosettes; cells	Grafted tumor cells (NU-78) showed neural figures including neurotubules, neurite-like cytoplasmic extensions, neurosecretory granules;	[13]

Continued on next page

Table 2 (continued)

Cell line	Original tumor	Xenograft	Ref
cytoplasmic processes; cells positive for NSE, synaptophysin, NFP, cytokeratin; after dbcAMP treatment: cells with long processes containing microtubules and neurosecretory granules and increase of neural markers	positive for NSE, synaptophysin, NFP, cytokeratin	cells positive for NFP, synaptophysin, cytokeratin	
KU-SN: polygonally shaped cells consisting of small cells and large flat cells; cells with unbranched cytoplasmic processes; positive for NSE, S100, NFP, CEA vimentin, cytokeratin, chromogranin, type 2-collagen	Small-round-cells separated by fibrovascular stroma; round hyperchromatic nuclei with basophilic cytoplasm; presence of Homer-Wright rosettes; positive for NSE, S100, NFP, CEA. Ultrastructurally: cells with dense core granules and neurite-like cytoplasmic processes	Small-round-cells with lobular pattern separated by fibrovascular stroma; absence of Homer-Wright rosettes; cells formed small amounts of hyaline cartilage. Ultrastructurally: principal cells showing scarce organelles, dark cells showing vacuolated ER and mitochondria and associated with cartilage (No cartilage formation found in xenografts of cell lines SK-N-MC, RD-ES, and SK-ES-1)	[20]
CADO-ES-1: major portion of cell population with long neurites and microtubules; minor portion of cell population with short processes lacking microtubules	Small-round-cells, undifferentiated, dented nuclei with faint chromatin	After differentiation of cells in vitro with dbcAMP tumorigenicity was depressed; cartilage formation by tumor cells at more than 7 passages in vitro and by dbcAMP-treated cells	[21]

Continued on next page

Table 2 (continued)

Cell line	Original tumor	Xenograft	Ref
N1000: cells with neuritic processes and microtubules; presence of neurosecretory granules; glycogen positive; NSE positive; same abnormal karyotype as original tumor	Small-round-cells with neural features; NSE positive; cells with translocation t(11;22)(q24;q12)	n.d.	[32]
TC-32: cells with neuritic processes and microtubules; presence of neurosecretory granules; glycogen positive; NSE positive; same abnormal karyotype as original tumor	Small-round-cells with neural features; NSE positive; cells with translocation t(11;22)(q24;q12)	n.d.	[32]
TC-131: cells with myogenic and neural traits; positive for MYF5, NFP, desmin, actin; negative for myogenin	rhabdomyoblastic cells with primitive neurites and dense core neurosecretory granules; positive for muscle-specific desmin and actin, and NFP	n.d.	[44]
TC-174: positive for MYF5, myogenin, NFP and muscle-specific desmin and actin	rhabdomyoblastic cells with myogenic and neural figures; positive for desmin, actin and NFP	n.d.	[44]
TC-206: positive for MYF5, myogenin, NFP, desmin, actin	rhabdomyoblastic cells with primitive neurites and neurosecretory granules; positive for desmin, actin, NFP	n.d.	[44]

Continued on next page

Table 2 (continued)

Cell line	Original tumor	Xenograft	Ref
TC-253: cells positive for MYF5, NFP, desmin, actin; negative for myogenin	rhabdomyoblastic cells with myogenic and neural traits; neurosecretory granules; positive for neurofilament, muscle-specific desmin and actin	n.d.	[44]
TTC-547: positive for MYF5, myogenin, NFP and muscle-specific desmin and actin	rhabdomyoblastic cells with primitive neurites and dense core granules; positive for desmin, actin and NFP	n.d.	[44]
W-ES: coexistence of two cell types: spindle-shaped cells, and polygonal epithelial-like cells; no neurite formation; cells positive for NSE, NFP; negative for S100	Small rounded, closely packed cells; glycogen positive; no Homer-Wright rosettes; nuclei rounded to oval with finely dispersed chromatin; cells were positive for NSE, cytokeratin, negative for NFP and S100	Cells showed almost identical features as original tumor but with somewhat increased pleomorphism; cells round/polygonal; positive for NSE,CEA,vimentin cytokeratin; negative for NFP, S100, desmin; no cytoplasmic processes; no secretory granules	[120]
SSCH-196: plump, bipolar cells; positive for Leu-7 and desmin; negative for vimentin, NFP, cytokeratin, chromogranin A; cells showed the same abnormal karyotype as the original tumor	Small-round-cells showing diffuse proliferation; neither proper structure nor Homer-Wright rosette formation; cells positive for NSE; negative for Leu-7, cytokeratin, NFP, desmin. chromogranin; cells with translocation t(11;22)	Histology similar to that of the original tumor	[120]
KU-9: small-round to spindle-shaped cells; cells contain	Poorly differentiated small-round-cell tumor arranged in clusters showing a	Similar features as original tumor although Homer-Wright rosettes not apparent; cells contain glycogen,	[125]

Continued on next page

Table 2 (continued)

Cell line	Original tumor	Xenograft	Ref
glycogen and are positive for NSE; negative for NFP, cytokeratin, muscle actin, myoglobin	variety of mitotic activity; presence of Homer-Wright rosettes; cells contain glycogen and are strongly positive for NSE; negative for NFP, cytokeratin, muscle actin, myoglobin	and are faintly positive for NSE; negative for NFP, cytokeratin, muscle actin, myoglobin; ultrastructurally: cells show cytoplasmic processes and neurosecretory granules	
JFEN: neuroblastic round cells with neuritic processes; positive for NFP, vimentin, keratin, synaptophysin, chromogranin A	Small-cell-tumor; cells positive for NSE, NFP; presence of dense core neurosecretory granules	n.d.	[128]
TC-268: epithelioid cells with neuritic processes; positive for NFP, vimentin, keratin, synaptophysin, chromogranin A	Small-cell-tumor; positive for NSE, NFP; presence of dense core neurosecretory granules	n.d.	[128]
WE-68: grow as loosely packed cells; firm adhesion to collagen and fibronectin; positive for Leu-7, HBA-71, vimentin, NFP, cytokeratin, chromogranin A, NGF-R; negative for NSE	Small-round-cell tumor without neural differentiation; positive for vimentin; negative for Leu-7, NFP, cytokeratin, chromogranin A, NGF-R, NSE	n.d.	[141]

Continued on next page

Table 2 (continued)

Cell line	Original tumor	Xenograft	Ref
RM-82: grow as loosely packed cells unless grown on collagen; positive for Leu-7, TNFα, NGF-R, NFP, vimentin, cytokeratin, synaptophysin chromogranin A; negative for S100, NSE	small-round-cells without neural differentiation; positive for vimentin; negative for Leu-7, NFP, cytokeratin, NGF-R, synaptophysin, chromogranin A, S100, NSE	Tumor from cell line shows similar features as cells in vitro; frequently, the tumor was lost after more than 15 subsequent passages in nude mouse; positive for cell-surface-associated human TNFα	[141]
JS-73: cells with unbranched cytoplasmic processes; positive for NSE, S100, vimentin, glycogen; negative for synaptophysin	Small rounded, closely packed cells; presence of Homer-Wright rosettes; positive for NSE, S100, vimentin, glycogen	n.d.	[141]
VH-64: grow as loosely packed cells unless grown on collagen or fibronectin; round to oval cells without cytoplasmic extensions; positive for Leu-7, vimentin, NFP, cytokeratin, NGF-R, synaptophysin, chromogranin A; negative for S100, NSE; after dbcAMP treatment: elongated neurite-like processes; strongly positive for cell-surface-associated fibronectin	Small-round-cells without neural differentiation; positive for vimentin; negative for Leu-7, NFP, cytokeratin, NGF-R, synaptophysin, chromogranin A, S100, NSE	n.d.	[141]

Continued on next page

Table 2 (continued)

Cell line	Original tumor	Xenograft	Ref
NT-68: round to oval, spindle-shaped cells without cytoplasmic extensions; positive for Leu-7, NFP, NSE, vimentin, NGF-R	Small-round-cells with mild pleomorphism; hyperchromatic nuclei intermingled with lobulated or indented nuclei; positive for glycogen; negative for NSE, S100, NFP	n.d.	[141]
GG-62: large cells with large nuclei; biphasic cell population: round cells without cytoplasmic extensions and oval to round cells with long neurites; positive for vimentin, Leu-7, NFP, NGF-R; negative for NSE, S100, HBA-71	Large-cell tumor with prominent polymorphic, hyperchromatic nuclei and broad cytoplasm zone; cells positive for NSE, Leu-7, S100, vimentin; negative for NFP, cytokeratin, NGF-R	n.d.	[141]
MC-77: grow as loosely packed cells; firm adhesion to collagen and fibronectin; positive for Leu-7, vimentin, NFP, S100, cytokeratin, chromogranin A, NSE, NGF-R	Small-round-cells with neurosecretory granules; positive for NSE, Leu-7, vimentin; negative for S100, NFP, cytokeratin, chromogranin A, NGF-R	n.d.	[141]

Continued on next page

Table 2 (continued)

Cell line	Original tumor	Xenograft	Ref
TC-83: small, round cells without cytoplasmic extensions coexisting with larger epithelioid cells with unbranched neurite-like processes; positive for S100, NSE, HBA-71, NFP, NGF-R	Small-round-cells; positive for NSE, HBA-71 and S100; occasional Homer-Wright rosettes	n.d.	(this paper)

NFP, neurofilament protein; NGF-R, nerve growth factor receptor; NSE, neuron-specific enolase; n.d., not determined

adjuvant to surgical pathology in evaluating the grade of tumor regression and chemotherapeutic response and may provide a new staging system for Ewing tumor therapy. Additionally, nested PCR amplification of the fusion transcripts is a sensitive procedure to screen bone marrow aspirates and peripheral blood samples for detection of occult Ewing cells [62–64]. Studies have indicated that EWS/FLI-1 and EWS/ERG proteins function as aberrant transcriptional activators and are capable of transforming NIH3T3 fibroblasts [65,66]. Targets for the EWS/FlI-1 fusion protein include the EAT-2 gene [67], the c-fos serum-responsive element [68] and the stromelysin-1 and cytokeratin-15 genes [69]. Inhibition of EWS/Fli-1 gene expression using antisense technology resulted in reduction of cell proliferation and cell viability *in vitro* and loss of tumorigenicity *in vivo* [70-74]. A variant reciprocal translocation t(7;22)(p22;q12) has been detected in one case of Ewing tumor [75]. As a result of this chromosomal rearrangement, EWS is fused to another member of the ETS family of transcription factors, ETV1 [76]. Attempts to establish a continuously growing cell line from the t(7;22)-containing tumor specimen were unsuccessful. In Ewing tumor samples having a t(17;22)(q12;q12) translocation and in cell line NCR-EW3, which expresses neither EWS/FLI-1 nor EWS/ERG chimeric genes, EWS/E1A-F fusion transcripts were identified [77,78]. The ETS gene E1A-F, which encodes the adenovirus E1A enhancer-binding protein, is known to activate transcription from matrix metalloproteinase genes that are associated with tumor invasion and metastasis [79,80]. In a subset of Ewing tumors carrying a t(2;22) chromosomal translocation EWS is fused to the ETS gene FEV [81]. These observations suggest that fusion of EWS to different ETS genes may play an important role in Ewing tumor genesis, proliferation and dissemination. The cell lines containing the main genetic aberrations associated with Ewing tumor are reported in Table 3.

4. BIOLOGICAL FEATURES

Studies on hormone receptor regulation of cellular transduction mechanisms have revealed that Ewing tumor cells (e.g. WE-68; VH-64; NT-68; RM-82; SK-N-MC) display β-adrenergic, D_1 dopaminergic and prostaglandin receptors and vasoactive intestinal peptide (VIP) and calcitonin gene-related peptide (CGRP) binding sites, which are coupled to glycogen hydrolysis through a cAMP-dependent pathway [82–86]. Ewing cells also express neuropeptide Y1 and α_2-adrenergic receptors, which are linked to inhibition of cAMP formation via a pertussis toxin-sensitive pathway [87-89]. SK-N-MC cells were reported to display endothelin and P_2 nucleotide binding sites and α1-adrenergic and M_1 muscarinic receptors, which are coupled to phosphoinositol metabolism [90]. Ewing tumor cell lines express neuropeptide genes

Table 3 Main genetic aberrations in Ewing tumor cells

Genetic changes	Cell line	Reference
t(11;22)(q24;q12)	IARC-EW13	[8]
	IARC-EW16	[8]
	IARC-EW17 [a]	[8]
	NCR-EW2	[12]
	H-15	[13]
	H-22	[13]
	H-825 [b]	[13]
	KU-SN	[20]
	N1043	[25]
	N1046	[25]
	N1050	[25]
	STA-ET-1	[28]
	STA-ET-3	[28]
	IARC-EW12	[31]
	SK-ES-1	[31]
	DES1	[31]
	TC-32	[32]
	TC-71	[32]
	TC-106	[32]
	A4573	[32]
	A9423	[32]
	5838	[32]
	6647	[32]
	N1000	[32]
	N1001 [c]	[32]
	N1002	[32]
	N1003	[32]
	N1007	[32]
	N1008	[32]
	N1016	[32]
	SK-N-MC [d]	[32]
	CHP100	[32]
	SK-PN-DW [e]	[37]
	TC-268	[42]
	JFEN	[42]
	TC-131	[44]
	TC-174	[44]
	TC-206	[44]
	TC-253	[44]
	TTC-547 [f]	[44]
	KP-EW-MS	[61]
	T50	[61]
	TC-135	[95]
	IARC-EW1	[119]
	IARC-EW2	[119]
	IARC-EW7	[119]
	IARC-EW11	[119]
	W-ES	[120]
	SCCH-196	[121]
	LAP-35	[122]

Continued on next page

Table 3 (continued)

Genetic changes	Cell line	Reference
	NUB-20	[123]
	STA-ET-6	[124]
	STA-ET-8.1	[124]
	STA-ET-8.2	[124]
	STA-ET-12	[124]
	ICB-112	[129]
	SCMC-ES1	[139]
	SCMC-ES2	[139]
	ES-1-OT [g]	[139]
	RD-ES [h]	[142]
	ES-1	Look/Ragsdale (pc)
	ES-2	Look/Ragsdale (pc)
	ES-3	Look/Ragsdale (pc)
	ES-4	Look/Ragsdale (pc)
	ES-5	Look/Ragsdale (pc)
	ES-6	Look/Ragsdale (pc)
	ES-7	Look/Ragsdale (pc)
	ES-8	Look/Ragsdale (pc)
	PNET-2	Look/Ragsdale (pc)
	VH-64	Ambros (pc)
	TC-83 [i]	Anderer (pc)
	A17/95	Anderer (pc)
EWS/FLI1 fusion	TC-106	[22]
	A4573	[22]
	TC-71	[37]
	6647	[37]
	LAP-35	[37]
	SK-PN-DW	[37]
	TC-268	[42]
	JFEN	[42]
	TC-131	[44]
	TC-174	[44]
	TC-206	[44]
	TC-253	[44]
	TTC-547	[44]
	STA-ET-1	[58]
	STA-ET-2.1	[58]
	STA-ET-2.2	[58]
	STA-ET-3	[58]
	A673	[58]
	SK-ES-1	[58]
	VH-64	[58]
	WE-68	[58]
	T50	[61]
	SCCH196	[61]
	KP-EW-Y1	[61]
	KP-EW-MS	[61]
	SK-N-LO	[61]
	IARC-EW24	[62]
	TC-135	[70]

Continued on next page

Table 3 (continued)

Genetic changes	Cell line	Reference
	TC-32	[143]
	IARC-EW11	[144]
	IARC-EW17	[144]
	RD-ES	[145]
	SK-N-MC	[145]
	NT-68	(this paper)
	TC-83	(this paper)
del(22)(q12)	STA-ET-2.1	[28]
	STA-ET-2.2	[28]
	SIM-1	[28]
	KU-9	[125]
	GG-62	Ambros (pc)
	RM-82	Ambros (pc)
	WE-68	Ambros (pc)
	MC-77	Ambros (pc)
t(21;22)(q22;q12)	SK-PN-LI	[37]
	TTC-466	[126]
	TTC-633	[126]
EWS/ERG fusion	SK-PN-LI	[37]
	SCMC-ES1 [j]	[61]
	SCMC-ES2 [j]	[61]
	IARC-EW3	[62]
	TTC-466	[126]
	TTC-633	[126]
	5838 [k]	[145]
	RM-82	(this paper)
p53 mutation	IARC-EW2	[138]
	STA-ET-2.1	[138]
	STA-ET-2.2	[138]
	STA-ET-7.1	[138]
	STA-ET-7.2	[138]
	RM-82	[138]
	SK-ES-1	[138]
	RD-ES	[138]
	SMB	[138]
	ES-1-OT	[139]
	SCMC-ES1	[139]
	SCMC-ES2	[139]
	W-ES	[139]

[a]a complex t(11;22;14); [b]a complex t(9;11;22;13); [c]a complex t(11;22;17); [d]a complex t(2;11;22;21); carries an additional t(15;22)(q21;q12) according to [131]; [e]carries no t(11;22) according to [84]; [f]a complex t(11;22;12); [g]carries no EWS/FLI1 fusion transcript according to [61]; [h]carries a del(22q) according to [37]; carries two masked t(11;22)(q24;q12) according to [131]; [i]a complex t(11;22;17); [j]carries a t(11;22) according to [139]; [k]microscopically no visible changes in chromosome #21; carries a t(11;22)(q24;q12) according to [32].

Secondary chromosomal changes involve trisomy chromosome #8 (A17/95; CADO-ES-1; DES1; IARC-EW2; IARC-EW3; IARC-EW12; KU-9; LAP-35; MC-77; N1000; N1002; NU78; NCR-EW2; SCCH-196; SK-N-MC; STA-ET-1; STA-ET-2.1; STA-ET-2.2; STA-ET-6; STA-ET-8.1; STA-ET-8.2; STA-ET-12; VH-64; WE-68) and der(16)t(1;16) (A17/95; A4573; LAP-35; N1001; STA-ET-8.1; STA-ET-8.2).

encoding CGRP (e.g. IARC-EW1; IARC-EW2; IARC-EW11; SK-ES-1), enkephalin, cholecystokinin and gastrin-releasing peptide (i.e. SK-N-MC) [91,92].

A survey on cytokine production by Ewing tumor cells showed that almost all cell lines constitutively release interleukin (IL)-8, monocyte chemotactic protein (MCP)-1, transforming growth factor (TGF)β1 and vascular endothelial growth factor/vascular permeability factor (VEGF/VPF). Some cell lines produced tumor necrosis factor (TNF)α, IL-6 and IL-10, and none produced IL-12 (Table 4). The potential role of VEGF/VPF, TGFβ1, MCP-1 and IL-8 in Ewing tumor angiogenesis and metastasis is currently under investigation.

Cell lines (e.g. WE-68; VH-64; RM-82) exhibit specific high-affinity insulin-like growth factor (IGF)-I receptors that upon activation with IGF-I or IGF-II stimulate proliferation, glucose transport and glycogen synthesis [93]. Anti-CD99 antibody HBA-71 inhibits Ewing tumor cell proliferation by interfering with the action of IGF-I [94]. Expression of IGF-I mRNA has been reported in several Ewing tumor cell lines (e.g. CHP-100; SK-N-MC; TC-71; LAP-35; SK-ES-1; RD-ES; IOR-EW4) [95,96]. These studies showed that blocking the IGF-I receptor with αIR3 antibody inhibited the *in vitro* growth of Ewing tumor cells, suggesting that IGF-I might be an autocrine growth factor in this group of tumors. It should be noted here that the amount of IGF-I secreted by Ewing tumor cells *in vitro* is inadequate to support long-term cell proliferation under serum-free conditions.

Many chemotherapeutic drugs have been found to have single-agent activity in the treatment of Ewing tumor. From the results of clinical studies,

Table 4 Cytokine production by Ewing tumor cells

Cell line	TNFα	IL-6	IL-8	IL-10	IL-12	TGFβ1	VEGF	MCP-1
WE-68	<	<	+	<	<	+++	++	+++
RM-82	+	<	++	<	<	++	+++	++
GG-62	<	<	+	<	<	++	++	<
VH-64	<	+	+++	+	<	+++	+++	+++
TC-83	<	<	++	<	<	n.d.	++	+++
RD-ES	+	<	+	<	<	++	+++	+++
SK-ES-1	+	<	+++	<	<	++	++	+
SK-N-MC	<	+	++	+	<	+++	++	+++
A673	<	<	+++	<	<	+++	+++	++
STA-ET-1	<	<	+++	<	<	++	+++	+++
STA-ET-2.1	<	<	+++	<	<	+++	++	+++
STA-ET-2.2	<	<	+	<	<	++	+++	++
STA-ET-3	<	+	+	+	<	++	+++	<

+ 1-100 pg/ml; ++ 101-500 pg/ml; +++ >500 pg/ml; <below threshold of sensitivity of enzyme-linked immunosorbent assay (ELISA) with 1 pg/ml for IL-8, 2 pg/ml for IL-6, 5 pg/ml for TNFα, IL-10, VEGF and MCP-1, 10 pg/ml for IL-12 and 50 pg/ml for TGFβ1; n.d., not determined

single-agent chemosensitivities of 60% in response to cyclophosphamide, 55% for doxorubicin, 60% for dactinomycin and 20% for vincristine have been reported [97]. The results from *in vitro* chemosensitivity tests confirmed the efficacy of cyclophosphamide, dactinomycin and doxorubicin in inhibiting the proliferation of Ewing tumor cells [98]. Also, the relative ineffectiveness of vinca alkaloids as demonstrated *in vitro* [98] correlated with the low response rate observed in vivo [97].

One of the main obstacles for therapeutic success in many tumors is resistance to multiple chemotherapeutic agents through the expression of a membrane-bound channel-forming transport protein, P-glycoprotein, expressed by the MDR1 gene [99]. Clinical data indicate that P-glycoprotein expression in Ewing tumors is associated with reduced responsiveness to chemotherapy [100]. Another study, however, revealed no difference in survival between patients who showed a high incidence of elevated P-glycoprotein expression and those who displayed no immunohistochemically detectable P-glycoprotein [101].

Plasminogen activators (PAs) are serine proteases that regulate a variety of processes associated with cell motility and tissue remodeling, including tumor invasion and metastasis. PA inhibitor (PAI)-1 regulates the activity of PAs by the formation of PA/PAI-1 complexes, which lead to inactivation of PAs. Measurement of tissue-type (t)PA and urokinase-type (u)PA in a series of Ewing tumor cell lines showed that tPA prevailed, and that IFNγ attenuated tPA activity in phorbol ester-treated cells by enhancing production of PAI-1 [102].

Conventionally, radiotherapy plays a major role in obtaining local control of Ewing tumor, due to its radiosensitivity. In radiation survival studies on Ewing tumor cells established from patients with recurrent disease, it was shown that they have a substantial capacity to repair sublethal radiation damage with a single dose of less than 400 cGy; no repair of potential lethal damage was shown in cells exposed to higher radiation doses. It was argued that these cell lines (A4573; 5838) might represent a resistant subpopulation of tumor cells that have escaped radiation and chemotherapeutic treatment *in vivo* [103]. Studies have indicated that Ewing tumor cells in early culture passage are more radiosensitive than those in late culture passage [104,105]. Cell lines (5838; A4573; RD-ES; TC-106) express enhanced activity in poly(ADP-ribose) polymerase (PARP), an enzyme involved in DNA-damage repair. This excess of PARP may contribute to the intrinsic radiosensitivity of Ewing tumor cells [105]. Irradiation-induced programmed cell death (apoptosis) in A4573 cells was found to result in proteolytic cleavage of PARP and fragmentation of DNA and to be associated with accumulation of ubiquinated proteins [106].

Ewing tumor cell lines (i.e. RM-82; RD-ES; SK-ES-1; SK-N-MC; STA-ET-2.1; STA-ET-2.2) release bioactive TNFα in response to γ irradiation following a priming step with interferon (IFN)γ through a protein kinase C-independent mechanism [107]. As Ewing tumor cells express binding sites for IFNα [108], IFNγ and TNFα [109], the latter cytokine may act as an autocrine regulator of Ewing cell function. Alone or synergistically with IFN, TNFα-mediated responses in Ewing tumor cells *in vitro* include: (i) inhibition of cell proliferation and subsequent apoptosis [109]; (ii) stimulation of β1 integrin-mediated cell-matrix interaction [110]; and (iii) modulation of immuno-regulatory antigens such as intercellular adhesion molecule (ICAM)-1, ICAM-2, leukocyte function antigen (LFA)-3, CD40 and major histocompatibility complex (MHC) class-I [111–113].

MHC-restricted and MHC-unrestricted effector mechanisms, such as lymphokine-activated killer (LAK) cell activities, may be of therapeutic value in the treatment of Ewing tumor. IL-2-induced LAK-type immune effector cells were found highly cytotoxic against cell lines A4573, 6647, 5838, SK-ES-1 and SK-N-MC [114, 115]. An increased susceptibility to lysis by LAK effector cells was observed in heat-shocked Ewing tumor 5838 cells, and appeared to correlate with induction of cell surface-associated heat-shock protein HSP72 [116]. IFNγ-treated Ewing tumor cells (RD-ES; 6647; TC-71; SK-N-MC) were found to be lysed by melanoma-specific HLA-A1- and HLA-A2-restricted cytotoxic T lymphocytes (CTLs) [117,118]. The lysis of Ewing tumor cells was strictly dependent on the expression of MAGE-1 and MAGE-3 genes, which code for antigenic peptides recognized by melanoma-derived CTL in the context of HLA-A1 and HLA-A2 [118]. Therefore, the possibility that IFN not only may directly inhibit the proliferation and metastatic propensity of Ewing tumor cells but also may contribute to the induction of a MHC-restricted antitumor immune response is worth considering.

5. SUMMARY

The properties of Ewing tumor cells *in vitro*, including chemotherapeutic drug and radiation sensitivity, chromosomal aberrations and EWS/FLI1 and EWS/ERG fusion transcripts are comparable, if not identical, to those of the tumors from which they were derived. Thus, most Ewing tumor cell lines are representative in terms of clinical presentation, cytogenetics and molecular genetics. However, the undifferentiated nature of Ewing's sarcoma cells generally is not maintained *in vitro*. In culture, differentiated features emerge, overlapping those of primitive neuroectodermal tumors. This observation favors the hypothesis that Ewing's sarcoma and possibly other members of the Ewing tumor family are of neural crest origin, although direct proof is lacking. The identification and cloning of chromosomal translocations in Ewing tumors is providing insight into the molecular pathogenesis. The fusion of EWS to different ETS genes seems to be a necessary step in tumorigenesis, but the presumptive tumor suppressor genes that are the targets of these chromosomal changes are yet to be identified. Tumor progression involves a series of changes (for example, in cytokine, enzyme and adhesion molecule receptor expression) culminating in metastasis. Although the published studies on Ewing tumor cells are scarce, a number of questions have been raised. For example, what role do VEGF, MCP-1 and IL-8 play in angiogenesis? Is Ewing tumor cell-derived tissue-type plasminogen activator involved in fibrinolysis and/or extracellular matrix degradation? Does $\alpha5\beta1$ and $\alpha6\beta1$ integrin expression correlate with invasive or metastatic potential? Ewing tumor cell lines provide us with model systems to address these questions and test new therapeutic approaches.

ACKNOWLEDGMENTS

I wish to thank my colleagues for supplying me with additional information on their cell lines;

Drs Peter Ambros and Heinrich Kovar
Children's Cancer Research Institute, St.Anna Kinderspital
Kinderspitalgasse 6, A-1090 Vienna, Austria
Phone: (43) 1-404-70421. FAX: (43) 1-408-7230

Dr Ursula Anderer
Institute for Pathology, Humboldt-University
Schumannstrasse 20/21, 10117 Berlin, Germany
Phone: (49) 30-2802-2261. FAX: (49) 30-2802-3407.
E-Mail: anderer@rz.charite.hu-berlin.de

Dr Alain Aurias
INSERM U434, Laboratoire de Génétique des Tumeurs, Institut Curie,
26, rue d`Ulm, 75231 Paris Cedex 05, France
FAX: (33) 14-234-6630.

Dr Nicola Baldini
Instituti Ortopedici Rizzoli, Laboratorio di Ricerca Oncologica
Via di Barbiano 1/10, I-40136 Bologna, Italy
Phone: (39) 51-636-6760. FAX: (39) 51-636-6761

Dr Yuji Fujii
Division of Pediatrics, Iwata General Hospital
57-1 Kounodai, Iwata, Shizuoka 438, Japan
FAX: (81) 538-37-6398.

Dr Gerhard Hamilton
University Clinic of Surgery, AKH, Waehringer Guertel 18-20
A-1090 Vienna, Austria.
Phone: (43) 1-40400-6959. FAX: (43) 1-40400-6982

Dr Chieko Homma
Saitama Cancer Center, 818 Komuro, Ina,
Saitama 362, Japan
Phone (81) 48-722-1111. FAX: (81) 48-722-1739

Dr Ken Kodama
Department of Thoracic Surgery, Osaka Medical Center for Cancer & Cardiovascular Diseases
3-1-1 Nakamichi, Higashinari-ku, Osaka 537, Japan
Phone: (81) 6-972-1181. FAX: (81) 6-981-8055

Dr Gerard Lenoir
International Agency for Research on Cancer, Claude-Bernard University
150, Cours Albert-Thomas, 69372 Lyon Cedex 08, France
Phone: (33) 4-7273-8485. FAX: (33) 4-7273-8575

Drs Antonio Llombart-Bosch and Rosa Noguera
Departamento de Patologia, Universidad de Valencia, Facultad de Medicina y Odontologia
Avda. Blasco Ibanez 17, 46010 Valencia, Spain
Phone: (96) 386-4146. FAX: (96) 386-4173

Drs Thomas Look, David Reardon and Susan Ragsdale
Department of Experimental Oncology, St. Jude Children's Research Hospital
P.O.Box 318, Memphis, TN 38101, USA
Phone: (1) 901-495-3500. FAX: (1) 901-495-2032.
E-Mail: thomas.look@stjude.org

Dr Kimihiko Sano
Department of Pediatrics, Kobe University School of Medicine
7-5-1 Kusunoki-cho, Chuo-ku, Kobe 650, Japan
Phone: (81) 78-341-7451. FAX (81) 78-371-6239.
E-Mail: sanoped@kobe-u.ac.jp

Dr Timothy J. Triche
Childrens Hospital Los Angeles, 4650 Sunset Boulevard, Mail Stop 43
Los Angeles, CA 90027, USA
Phone: (1) 213-669-4516. FAX: (1) 213-667-1123.

Dr Herman Yeger
Department of Pediatric Laboratory Medicine, Division of Pathology,
Hospital for Sick Children
555 University Avenue, Toronto, Ontario, Canada M5G 1X8, USA
Phone: (1) 416-813-5967. FAX: (1) 416-813-5974

REFERENCES

1. Ewing J. *Proc NY Pathol Soc* 21: 17, 1921.
2. Friedman B, Gold H. *Cancer* 22: 307, 1968.
3. Kadin ME, Bensch KG. *Cancer* 27: 257, 1971.
4. Takahashi K, Sato T, Kojima M. *Acta Pathol Jpn* 26: 167, 1976.
5. Dickman PS, Liotta LA, Triche TJ. *Lab Invest* 47: 375, 1982.
6. Miettinen M, Lehto VP, Virtanen I. *Virchows Arch [Cell Pathol]* 41: 277, 1982.
7. Schmidt D, Mackay B, Ayala AG. *Ultrastruct Pathol* 3: 143, 1982.
8. Lipinski M, Braham K, Philip I et al. *Cancer Res* 47: 183, 1987.
9. Cavazzana AO, Miser JS, Jefferson J et al. *Am J Pathol* 127: 507, 1987.
10. van Valen F, Prior R, Wechsler W et al. In: Heuck FHW, Keck E (eds), Osteologia vol.3, p 341, Springer, Berlin, 1988.
11. Navarro S, Gonzales-Devesa M, Ferrandez-Izquierdo A et al. *Virchows Arch [Pathol Anat]* 416: 383, 1990.
12. Hara S, Adachi Y, Kaneko Y et al *Br J Cancer* 64: 1025, 1991.
13. Noguera R, Navarro S, Peydro-Olaya A et al. *Cancer* 73: 616, 1993.
14. Stout AP. *Proc NY Pathol Soc* 18: 2, 1918.
15. Dehner LP. *Am J Surg Pathol* 17: 1, 1993.
16. Dellagi K, Lipinski M, Paulin D et al. *Cancer Res* 47: 1170, 1987.
17. Moll R, Lee I, Gould VE et al. *Am J Pathol* 127: 288, 1987.
18. Siegal GP, Miller EW, Mirmiran SA et al. *Pediatr Pathol* 5: 485, 1986.
19. Scarpa S, Modesti A, Triche TJ. *Am J Pathol* 129: 74, 1987.
20. Goji J, Sano K, Nakamura H et al. *Cancer Res* 52: 4214, 1992.
21. Kodama K, Doi O, Higashiyama M et al. *Jpn J Cancer Res* 85: 335, 1994.
22. Pagani A, Fischer-Colbrie R, Eder U et al. *Int J Cancer* 63: 738, 1995.
23. Horstadius S. The Neural Crest: Its properties and derivatives in the light of experimental research, p 1, Hafner Publ. Comp., New York, 1969.
24. Thiele C, McKeon C, Triche TJ et al. *J Clin Invest* 80: 804, 1987.
25. McKeon C, Thiele CJ, Ross RA et al. *Cancer Res* 48, 4307, 1988.
26. Hamilton G, Fellinger EJ, Schratter I et al. *Cancer Res* 48: 6127, 1988.
27. Kovar H, Dworzak M, Strehl S et al. *Oncogene* 5: 1067, 1990.
28. Ambros IM, Ambros PF, Strehl S et al. *Cancer* 67: 1886, 1991.
29. Fellinger EJ, Garin-Chesa P, Triche TJ et al. *Am J Path* 139: 317, 1991.
30. Aurias A, Rimbaut C, Buffe D et al. *N Engl J Med* 309: 496, 1983.
31. Turc-Carel C, Aurias A, Mugneret F et al. *Cancer Genet Cytogenet* 32: 229, 1988.
32. Whang-Peng J, Triche TJ, Knutsen T et al. *Cancer Genet Cytogenet* 21: 185, 1986.
33. Delattre O, Zucman J, Plougastel B et al. *Nature* 359: 162, 1992.
34. Delattre O, Zucman J, Melot T et al. *N Eng J Med* 331: 294, 1994.
35. Zucman J, Delattre O, Desmaze C et al. *Genes Chromosom Cancer* 5: 271, 1992.
36. Zucman J, Melot T, Desmaze C et al. *EMBO J* 12: 4481, 1993.
37. Giovannini M, Biegel JA, Serra M et al. *J Clin Invest* 94: 489, 1994.

38. Horowitz ME, Malawer MM, Delaney TF et al. In: Pizzo PA, Poplack DG (eds), *Principles and Practice of Paediatric Oncology*, p.795, Lippincott, Philadelphia, 1993.
39. Askin FB, Rosai J, Sibley RK et al. *Cancer* 43: 2438, 1979.
40. Seemayer TA, Vekemans M, de Chadarevian JP. *Virchows Arch [Pathol Anat]* 408: 289, 1985.
41. Whang-Peng J, Freter CE, Knutsen T et al. *Cancer Genet Cytogenet* 29: 155, 1987.
42. Sorensen PHB, Wu JK, Berean KW. *Proc Natl Acad Sci USA* 93: 1038, 1996.
43. Whang-Peng J, Knutsen T, Thiel K et al. *Genes Chromosom Cancer* 5: 299, 1992.
44. Sorensen PHB, Shimada H, Liu XF et al. *Cancer Res* 55: 1385, 1995.
45. Thorner P, Squire J, Chilton-MacNeill S et al. *Am J Pathol* 148: 1125, 1996.
46. Nelson RS, Perlman EJ, Askin FB. *Human Pathol* 26: 639, 1995.
47. Swanson PE, Wick MR, Garin-Chesa P et al. *Lab Invest* 70: 11A, 1994.
48. van Valen F, Hötte A, Zardi L et al. *Med. Ped. Oncology* 23: 187, 1994.
49. Jürgens H, Donaldson SS, Göbel U. In: Voûte PA, Barrett A, Lemerle J (eds), *Cancer in Children: clinical management*, p 295, Springer, Berlin, 1992.
50. Llombart-Bosch A, Blache R, Peydro-Olaya A. *Cancer* 41: 1362, 1978.
51. Nascimento AG, Unni KK, Pritchard DJ et al. *Am J Surg Pathol* 4: 29, 1980.
52. Hou-Jensen K, Priori E, Dmochowski L. *Cancer* 29: 280, 1972.
53. Mahoney JP, Alexander RW. *Am J Surg Pathol* 2: 283, 1978.
54. Llombart-Bosch A, Peydro-Olaya A. *Virchows Arch [Pathol Anat]* 398: 329, 1983.
55. Schmidt D, Hermann C, Jürgens H et al. *Cancer* 68: 2251, 1991.
56. Enzinger FM, Weiss SW. Soft Tissue Tumors. 2nd Edition, p.1, The C.V.Mosby Co., St Louis, 1988.
57. Downing JR, Head DR, Parham DM et al. *Am J Pathol* 143: 1294, 1993.
58. Dockhorn-Dworniczak B, Schäfer KL, Dantcheva R et al. *Virchows Arch [Pathol Anat]* 425: 107, 1994.
59. Pellin A, Boix J, Blesa JR et al. *Eur J Cancer* 30A: 827, 1994.
60. Zoubek A, Pfleiderer C, Salzer-Kuntschik M et al. *Br J Cancer* 70: 908, 1994.
61. Ida K, Kobayashi S, Taki T et al. *Int J Cancer* 63: 500, 1995.
62. Peter M, Magdelenat H, Michon J et al. *Br J Cancer* 72: 96, 1995.
63. Pfleiderer C, Zoubek A, Gruber B et al. *Int J Cancer* 64: 135, 1995.
64. West DC, Grier HE, Swallow MM et al. *J Clin Oncol* 15: 583, 1997.
65. May WA, Gishizky ML, Lessnick SL et al. *Proc Natl Acad Sci USA* 90: 5752, 1993.
66. Ohno T, Ouchida M, Lee L et al. *Oncogene* 9: 3087, 1994.
67. Thompson AD, Braun BS, Arvand A et al. *Oncogene* 13: 2649, 1996.
68. Magnaghi-Jaulin L, Masutani H, Robin P et al. *Nucleic Acids Res* 24: 1052, 1996.
69. Braun BS, Frieden R, Lessnick SL et al. *Mol Cell Biol* 15: 4623, 1995.
70. Ouchida M, Ohno T, Fujimura Y et al. *Oncogene* 11: 1049, 1995.
71. Kovar H, Aryee DN, Jug G et al. *Cell Growth Differ* 7: 429, 1996.
72. Tanaka K, Iwakuma T, Harimaya K et al. *J Clin Invest* 99: 239, 1997.
73. Toretsky JA, Connell Y, Neckers L et al. *J Neuro-oncol* 31: 9, 1997.
74. Yi H, Fujimara Y, Ouchida M et al. *Oncogene* 14: 1259, 1997.
75. Squire J, Zielenska M, Thorner P et al. *Genes Chromosom Cancer* 8: 190, 1993.
76. Jeon I-S, Davis JN, Braun BS et al. *Oncogene* 10: 1229, 1995.
77. Urano F, Umezawa A, Hong W et al. *Biochem Biophys Res Commun* 219: 608, 1996.
78. Kaneko Y, Yoshida K, Handa M et al. *Genes Chromosomes Cancer* 15: 115, 1996.
79. Higashino F, Yoshida K, Fujinaga Y et al. *Nucleic Acids Res* 21: 547, 1993.
80. Higashino F, Yoshida K, Noumi T et al. *Oncogene* 10: 1461, 1995.
81. Peter M, Couturier J, Pacquement H et al. *Oncogene* 14: 1159, 1997.
82. van Valen F, Jürgens H, Winkelmann W et al. *Biochem Biophys Res Commun* 146: 685, 1987.

83. van Valen F, Keck E. *J Cancer Res Clin Oncol* 114: 266, 1988.
84. van Valen F, Keck E, Jürgens H. *FEBS Lett* 256: 170, 1989.
85. van Valen F, Jürgens H, Winkelmann W et al. *Cell Signal* 1: 435, 1989.
86. van Valen F, Piechot G, Jürgens H. *Neurosci Lett* 119: 195, 1990.
87. van Valen F, Keck E, Jürgens H. *FEBS Lett* 249: 271, 1989.
88. van Valen F, Winkelmann W, Jürgens H. *J Cancer Res Clin Oncol* 118: 529, 1992.
89. Schaak S, Cayla C, Blaise R et al. *J Pharmacol Exp Ther* 281: 983, 1997.
90. Fisher SK, Landon RE. *J Neurochem* 57: 1599, 1991.
91. Höppener JWM, Steenbergh PH, Slebos RJC et al. *J Clin Endocrinol Metab* 64: 809, 1987.
92. Verbeek MA, Draaijer M, Burbach JP. *J Biol Chem* 265: 18087, 1990.
93. van Valen F, Winkelmann W, Jürgens H. *J Cancer Res Clin Oncol* 118: 269, 1992.
94. Hamilton G, Mallinger R, Hofbauer S et al. *Thymus* 18: 33, 1991.
95. Yee D, Favoni RE, Lebovic GS et al. *J Clin Invest* 86: 1806, 1990.
96. Scotlandi K, Benini S, Sarti M et al. *Cancer Res* 56: 4570, 1996.
97. Donaldson SS. *Cancer* 55: 2184, 1985.
98. Hofbauer S, Hamilton G, Theyer G et al. *Eur J Cancer* 29A: 241, 1993.
99. Endicott JA, Ling V. *Annu Rev Biochem* 58: 137, 1989.
100. Roessner A, Ueda Y, Blasius S et al. *J Cancer Res Clin Oncol* 119: 185, 1993.
101. Hijazi YM, Axiotis CA, Navarro S et al. *Am J Clin Pathol* 102: 61, 1994.
102. van Valen F, Kentrup-Lardong V, Truckenbrod B et al. *Acta Orthop Scand* [suppl.276]: 68, 1997.
103. Kinsella TJ, Mitchell JB, McPherson S et al. *Int J Rad Oncol Biol Phys* 10: 1005, 1984.
104. Weichselbaum RR, Beckett MA, Simon MA et al. *Int J Rad Oncol Biol Phys* 15: 937, 1988.
105. Prasad SC, Thraves PJ, Bhatia KG et al. *Cancer Res* 50: 38, 1990.
106. Soldatenkov VA, Dritschilo A. *Cancer Res* 57: 3881, 1997.
107. van Valen F, Kentrup-Lardong V, Truckenbrod B et al. *J Cancer Res Clin Oncol* 123: 245, 1997.
108. Rosolen A, Todesco A, Colamonici OR et al. *Mod Pathol* 10: 55, 1997.
109. van Valen F, Winkelmann W, Burdach S et al. *J Cancer Res Clin Oncol* 119: 615, 1993.
110. van Valen F, Hanenberg H, Jürgens H. *Eur J Cancer* 30A: 2119, 1994.
111. van Valen F, Hanenberg H, Fried C et al. *Med Ped Oncology* 19: 342, 1991.
112. van Valen F, Winkelmann W, Jürgens H. *Eur J Cell Biology* 60 [Suppl.37]: 21, 1993.
113. van Valen F, Truckenbrod B, Kentrup-Lardong V et al. *Med Ped Oncol* 29: 327, 1997.
114. Atzpodien J, Gulati SC, Shimazaki C et al. *Oncology* 45: 437, 1988.
115. Chin T, Toy C, Vandeven C et al. *Pediatr Res* 25: 156, 1989.
116. Multhoff G, Botzler C, Wiesnet M et al. *Blood* 86: 1374, 1995.
117. Shamamian P, Mancini M, Kawakami Y et al. *Cancer Immunol Immunother* 39: 73, 1994.
118. Ishida H, Matsumura T, Salgaller ML et al. *Int J Cancer* 69: 375, 1996.
119. Turc-Carel C, Philip I, Berger MP et al. *Cancer Genet Cytogenet* 12: 1, 1984.
120. Fujii Y, Hongo T, Nakagawa Y et al. *Cancer* 64: 43, 1989.
121. Homma C, Kaneko Y, Sekine K et al. *Jpn J Cancer Res* 80: 861, 1989.
122. Bagnara GP, Serra M, Giovannini M et al. *Int J Cell Cloning* 8: 409, 1990.
123. Yeger H, Mor O, Pawlin G et al. *Cancer Res* 50: 2794, 1990.
124. Hattinger CM, Rumpler S, Ambros IM et al. *Genes Chromosom Cancer* 17: 141, 1996.
125. Nakashima J, Horiguchi Y, Ueno M et al. *Jpn J Cancer Res* 86: 1172, 1995.
126. Sorensen PHB, Lessnick SL, Lopez-Terrada D et al. *Nature* Gen 6: 146, 1994.
127. Hay R, Caputo J, Chen TR et al. (eds) American Tissue Type Culture Collection. Catalogue, 7th Edition, Rockville MD, USA, 1992.
128. Cavazzana AO, Navarro S, Noguera R et al. *Adv Neuroblastoma Res* 2: 463, 1988.

129. O'Regan S, Diebler M-F, Meunier F-M et al. *J Neurochem* 64: 69, 1995.
130. Llombart-Bosch A, Carda C, Peydro-Olaya A et al. *Cancer* 66: 2589, 1990.
131. Drexler HG, Dirks W, MacLeod RAF et al. (eds) DSM Catalogue of Human and Animal Cell Lines, 5th Edition, Braunschweig, 1995.
132. Biedler JL, Helson L, Spengler B. *Cancer Res* 33: 2643, 1973.
133. Schlesinger HR, Gerson JM, Moorhead PS et al. *Cancer Res* 36: 3094, 1976.
134. Bloom ET. Cancer Res 32: 960, 1972.
135. Giard DJ, Aaronson SA, Todaro GJ et al. *J Natl Cancer Inst* 51: 1417, 1973.
136. Reynolds CP, Tomayko MM, Donner L et al. *Adv Neuroblastoma Res* 2: 291, 1988.
137. Roberts WM, Douglass EC, Peiper SC et al. *Cancer Res* 49: 5407, 1989.
138. Kovar H, Auinger A, Jug G et al. *Oncogene* 8: 2683, 1993.
139. Komuro H, Hayashi Y, Kawamura M et al. *Cancer Res* 53: 5284, 1993.
140. Potluri VR, Gilbert F, Helson C et al. *Cancer Genet Cytogenet* 24: 75, 1987.
141. Hoffmann Ch. In: Thesis, Heinrich-Heine-University of Düsseldorf, Germany, 1993.
142. Taylor C, Patel K, Jones T et al. *Br J Cancer* 67: 128, 1993.
143. May WA, Lessnick SL, Braun BS et al. *Mol Cell Biol* 13: 7393, 1993.
144. Bonin G, Scamps C, Turc-Carel C et al. *Cancer Res* 53: 3655, 1993.
145. Dunn T, Praissman L, Hagag N et al. *Cancer Genet Cytogenet* 76: 19, 1994.

Chapter 4

Mesothelioma

Marjan A. Versnel
Dept. of Immunology, Erasmus University Rotterdam, PO Box 1738, 3000 DR Rotterdam, The Netherlands. Tel 0031-10-408-8086. Fax 0031-10-436-7601. E-mail versnel @ immu.fgg.eur.nl

Human malignant mesothelioma is a tumor originating from mesodermally derived tissues of the coelomic cavities. Most malignant mesotheliomas are found in the pleura, and less frequently in the peritoneum, the pericardium and the tunica vaginalis testis. Establishment of the diagnosis malignant mesothelioma is difficult, because it is difficult to differentiate from reactive mesothelial cells and adenocarcinoma cells.

Most patients with malignant mesothelioma die within one year of diagnosis. Although a variety of different therapeutic approaches have been explored, the results are poor.

The development of malignant mesothelioma is associated with asbestos exposure. Wagner et al. described a series of malignant mesotheliomas in South African asbestos miners in 1960 (1). Thereafter a number of reports confirmed the association between asbestos exposure in asbestos mines or industries using asbestos and the occurrence of malignant mesotheliomas. Asbestos has been used in many industries, including ship building, the aircraft industry, construction work and arms manufacture. Although asbestos usage has been greatly reduced in recent years, the incidence of malignant mesotheliomas is still increasing. This delay is due to the long latency period of 15 to over 40 years between exposure to asbestos and clinical detection. Furthermore, the millions of tons of asbestos incorporated into buildings will continue to cause asbestos exposure in the future. Some investigators expect a rise in asbestos related disease until 2010, while others expect a significant decrease in the incidence of malignant mesothelioma before 2000. As malignant mesothelioma is a relatively rare tumor, the establishment of cell lines from this malignancy is important for studies on the mechanism of transformation.

J.R.W. Masters and B. Palsson (eds.), Human Cell Culture Vol. 1, 87–105.

1. CELL CULTURE

The differential diagnosis between malignant mesothelioma and adenocarcinoma and reactive mesothelial cells is difficult, and a definitive diagnosis is important. Therefore, we have used material diagnosed as malignant mesothelioma by at least two different techniques. Most samples collected for the establishment of cell lines in Rotterdam were pleural effusions, and therefore the diagnosis was based on cytology and confirmed by electron microscopy or histology on biopsies or autopsy material (2).

Immediate transportation to the laboratory and processing of the samples is a prerequisite for successful isolation of a cell line (2,3).

The method used for the establishment of malignant mesothelioma cell lines from pleural effusions is similar to that described (2). Briefly, effusions were collected and after centrifugation the cells were seeded in culture medium. We collected the material in tubes with heparin and always changed the medium after overnight incubation in order to eliminate erythrocytes and floating cells. Modifications used by others include Ficoll gradient centrifugation and lysis with buffered ammonium chloride if large numbers of erythrocytes are present (4).

For the establishment of cell lines from solid tissues, the material is minced with scissors into small pieces and in most studies placed directly into culture. We incubate the tissues overnight in collagenase (8 mg/ml) (2).

The medium most frequently used is RPMI 1640 with 5 or 10% fetal calf serum. The cell lines established in Rotterdam were all grown in Ham's F10 medium, which was originally developed for adherent cells. Pass and coworkers (4) have compared various media with different supplements, including epidermal growth factor (EGF), HEPES, triiodothyronine, and ethanolamine. However, the results were unpredictable and therefore the use of multiple media, harvesting methods and cloning techniques was advised (4). EGF and hydrocortisone, or EGF alone as culture supplements, increased the success rate for the isolation of a cell line. However, the effect of these supplements on individual samples again was unpredictable (2). The success rate for the establishment of malignant mesothelioma cell lines ranges from 28 to 50 % (2,3,4).

2. ESTABLISHED CELL LINES

Table 1 lists established malignant mesothelioma cell lines that are well-characterized. Due to the relationship with asbestos exposure and the long latency period the majority of patients are elderly men with a pleural mesothelioma. Therefore, the established cell lines are representative of this patient group. However, as the latency period is so long, it is questionable

Table 1 Overview of established malignant mesothelioma cell lines reported in the literature*

Cell line	Patient age/sex	Asbestos exposure	Primary site	Specimen/ site	Culture method	Reference
JMN	57/M	–	Pe	E/A	S	5
SMC-1	43/M	ND	P	B/P	D	6
H-MESO-1	35/M	+	P	T/P	D[1]	7
DND	59/M	+	?	E/A	D	8
MS-1	ND	ND	ND	E/P	S	9
MS-2	ND	ND	ND	E/P	S	9
Mero-14	48/M	+	P	E/P	S	2
Mero-25	78/M	+	P	A/P	D	2
Mero-41	61/M	++	P	E/P	S	2
Mero-48a**	62/M	++	P	A/P	D	2
Mero-48b**				A/omentum	D	2
Mero-48c**				A/liver	D	2
Mero-48d**				A/Pc	D	2
Mero-72	66/M	+	P	E/P	S	2
Mero-82	63/F	+	P	E/P	S[2]	2
Mero-83	52/M	++	P	E/P	S	2
Mero-84	70/M	+	P	E/P	S	2
Mero-95	57/M	++	P	E/P	S	2
Mero-96	46/M	+	P	B/P	D	2
Mero-105	67/M	++	P	E/P	S	10
Mero-134	57/M	ND	P	E/P	S	this paper
?	59/M	?	P	E/P	S	11
M9K	ND/M	+	P	T/P	D	3
M10K	ND/M	+	P	T/P	D	3
M14K	ND/M	+	P	T/P	D	3
M14	ND/M	+	P	T/P	D	3
M20	ND/M	+	P	T/P	D	3
M33K	ND/M	+	P	T/P	D	3
M38K	ND/M	+	P	T/P	D	3
M22K	ND/M	ND	P	T/P	D	3
M19	ND/M	ND	P	T/P	D	3
NO36	ND/M	+	P	E/P	S	12
JU77	ND/M	+	P	E/P	S	12
LO68	ND/M	+	P	E/P	S	12
ONE58	ND/M	+	P	E/P	S	12
DeH128	ND/M	+	P	E/P	S	12
Ma-18	43/M	ND	ND	E/P	S	13
ZL5	43/M	+	P	E/P	S	14
ZL34	53/M	+	P	T/P	D	14
ZL55	52/M	+	P	T/P	D	14
SPC78	61/M	+	P	T/P	D	14
ZL92	60/M	+	P	E/P	S	14
SPC111	55/M	+	P	E/P	S	14
SPC212	47/F	+	P	T/P	D	14
HMCL1	65/M	2	ND	E/Pe	S	15
HMCL2	87/F	+	ND	E/P	S	15
HMCL3	59/M	–	ND	B/P	D	15
HMCL4	71/M	+	ND	B/P	D	15
HMCL5	79/M	+	ND	B/P	D	15

Continued on next page

Table 1 (continued)

Cell line	Patient age/sex	Asbestos exposure	Primary site	Specimen/ site	Culture method	Reference
HMCL6	65/M	ND	ND	E/P	S	15
HMCL7	78/M	+	ND	B/P	D	15
HMCL8*	47/M	–	ND	E/Pc	S	15
HMCL9*				E/Pc	S	15
HMCL10*				E/Pe	S	15
HMCL11*				E/Pe	S	15
HMCL12	68/M	–	ND	B/P	D	15
HMCL13	68/M	–	ND	B/P	D	15
HMCL14	66/M	+	ND	E/Pc	S	15
HMCL15	68/M	–	ND	E/P	S	15
HMCL16	65/F	+	ND	E/P	S	15
HMCL17	60/M	ND	ND	B/P	D	15
HMCL18	65/M	+	ND	B/P	D	15
HMCL19*	71/M	+	ND	B/P	D	15
HMCL20*				E/P	S	15
HMCL21**				E/P	S	15
HMCL22	66/M	+	ND	B/P	D	15
HMCL23	64/F	–	ND	E/P	S	15
HMCL24**	49/M	–	ND	B/P	D	15
HMCL25**				B/P	D	15
HMCL26	70/M	+	ND	E/P	S	15
HMCL27**	69/M	+	ND	B/P	D	15
HMCL28**				B/P	D	15
HMCL29**				B/P	D	15
HMCL30	65/M	3	ND	B/P	D	15
H2596	ND	ND	P	B/P	D	4
H2591	ND	ND	P	B/P	D	4
H2461	ND	ND	P	B/P	D	4
HP-1	ND	ND	P	E/P	S	4
HP-2	ND	ND	P	E/P	S	4
H2373	ND	ND	P	B/P	D	4
H2452	ND	ND	P	B/P	D	4
HP-3	ND	ND	P	E/P	S	4
H2593	ND	ND	P	B/P	D	4
T85	55/M	ND	Pe	A/Pe	S	16

* well-characterized cell lines
** derived from a single patient
1 isolated from a human tumor after s.c. transplantation in nude mice
2 in soft agar
3 possible
ND no data available

P, pleura; Pe, peritoneum; Pc, pericardium; A, autopsy; B, biopsy; E, effusion; T, tumor tissue; D, dissociation; S, Seeding

whether someone with a malignant mesothelioma has had no asbestos contact, and some pathologists believe that all malignant mesothelioma cases have had asbestos exposure. Recently a new TNM staging for malignant mesothelioma was proposed by members of the International Mesothelioma Interest Group (17), but no data on clinical staging were available for the malignant mesothelioma patients from which the cell lines listed in Table 1 were established.

3. PATHOLOGICAL FEATURES

Histologically, three subtypes of malignant mesothelioma are distinguished: epithelial, fibromatous and biphasic. The epithelial type mainly has an epithelial morphology. In fibromatous or sarcomatous mesotheliomas the predominant cell type is fibroblastoid. Biphasic or mixed mesotheliomas contain areas with both epithelial and fibroblastoid features. In the literature the frequencies of the occurrence of the various subtypes differ considerably. This variability might be due to the small sample sizes (18). Correlations between prognosis and the different histological subtypes were found in several studies (19,20,21). However, others found no evidence for such a relationship (22,23,24). The morphology of the malignant mesothelioma cell lines listed in Table 2 covers the various subtypes. The majority of the cell lines are either epithelial or fibrous, but some cell lines exhibit a mixture of these morphologies, reflecting the biphasic subtype. Interestingly, some cell lines with a fibrous morphology are derived from tumors with an epithelial morphology e.g. Mero-41 (2). This observation might favor the hypothesis of the 'multipotential subserosal precursor', which can differentiate to cells with an epithelial or fibroblastoid morphology (25). However, it cannot be excluded that the limited size of the specimen used for diagnosis only exhibited the epithelial features of a biphasic tumor or that the morphology is determined by the culture conditions.

For the characterization of malignant mesothelioma cell lines several markers have been used. However, there is no single marker specific for malignant mesothelioma cells which is not also found in reactive mesothelial cells and adenocarcinoma cells. Coexpression of vimentin and keratin is useful for differentiation between mesothelioma and adenocarcinoma cells, but cannot differentiate between normal and malignant mesothelial cells. Furthermore, vimentin expression can be induced in culture. EMA (epithelial membrane antigen) is a marker for malignant and normal epithelial cells and is frequently expressed by malignant mesothelioma cells. However, normal mesothelial cells in pleural effusions can also exhibit a faint EMA expression. EMA was expressed in 50 and 30% of the malignant mesothelioma cell lines (15, 26). The most consistent marker is ME1, which is expressed by normal and malignant mesothelial cells (26). In all the cell lines investigated by Zeng et al. (15) ME1 expression was found, irrespective of the morphological phenotype of the cell line. We found ME1 expression in most malignant mesothelioma cell lines with an epithelial phenotype, while the fibrous cell lines were negative for ME1 (27). Markers specific for adenocarcinoma cells like CEA (carcino embryonic antigen), BerEp4 and B72.3 are not expressed (4,15).

Only a limited number of cell lines have been cytogenetically characterized (2,3) (Table 2). In addition to the studies on the genetics of mesothelioma (see next paragraph), frequent cytogenetic characterization of cell lines can be helpful to detect cross-contamination between cell lines.

Xenograft experiments were performed on a limited set of the established malignant mesothelioma cell lines. In our panel of malignant mesothelioma cell lines the incidence of tumors following intraperitoneal or subcutaneaous inoculation was extremely low (2). Higher tumorigenicity was found in the cell lines from Pelin-Enlund et al. (3). However, the cell lines from Pelin-Enlund et al. (3) were mostly established from tumor tissue, while most of the cell lines we tested were derived from pleural effusions.

In conclusion, the best marker to establish the mesothelial nature of a cell line is ME1, although cell lines with a fibrous morphology can be ME1 negative. Further studies using markers for adenocarcinoma cells may help. Discrimination between normal mesothelial cells and malignant mesothelioma cells in vitro is possible by the limited lifespan and the normal karyotype of normal mesothelial cells.

4. MOLECULAR AND GENETIC CHARACTERISTICS

The cytogenetic characteristics of malignant mesothelioma are complex and heterogeneous (10, 18, 28). Malignant mesothelioma cell lines with a complete cytogenetic characterization are useful tools for genetic studies (Table 2). Chromosomal regions frequently deleted and known to contain tumor suppressor genes were studied in more detail. The p16 gene mapped to chromosome 9p21 was deleted in 85% of the malignant mesothelioma cell lines investigated by Cheng et al. (29). However, these cell lines have not been characterised and therefore these data are not included in Table 3. In another study and in our own panel of malignant mesothelioma cell lines all lines investigated had a deletion of p16 (see Table 3) (30,31). The p16 gene encodes a cyclin dependent kinase inhibitor. Interestingly, the incidence of homozygous p16 deletion in fresh tumor tissue from malignant mesothelioma patients is lower than the incidence in cell lines.

Another tumor suppressor gene altered in malignant mesothelioma cell lines is the NF2 (Neurofibromatosis type 2) gene. This gene encodes a protein

Table 2 Pathological features of malignant mesothelioma cell lines

Cell line	Pathology original tumor	Morphology cell line detected markers	Immunohistochemical characterization	Cytogenetic characterization	Xenograft
JMN	B	spindle-shaped cells with cytoplasmic processes fine cytoplasmic vacuoles EM microvilli	ND	Y[1]	mesenchymal mesothelioma
SMC-1	B	spindle-like, polygon-like cells with cytoplasmic projections	ND	Y[1]	resemblance to resected specimen from patient
H-MESO-1	E	the morphology of the transplanted cell line is described in detail(7)	ND	Y[1]	serially transplantable, can grow as a biphasic solid and ascitic tumor
DND	E	ND	ND	Y	ND
MS-1		spindle-shaped cells with microvilli	keratin and vimentin	Y	i.p., cytokeratin+ and vimentin+
MS-2		spindle-shaped cells with microvilli	keratin and vimentin		-
Mero-14	E	fibrous morphology, spindle-shaped cells with polymorphic and hyperchromatic nuclei, no villi	keratin 18,8,9,19 vimentin EMA^-, $PDGF^+$, $PDGF\beta R^+$, $PDGF\alpha R^+$	Y	-
Mero-25	E	epithelial morphology, polymorphic nuclei with coarse chromatin, villi and some tight junctions	keratin 18,8,7,19 vimentin EMA^+ (4%), $ME1^+$,	Y	s.c., histological features similar to original tumor

Continued on next page

Table 2 (continued)

Cell line	Pathology original tumor	Morphology cell line detected markers	Immunohistochemical characterization	Cytogenetic characterization	Xenograft
			$PDGF^{+}$, $PDGF\alpha R^{+}$ (66%), $PDGF\beta R^{-}$		
Mero-41	E[2]	epithelial like cells with polymorphic nuclei	keratin18,8,7,19 vimentin EMA^{+} (69%), $ME1^{+}$, $PDGF^{+}$, $PDGF\alpha$+, $PDGF\beta R^{+}$	Y	- i.p. and s.c unsuccessful
Mero-48a	B	epithelial morphology (fibrous morphology)[3]	keratin 18^{+}, $vimentin^{+}$,EMA^{-} $ME1^{+}$, $PDGF^{+}$, $PDGF\alpha R^{+}$ (54%), $PDGF\beta R^{-}$	Y	- i.p. and s.c inoculation was unsuccessful
Mero-48b	E	fibrous (epithelial)3	keratin 18^{+}, $vimentin^{+}$, EMA^{+}, $ME1^{+}$, $PDGF^{+}$, $PDGF\alpha R^{+}$, $PDGF\beta R^{+}$	Y	ND
Mero-48c	B	fibrous (epithelial)	keratin 18^{+}, vimentin+, EMA^{+} (77%), $ME1^{+}$, $PDGF^{+}$, $PDGF\alpha R^{+}$, $PDGF\beta R^{+}$	Y	ND
Mero-48d	B	fibrous	keratin 18^{+}, $vimentin^{+}$	Y	ND
Mero-72	B	fibrous	keratin 18^{+}, $vimentin^{+}$, EMA^{+}, $ME1^{+}$, $PDGF^{+}$, $PDGF\alpha R^{+}$, $PDGF\beta R^{+}$	Y	ND

Continued on next page

Table 2 (continued)

Cell line	Pathology original tumor	Morphology cell line detected markers	Immunohistochemical characterization	Cytogenetic characterization	Xenograft
Mero-82	F	fibrous (epithelial)	keratin 18^+, vimentin$^+$, EMA$^+$, ME1$^+$ (4%), PDGF$^+$, PDGFαR$^+$, PDGFβR$^+$ (4%)	Y	ND
Mero-83	B	fibrous	keratin 18^+, vimentin$^+$, EMA$^+$(3%), ME1$^+$, PDGF$^+$, PDGFαR$^+$ (68%), PDGFβR$^+$ (3%)	Y	ND
Mero-84	E	fibrous (epithelial)	keratin 18^+, vimentin$^+$, EMA$^+$, ME1$^+$, PDGF$^+$, PDGFαR$^+$, PDGFβR$^+$	Y	ND
Mero-95	E[1]	fibrous (epithelial)	keratin 18^+, vimentin$^+$, EMA$^+$ (4%), ME1$^+$, PDGF$^+$, PDGFαR$^+$ (69%), PDGFβR$^+$ (1%)	Y	ND
Mero-96	F	biphasic: epithelial and fibrous cells	keratin 18^+ (6%), vimentin$^+$, EMA$^+$ (4%), ME1$^+$, PDGF$^+$, PDGFαR$^+$, PDGFβR$^+$	Y	ND
Mero-123	F	fibrous	keratin 18^+, vimentin$^+$, EMA$^-$, ME1$^-$, PDGF$^+$, PDGFαR$^+$, PDGFβR$^+$	Y	ND
Mero-134	E[1]	ND	keratin 18^+, ME1$^-$, EMA$^-$, PDGF$^+$, PDGFαR$^+$, PDGFβR$^-$	Y	ND

Continued on next page

Table 2 (continued)

Cell line	Pathology original tumor	Morphology cell line detected markers	Immunohistochemical characterization	Cytogenetic characterization	Xenograft
M9K	B	multinucleation and microvilli	keratin+	Y	3/5 mice, transplantable
M10K	B	multinucleation and microvilli	keratin+	Y	2/5 mice, transplantable
M14K	E	multinucleation and microvilli	keratin+	Y	2/5 mice, transplantable
M14	E	multinucleation and microvilli	keratin+	Y	
M20	E	multinucleation and microvilli	keratin+	Y	2/8 mice, transplantable
M33K	B	multinucleation and microvilli	keratin+	Y	2/5 mice, transplantable
M38K	B	multinucleation and microvilli	keratin+	Y	ND
M22K	B	multinucleation and microvilli	keratin+	Y	ND
M19	B	multinucleation and microvilli	keratin+	Y	ND

Continued on next page

Table 2 (continued)

Cell line	Pathology original tumor	Morphology cell line detected markers	Immunohistochemical characterization	Cytogenetic characterization	Xenograft
NO36	E^2	stellate-shaped cells with vacuoles, microvilli	EMA^+, $cytokeratin^+$	Y	-
JU77	E^2	spindle-shapes with few vacuoles, microvilli	EMA^+, $cytokeratin^+$	Y	-
LO68	E^2	spindle-shapes with few vacuoles, microvilli	EMA^+, $cytokeratin^+$	Y	-
ONE58	E^2	spindle-shapes with few vacuoles, microvilli	EMA^+, $cytokeratin^+$	Y	+
DeH128	E^2	pleomorphic with numerous vacuoles, microvilli	EMA^+, $cytokeratin^+$	Y	+
ZL5	E^2	adherent and floating cells	cytokeratin 8^+, 18+, $ME1^+$ CEA^-	ND	ND
ZL34	F	monolayer with contact inhibition, cells with thin cytoplasmic processes	cytokeratin 8^+, 18^+, $ME1^+$ CEA^-	ND	ND
ZL55	E	contact inhibition	cytokeratin 8^+, 18^+, $ME1^+$ CEA^-	ND	ND
SPC78	F	adherent and floating cells	cytokeratin 8^+, 18^+, $ME1^+$ CEA^-	ND	ND

Continued on next page

Table 2 (continued)

Cell line	Pathology original tumor	Morphology cell line detected markers	Immunohistochemical characterization	Cytogenetic characterization	Xenograft
ZL92	ND	small fibroblast-like cells	cytokeratin 8^+, 18^+, $ME1^+$ CEA-	ND	ND
SPC111	B	compact cells with a tendency to form 3 dimensional structures	cytokeratin 8^+, 18^+, $ME1^+$ CEA^-	ND	ND
SPC212	B	monolayer with contact inhibition, cells made thin cytoplasmic processes	cytokeratin 8^+, 18^+, $ME1^+$ CEA^-	ND	ND
HMCL1	E	E	$cytokeratin^+$, $vimentin^+$, CEA^- $LeuM1^-$· Factor VIII^-	ND	ND
HMCL2	E	E	$cytokeratin^+$, $vimentin^+$, CEA^- $LeuM1^-$, Factor VIII^-	ND	ND
HMCL3	E	E	$cytokeratin^+$, $vimentin^+$, CEA^- $LeuM1^-$, Factor VIII^-	ND	ND
HMCL4	E	E	$cytokeratin^+$, $vimentin^+$, CEA^- $LeuM1^-$, Factor VIII^-	ND	ND
HMCL5	E	E	$cytokeratin^+$, $vimentin^+$, CEA^- $LeuM1^-$, Factor VIII^-	ND	ND
HMCL6	E	E	$cytokeratin^+$, $vimentin^+$, CEA^- $LeuM1^-$, Factor VIII^-	ND	ND

Continued on next page

Table 2 (continued)

Cell line	Pathology original tumor	Morphology cell line detected markers	Immunohistochemical characterization	Cytogenetic characterization	Xenograft
HMCL7	E	B	cytokeratin+, vimentin+, CEA- LeuM1-, Factor VIII-	ND	ND
HMCL8	B	F	cytokeratin$^+$, vimentin$^+$, CEA$^-$ LeuM1$^-$, Factor VIII$^-$	ND	ND
HMCL9		B	cytokeratin$^+$, vimentin$^+$, CEA$^-$ LeuM1$^-$, Factor VIII-	ND	ND
HMCL10		E	cytokeratin$^+$, vimentin$^+$, CEA$^-$ LeuM1$^-$, Factor VIII$^-$	ND	ND
HMCL11		E	cytokeratin$^+$, vimentin$^+$, CEA$^-$ LeuM1$^-$, Factor VIII$^-$	ND	ND
HMCL12	E	E	cytokeratin$^+$, vimentin$^+$, CEA$^-$ LeuM1$^-$, Factor VIII$^-$	ND	ND
HMCL13		E	cytokeratin$^+$, vimentin$^+$, CEA$^-$ LeuM1$^-$, Factor VIII$^-$	ND	ND
HMCL14	E	E	cytokeratin$^+$, vimentin$^+$, CEA$^-$ LeuM1$^-$, Factor VIII$^-$	ND	ND
HMCL15	E	E	cytokeratin$^+$, vimentin$^+$, CEA$^-$ LeuM1$^-$, Factor VIII$^-$	ND	ND
HMCL16	E	E	cytokeratin$^+$, vimentin$^+$, CEA$^-$ LeuM1$^-$, Factor VIII	ND	ND

Continued on next page

Table 2 (continued)

Cell line	Pathology original tumor	Morphology cell line detected markers	Immunohistochemical characterization	Cytogenetic characterization	Xenograft
HMCL17	E	E	cytokeratin+, vimentin+, CEA-LeuM1-, Factor VIII-	ND	ND
HMCL18	E	F	cytokeratin+, vimentin+, CEA-LeuM1-, Factor VIII-	ND	ND
HMCL19	E	E	cytokeratin+, vimentin+, CEA-LeuM1-, Factor VIII-	ND	ND
HMCL20		E	cytokeratin+, vimentin+, CEA-LeuM1-, Factor VIII-	ND	ND
HMCL21		E	cytokeratin+, vimentin+, CEA-LeuM1-, Factor VIII-	ND	ND
HMCL22	E	E	cytokeratin+; vimentin+, CEA-LeuM1-, Factor VIII-	ND	ND
HMCL23	E	F	cytokeratin+, vimentin+, CEA-LeuM1-, Factor VIII-	ND	ND
HMCL24	E	E	cytokeratin+, vimentin+, CEA-LeuM1-, Factor VIII-	ND	ND
HMCL25		E	cytokeratin+, vimentin+, CEA-LeuM1-, Factor VIII-	ND	ND
HMCL26	E	E	cytokeratin+, vimentin+, CEA-LeuM1-, Factor VIII-	ND	ND

Continued on next page

Table 2 (continued)

Cell line	Pathology original tumor	Morphology cell line detected markers	Immunohistochemical characterization	Cytogenetic characterization	Xenograft
HMCL27	E	E	cytokeratin$^+$, vimentin$^+$, CEA$^-$ LeuM1$^-$, Factor VIII$^-$	ND	ND
HMCL28		E	cytokeratin$^+$, vimentin$^+$, CEA$^-$ LeuM1$^-$, Factor VIII$^-$	ND	ND
HMCL29		E	cytokeratin$^+$, vimentin$^+$, CEA$^-$ LeuM1$^-$, Factor VIII$^-$	ND	ND
HMCL30	E	E	cytokeratin$^+$, vimentin$^+$, CEA$^-$ LeuM1$^-$ Factor VIII$^-$	ND	ND
H2592	F	EM: villi, filaments, glycogen	keratin$^+$ B72.3$^-$, CEA$^-$ LeuM1$^-$, BerEp4$^-$	Y	ND
H2591	E	EM: villi, filaments, glycogen	keratin$^+$ B72.3$^-$, CEA$^-$ LeuM1$^-$, BerEp4$^-$	Y	ND
H2461	E	EM: villi, filaments, glycogen	keratin$^+$ B72.3$^-$, CEA$^-$ LeuM1$^-$, BerEp4$^-$	Y	ND
HP-1	B	EM: villi, filaments, glycogen	keratin$^+$ B72.3$^-$, CEA$^-$ LeuM1$^+$, BerEp4$^-$	Y	ND

Continued on next page

Table 2 (continued)

Cell line	Pathology original tumor	Morphology cell line detected markers	Immunohistochemical characterization	Cytogenetic characterization	Xenograft
HP-2	E	EM: villi, filaments, glycogen	keratin$^+$ B72.3$^-$, CEA$^-$ LeuM1$^-$, BerEp4$^-$	Y	ND
H2373	F	EM: villi, filaments, glycogen	keratin$^+$ B72.3$^-$, CEA$^-$ LeuM1$^-$, BerEp4$^-$	Y	ND
H2452	B	EM: villi, filaments, glycogen	keratin$^+$ B72.3$^-$, CEA$^-$ LeuM1$^-$, BerEp4$^-$	Y	ND
HP-3	E	EM: villi, filaments, glycogen	keratin$^+$ B72.3$^-$, CEA$^-$ LeuM1$^-$, BerEp4$^-$	Y	ND
H2595	E	EM: villi, filaments, glycogen	keratin$^+$ B72.3$^-$, CEA$^-$ LeuM1$^-$, BerEp4$^-$	Y	ND
T85	E	ND	cytokeratin$^+$, collagen type IV$^+$, laminin$^+$, vimentin$^+$, fibronectin$^+$	ND	ND

Y: Yes; 1: limited; 2: by cytology only; 3: minority of the cells; ND: no data available

Table 3 Modified tumor suppressor genes in malignant mesothelioma cell lines

Genetic changes	Cell line	Reference
p16 deletion	M9K	30
	M24K	30
	M14M	30
	M19	30
	M33K	30
	Mero-14	31
	Mero-25	31
	Mero-41	31
	Mero-48	31
	Mero-72	31
	Mero-82	31
	Mero-83	31
	Mero-84	31
	Mero-95	31
	Mero-96	31
	Mero-123	31
deletion in NF2	H290	32
	H2591	32
	H2461	32
	H2373	32
nonsense mutation in NF2	HP-3	32
nonsense mutation in NF2	H2052	32

called merlin or schwannomin, which plays a role in cell surface dynamics and structure. Inactivation of NF2 by deletion or mutation has been found in about half the cell lines investigated (32,33). In the matched primary tumors from which the cell lines originated similar results were found in one study (33) while Sekido et al. (32) reported a lower proportion of NF2 alterations in fresh tumor tissue. Studies on the retinoblastoma gene Rb and p53 indicate that inactivation of these genes is not common in malignant mesothelioma (34–37). Expression of the Wilms' tumor gene WT1 has been found in 75% of the investigated malignant mesothelioma cell lines, but no genetic changes have been identified in the region encoding WT1 (38,39). No consistent changes in other tumor suppressor genes have been reported.

In conclusion, the data on the genetics of malignant mesothelioma demonstrate that correlation of cytogenetic characteristics with the location of tumor suppressor genes has been the most successful approach. Good cytogenetic characterization of cell lines is essential for this approach. However, the p16 results indicate that the established malignant mesothelioma cell lines may either represent a subpopulation of the original tumor or have acquired such changes during culture.

5. GROWTH FACTOR PRODUCTION

Malignant mesothelioma cell lines produce a number of cytokines and growth factors. These mesothelioma-derived cytokines might function as autocrine growth factors, or by a paracrine effect contribute to the spread of the tumor.

Expression of the genes for colony-stimulating factors G-CSF, M-CSF and GM-CSF has been described (4,40,41). Only in the study of Pass and co-workers (4) were all the cell lines investigated found to produce GM-CSF. IL-6 production was found in all mesothelioma cell lines investigated (4,27). Malignant mesothelioma cell lines were also found to express TGF-β1 and TGF-β2 mRNA and protein (41,42). Transfection studies in murine malignant mesothelioma indicate that TGF-β may play an important role in immunosuppression (42).

Malignant mesothelioma cell lines were found to express PDGF (platelet-derived growth factor) A- and B-chains and PDGF β-receptors (41,43,44). The coexpression of PDGF and the appropriate receptor on malignant mesothelioma cell lines led to the hypothesis that PDGF is an autocrine growth factor in malignant mesothelioma cells. Data on the autocrine role of PDGF in malignant mesothelioma cell lines are controversial and further studies are needed.

Two other growth factors were identified as potential autocrine growth factors for malignant mesothelioma cells. Lee et al. (45) described the expression of IGF-1 (insulin growth factor) by normal and malignant mesothelioma cells and reported a stimulation of the proliferation of normal, but not malignant, mesothelial cells by exogenous IGF-1. Hepatocyte growth factor (HGF) and its receptor were recently demonstrated on malignant mesothelioma cells in vivo and in vitro and a possible autocrine stimulating loop is under investigation (46 and Dr. R. Warn personal communication).

ACKNOWLEDGMENTS

I am grateful to Professor Dr. A. Hagemeijer and Professor Dr. R. Benner for continuous support and Mrs. C.A.J. van der Linden and E. Postma for technical assistance. Professor H.C. Hoogsteden and the chest physicians in Rotterdam and Vlissingen are acknowledged for sending patient material. Ms. P.C. Assems is acknowledged for secretarial assistance.

REFERENCES

1. Wagner J.C. et al. *Br J Ind Med* 17,260, 1960.
2. Versnel M.A. et al. *Int J Cancer* 44, 256, 1989.
3. Pelin-Enlund K. et al. *Carcinogenesis* 11, 673, 1990.
4. Pass H.I. et al. *Ann Thorac Surg* 59, 829, 1995.
5. Behbehani A.M. et al. *Hum Pathol* 13, 862, 1982.
6. Shang-Fang W. et al. *Scientia Sinica* 3, 281, 1985.
7. Reale F.R. et al. *Cancer Res* 47, 3199, 1987.
8. Popescu N.C. et al. *Cancer Res* 48, 142, 1988.
9. Hsu S.M. et al. *Cancer Res* 48, 5228, 1988.
10. Hagemeijer A. et al. *Cancer Genet Cytogenet* 47, 1, 1990.
11. Klominek J. et al. *Cancer Res* 49, 6118, 1989.
12. Manning L.S. et al. *Int J Cancer* 47, 285, 1991.
13. Masuda N. et al. *Chest* 100, 429, 1991.
14. Schmitter D. et al. *Int J Cancer* 51, 296, 1992.
15. Zeng L. et al. *Human Pathol* 25, 227, 1994.
16. Tange T. et al. *Pathol Internat* 45, 791, 1995.
17. Rusch V.W. et al. *Chest* 108, 1123, 1995.
18. Van Gelder T. et al. *Virchows Archiv A (Pathol Anat)* 418, 315, 1991.
19. Adams V.I. et al. *Cancer* 58, 1540, 1986.
21. Huncharek M. et al. *Thorax* 42, 897, 1987.
22. Antman K. et al. *J Clin Oncol* 6, 147, 1988.
23. Chailleux E. et al. *Chest* 93, 159, 1988.
24. Van Gelder T. et al. *Eur Resp J* 7, 1035, 1994.
25. Bolen et al. *Ultrastruct Pathol* 11, 251, 1987.
26. Stahel R.A. et al. *Int J Cancer* 41, 218, 1988.
27. Langerak A.W. et al. *J Pathol* 178, 151, 1996.
28. Tiainen M. et al. *Br J Cancer* 60, 618, 1989.
29. Cheng J.Q. et al. *Cancer Res* 54, 5547, 1994.
30. Okamoto A. et al. *Proc Natl Acad Sci USA* 91, 11045, 1994.
31. Prins J-B. et al. *Int J Cancer* 75, 649, 1998.
32. Sekido Y. et al. *Cancer Res* 55, 1227, 1995.
33. Bianchi A.B. et al. *Proc Natl Acad Sci USA* 92, 10854, 1995.
34. Van der Meeren A. et al. *Eur Resp Rev* 3, 177, 1993.
35. Shimizu E. et al. *Oncogene* 9, 2441, 1994.
36. Cote R.J. et al. *Cancer Res* 51, 5410, 1991.
37. Metcalf R.A. et al. *Cancer Res* 52, 2610, 1992.
38. Langerak A.W. et al. *Genes Chromosom Cancer* 12, 87, 1995.
39. Amin K.M. et al. *Am J Pathol* 146, 344, 1995.
40. Demetri G.D. et al. *Blood* 74, 940, 1989.
41. Morocz I.A. et al. *Br J Cancer* 70, 850, 1994.
42. Gerwin B.I. et al. *Cancer Res* 47, 6180, 1987.
43. Fitzpatrick D.R. et al. *Growth Factors* 11, 29, 1994.
44. Versnel M.A. et al. *Oncogene* 2, 601, 1988.
45. Versnel M.A. et al. *Oncogene* 6, 2005, 1991.
46. Lee T.C. et al. *Cancer Res* 53, 2858, 1993.
47. Harvey P. et al. *J Pathol* 180, 389, 1996.

Chapter 5

Pancreatic Tumors

Takeshi Iwamura[1] and Michael A. Hollingsworth[2]
[1]*Department of Surgery I, Miyazaki Medical College, 5200 Kihara, Kiyotake, Miyazaki 889-16, Japan. (current address: Department of Surgery Miyazaki Prefectural General Hospital) Tel: 0081-985-85-9284; Fax: 0081-985-85-5814; E-mail: tiwamura@post1.miyazaki-med.ac.jp;* [2]*The Eppley Institute for Research in Cancer and Allied Diseases, University of Nebraska Medical Center, 600 South 42nd Street, Omaha NE 68198-6805. Tel: 001-402-559-8343; Fax: 001-402-559-4651; E-mail: mahollin@unmc.edu*

Pancreatic cancer is among the most lethal and difficult diseases to study. Only 2% to 5% of patients with pancreatic adenocarcinoma survive 2 years. The anatomic location of the pancreas makes diagnosis difficult using conventional methods. Patients with pancreatic tumors often present with vague complaints of gastrointestinal problems late in the disease. Surgery on late stage patients is generally ineffective. It is difficult to obtain many clinical specimens of pancreatic tumors for study and it is nearly impossible to obtain clinical samples from patients with early stage premalignant or malignant disease (1).

Over 90% of human pancreatic cancers have morphological and biological properties consistent with their being classified in the ductal cell lineage of the exocrine pancreas. However, the processes that lead to malignant alterations in the pancreas and the cell types that are affected are poorly understood. Much of our current knowledge of the biological properties of pancreatic adenocarcinoma is derived from *in vitro* studies of pancreatic tumor cell lines. This chapter contains references for many of the publications that describe the establishment of human pancreatic adenocarcinoma cell lines.

1. PRIMARY CULTURE

The first report of successful culture of a pancreatic cancer cell line was by Dobrynin et al. in 1963 (2), 12 years after HeLa cells were established (3).

J.R.W. Masters and B. Palsson (eds.), Human Cell Culture Vol. 1, 107–122.

Establishment of cell lines from primary pancreatic adenocarcinoma is difficult because the primary tumor is usually small and fibrotic, and the disease is sometimes exacerbated by pancreatitis. The majority of investigators have produced pancreatic tumor cell lines from metastatic lesions or from ascites.

Tumor tissue should be explanted quickly following surgical resection or harvest of ascites. Resected tissue can be kept in medium for a short time on ice (within 2 hr), but longer storage may reduce the number of viable tumor cells. Antibiotics (penicillin, streptomycin and occasionally Fungizone) are often included in the collection fluid and growth medium until the second subculture. Cell cultures are initiated by first mincing the tumor fragments with scissors to about 1-2 mm^3. Tumor cells may be dispersed using collagenase, streptokinase, dispase, or trypsin. Tumor explants are incubated in culture dishes with growth medium (in our hands, a 50/50 mixture of Dulbecco's modified minimal essential medium and Ham's F12, supplemented with 10% fetal bovine serum) for 2 to 3 days.

Tumor cells are generally obvious in primary culture; however they are often overgrown in subsequent subcultures by fibroblasts. Fibroblasts can be removed directly by preparing small glass rods (use a flame to seal the ends of Pasteur pipettes) and removing unwanted cells from cultures. Alternatively, enzymatic removal of fibroblasts (*e.g.* by differential sensitivity to treatment with trypsin) can be achieved by using a solution of 0.05% trypsin and 0.01% EDTA. The cultures should be observed carefully under phase contrast microscopy and the reaction should be halted by adding serum-containing growth medium when fibroblasts are released from the culture flask or dish. Released cells are removed by gentle washing of the cultures in growth media. This procedure can be repeated several times, minimally at weekly intervals, until a monolayer of epithelial cells is formed that is free of fibroblasts.

It is recommended that cell lines be frozen early and often during subculture since alterations in the morphology and biological properties of the tumor cells occur during progressive growth of the cultures. Antibiotics are usually unnecessary during later stages of culture. Cell lines should be evaluated for growth properties in different defined media and cultures should be frequently evaluated for contamination with mycoplasma. As a rule, cultures of pancreatic tumor cell lines should be split at high density, generally 1:2 to 1:4, and seldom more than 1:10.

Many pancreatic tumor cells can be adapted to growth in reduced serum (0.5% or less) or serum free conditions (4,5). To establish cell lines that are stable in serum free conditions, it is recommended that cells be grown in a 50/50 mixture of Dulbecco's modified minimal essential medium and Ham's F12 (DMEM/F-12) and that the concentration of serum in media be reduced by 50% per week at progressive stages of subculture, until the desired growth conditions are established.

2. CLASSIFICATION OF PANCREATIC TUMORS AND CELL LINES

There has been considerable debate regarding the classification of pancreatic tumors. The most recent and extensive treatment of this subject is found in the *Atlas of Exocrine Pancreatic Tumors*, edited by Pour, Konishi, Klöppel and Longnecker, Springer-Verlag, Tokyo (1994). The vast majority (>90%) of pancreatic tumors are ductal adenocarcinoma or cystadenocarcinoma, which show morphological and biological evidence of being similar in differentiation status to the ductal epithelial lineage of pancreatic cells. Acinar cell carcinomas comprise less than 1% of pancreatic tumors, and about 5% of the tumors are unusual carcinomas (adenosquamous, squamous, anaplastic). The remainder (4%) are endocrine tumors of islet cells (insulinoma, glucagonoma, gastrinoma).

Among the ductal adenocarcinomas, pancreatic tumors are generally categorized according to morphological criteria as exhibiting grades of differentiation from undifferentiated to well differentiated. Many classification systems include the following descriptors: well differentiated (WD) tumors exhibit extensive glandular formation comprised of cuboidal or cylindrical tumor cells with well defined epithelial-like polarity that often show evidence of secretory activity; moderately differentiated (MD) tumors have less differentiated ductular structures, and tumor cells that show greater variation in size and shape and less evidence of the cellular polarity that is typical of ductal cells; poorly differentiated (PD) tumors show a diffuse proliferation, little or no evidence of ductal structures, and little cellular polarity; undifferentiated tumors (UD) are anaplastic and show great variation in size, shape and growth characteristics, and may not resemble ductal epithelia. It is common to find areas of more than one grade within the same tumor. Among ductal adenocarcinomas, there is no proven association between degree of differentiation and tumor aggressiveness: these tumors are almost invariably metastatic and lethal. A small percentage of pancreatic tumors, including the intraductal non-invasive adenomas, mucinous cystic tumors, and papillary cystic tumors have less malignant potential and may represent relatively benign adenomas that are distinct from the ductal adenocarcinomas.

Table 1 contains a list of many of the pancreatic cancer cell lines that have been reported in the literature. With two exceptions, all cell lines are believed to be derived from ductal adenocarcinoma. The two exceptions are QGP-1, which is reportedly derived from a non-typical islet cell carcinoma, and HPC-Y0, which is reported to be derived from an acinar cell carcinoma. The cell line MDAPanc-28 is reported to have features of both ductal cells and acinar cells. Fourteen cell lines are listed in the American Type Culture Collection (ATCC) and are available.

Table 1 Human pancreatic tumor cell lines

Cell line	Patient age(y)/sex	TNM	Path.stage histolo.grade	Primary site	Specimen site	Culture method	Authentication	Availability	Primary reference
CaPa	52/n.r.	TxN1M1	IV G2-G1	head	l. n. met.	medium 199 with human (3–5%) and bovine (5–7%) serum	n.r.	B	Dobrynin (1963) (2)
PANC-1	56/m	T2N1MO	III G4	head	primary tumor	D-MEM with 10% FCS	G6PDH, chromosome	A	Lieber (1975) (6)
Hs 700T	61/m	TxNxM1	IV (colon, intestine, or panc metastatic to hip)	n.r.	metastasis	D-MEM with 10% FCS	chromosome, G6PDH	A	Owens (1976) (7)
Hs 766T	64/m	TxNxM1	IV G4	n.r.	l. n. met.	D-MEM with 10% FCS	chromosome, G6PDH	A	Owens (1976) (7)
HGC 25	64/m	TXNxM1	IV G4	whole	ascites	RPMI 1640 with 10% FCS	ALP, G6Pase LDH, G6PDH	B	Akagi (1977) (8)
MIA PaCa-2	65/m	T3NxMx	n.r. G4	body	primary tumor	D-MEM with 10% FCS	chromosome	A	Yunis (1977) (9)
Capan-1	40/m	TxN1M1	IV	head	liver met.	RPMI 1640 15% FCS	chromosome, G6PDH	A	Fogh (1977) (10)
SW-850	n.r.	TxNxMx	n.r.	n.r.	n.r.	n.r.	n.r.	B	Fogh (1977) (10)
SW-979	n.r.	TxNxMx	n.r.	n.r.	n.r.	n.r.	n.r.	B	Fogh (1977) (10)
GER	56/f	TxNxMx	n.r. G3	head	primary tumor	Ham's F12 with supplement	G6PDH	B	Grant (1979) (11)
QGP-1	61/m	T2N1M1	IV islet cell Ca.	tail	primary tumor	Medium 199 10% FCS	chromosome, CEA	B	Kaku (1980) (12)
COLO 357	77/f	T2N1Ma	IV G1	head	l. n. met.	RPMI 1640 10% FBS	G6PDH	B	Morgan (1980) (13)

Continued on next page

Table 1 (continued)

Cell line	Patient age(y)/sex	TNM	Path.stage histolo.grade	Primary site	Specimen site	Culture method	Authentication	Availability	Primary reference
AsPC-1	62/f	TxNxM1	IV G2	head	ascites	RPMI 1640 10% FBS	chromosome, G6PDH	A	Chen (1982) (14)
RWP-1	57/f	TxNxM1	IV G2-G1	head	liver met.	RPMI 1640 20% FCS	chromosome	B	Dexter (1982) (15)
RWP-2	40/f	TxNxM1	IV G2-G1	n.r.	liver met.	RPMI 1640 20% FCS	chromosome	B	Dexter (1982) (15)
HPAF	n.r.	TxNxM1	IV G2-G1	n.r.	ascites	E-MEM 10% FCS	n.r.	A	Metzgar (1982) (16)
HPC-Y1	52/f	TxN1Mx	n.r. G1	head & pancreas	l. n. met.	RPMI 1640 10% FCS	chromosome, LDH	B	Yamaguchi (1983) (17)
T3M-4	64/m	TxN1M1	IV n.r.	head	l. n. met.	F-10 10% FCS	chromosome, CEA	B	Okabe (1983) (19)
SW-1990	56/m	TxNxMx	n.r. G1	tail	primary tumor	AL-15 15% FBS	chromosome, CEA, LDH	A	Kyiazis (1983) (19)
PK-1	n.r.	T3N1M1	IV	body	liver met.	RPMI 1640 10% FBS	chromosome, CEA G6PDH	B	Kobari (1984) (20)
BXPC-3	61/f	TxNxMx	n.r. G2-G3	body	primary tumor	RPMI 1640 20% FBS	chromosome	A	Tan (1986) (21)
PSN-1	n.r.	TxN1Mx	n.r. G3	n.r.	primary tumor l. n . met.	RPMI 1640 10% FBS	c-myc, c-Ki-ras	B	Yamada (1986) (22)
HPC-YT	n.r.	TxNxMx	n.r. G2	n.r.	primary tumor	Ham's F12 w/o FBS	chromosome	B	Yamaguchi (1986) (23)
PK-8	n.r.	TxNxM1	IV G2	head	n. m. xenograft (liver, met.)	RPMI 1640 10% FBS	chromosome, CEA, G6PDH	B	Kobari (1986) (24)
PK-9	n.r.	TxNxMx	n.r. G2	head	primary tumor	RPMI 1640 10% FBS	chromosome, CEA, G6PDH	B	Kobari (1986) (24)

Continued on next page

Table 1 (continued)

Cell line	Patient age(y)/sex	TNM	Path.stage histolo.grade	Primary site	Specimen site	Culture method	Authentication	Availability	Primary reference
PK-12	n.r.	TxNxM1	IV G2	head	n. m. xenograft (liver met.)	RPMI 1640 10% FBS	chromosome, CEA, G6PDH	B	Kobari (1986) (24)
PK-14	n.r.	TxNxMx	n.r. G4	head	n. m. xenograft (primary tumor)	RPMI 1640 10% FBS	chromosome, CEA, G6PDH	B	Kobari (1986) (24)
PK-16	n.r.	TxNxMx	IV G2	body	n. m. xenograft (perit. dissem.)	RPMI 1640 10% FBS	chromosome, CEA, G6PDH	B	Kobari (1986) (24)
HPC-3	n.r.	TxNxM1	IV n.r.	n.r.	ascites	RPMI 1640 10% FCS	n.r.	B	Sato (1986) (25)
HPC-4	n.r.	TxNxM1	IV n.r.	n.r.	ascites	RPMI 1640 10% FCS	n.r.	B	Sato (1986) (26)
SUIT-2	73/m	T3N1M1	IV G2	whole	liver met.	D-MEM 5% FBS or w/o FBS	chromosome, CEA, Ca19-9	B	Iwamura (1987) (27)
SU.86	57/f	TxNxM1	IV G2-G3	head	primary tumor	RPMI 1640 20% FBS	chromosome, CEA	A	Drucker (1988) (27)
KP-1N	69/m	TxNxM1	IV G2	n.r.	liver met.	Daigo's T medium 10% FCS	chromosome, CEA, Ca19-9, SLX	B	Ikeda (1990) (28)
KP-2	65/f	TxNxMx	n.r. G2	n.r.	primary tumor	n.r.	n.r.	B	Ikeda (1990) (28)
KP-3	75/m	TxNxM1	IV adenosquamous	n.r.	n. m. xenograft (liver met.)	n.r.	n.r.	B	Ikeda (1990) (28)
PC-1	54/f	TxNxMx	IV G3	head	l.n. met	Ham's F12 15% FCS	chromosome, CEA	B	Jie (1990) (29)
PC-2	56/m	TxNxMx	IV G2	whole	omental mass	Ham's F12 15% FCS	chromosome, CEA	B	Jie (1990) (29)

Continued on next page

Table 1 (continued)

Cell line	Patient age(y)/sex	TNM	Path.stage histolo.grade	Primary site	Specimen site	Culture method	Authentication	Availability	Primary reference
HPC-Y0	n.r.	TxNxMx	Acinar cell carcinoma	n.r.	primary tumor	Ham's F10	CEA, CA19-9,	B	Yamaguchi (1990) (30)
HPC-Y1	n.r.	TxNxMx	n.r. n.r.	n.r.	primary tumor	Ham's F10	CEA, POA, CA19-9	B	Yamaguchi (1990) (30)
HPC-Y5	n.r.	TxNxMx	n.r. n.r.	n.r.	l.n. met	Ham's F10	CEA, POA, Ca19-9, SLX	B	Yamaguchi (1990) (30)
HPC-Y9	n.r.	TxNxMx	n.r. n.r.	n.r.	n. m. xenograft (primary tumor)	Ham's F10	CEA, POA, Ca19-9, SLX	B	Yamaguchi (1990) (30)
HPC-Y11	n.r.	TxNxMx	n.r. n.r.	n.r.	primary tumor	Ham's F10	CEA, POA, Ca19-9,	B	Yamaguchi (1990) (30)
HPC-Y15	n.r.	TxNxMx	n.r. n.r.	n.r.	primary tumor	Ham's F10	CEA, POA, Ca19-9,	B	Yamaguchi (1990) (30)
HPC-Y19	n.r.	TxNxMx	n.r. n.r.	n.r.	primary tumor	Ham's F10	CEA, POA, Ca19-9, SLX	B	Yamaguchi (1990) (30)
HPC-YP	n.r.	TxNxMx	n.r. n.r.	n.r.	primary tumor	Ham's F10	CEA, POA, Ca19-9,	B	Yamaguchi (1990) (30)
HPC-YS	n.r.	TxNxMx	n.r. n.r.	n.r.	primary tumor	Ham's F10	CEA, POA, Ca19-9, SLX	B	Yamaguchi (1990) (30)
HPC-Y25	n.r.	TxNxMx	n.r. n.r.	n.r.	primary tumor	Ham's F10	CEA, POA, Ca19-9,	B	Yamaguchi (1990) (30)
A-818	n.r.	n.r.	n.r. G1	n.r.	ascites	n.r.	CEA, POA, CA19-9	B	Schmigel (1990) (31)
CFPAC-1	26/m	TxNxM1	IV G1	head	primary tumor	Iscove's DMEM 10% FBS	CEA, POA, CA19-9, CF gene	A	Schoumacher (90) (32)
Capan-2	56/m	n.r.	n.r. G1	n.r.	n.r.	McCoy's 5a 10% FBS	G6PDH	A	Fogh (1993) (33)

Continued on next page

Table 1 (continued)

Cell line	Patient age(y)/sex	TNM	Path.stage histolo.grade	Primary site	Specimen site	Culture method	Authentication	Availability	Primary reference
JF305	n.r.	TxNxMx	n.r. n.r.	n.r.	n. m. xenograft	RPMI 1640 20% FBS	chromosome	B	Li (1994) (34)
HPAC	64/f	TxNxMx	n.r. G1-G2	head	n. m. xenograft (primary tumor)	D-MEM/F-12 5% FBS	chromosome	A	Gower (1994) (35)
COLO 587	55/f	TxNxM1	IV n.r.	n.r.	perit. dissem.	RPMI 1640 10% FBS	n.r.	A	Moore (n.r.)
MDAPanc-28	69/f	n.r.	n.r. n.r.	whole	primary	Leibovitz L-15 10% FBS	chromosome	B	Frazier (36)

TNM: TNM category of the pancreatic cancer; path.: pathological; m: male; f: female; n.r.: not reported; G1: well differentiated adenocarcinoma; G2: moderately differentiated adenocarcinoma; G3: poorly differentiated adenocarcinoma; G4: undifferentiated adenocarcinoma; l. n. met: lymph node metastasis; liver met: liver metastasis; n. m. xenograft: nude mouse xenograft; perit. dissem.: peritonial dissemination; D-MEM: Dulbecco's modification of minimal essential medium; FCS: fetal calf serum; FBS: fetal bovine serum; G6PDH: glucose-6-phosphate dehydrogenase; ALP: alkaline phosphatase; G6Pase: glucose-6-phosphatase; LDH: lactate dehydrogenase; CEA: carcinoembryonic antigen; CA19-9: carbohydrate antigen 19-9; SLX: sialyl SSEA-1 antigen; POA: pancreatic oncofetal antigen; CF: cystic fibrosis; A: available from ATCC; B: contact investigator

Many pancreatic tumor cell lines exhibit morphological characteristics that are consistent with the grades of differentiation reported for the tumor from which the cell line was derived, and many of these maintain those features when grown in culture or as xenografts in nude mice. The differentiation status reported for different pancreatic tumor cells following *in vitro* culture or growth as a xenograft in nude mice is summarized in Table 2. Cell lines such as PANC-1, HS766T, and HGC 25 retain many of the properties of poorly differentiated adenocarcinomas, and well differentiated cell lines such as HPAF, HPAC, and COLO-357 likewise retain many of the morphological features of well differentiated tumors. It is also common for the cells to undergo selection or alteration in culture so that sublines with different properties arise.

Several pancreatic tumor cell lines were used as immunogens to derive monoclonal antibodies (MAb) with the hope of improving early detection and treatment of pancreatic cancer. Examples of MAbs and cell lines, respectively, include: DU-PAN-2 against HPAF (16); AR2-20 and AR1-28 against RWP-1 and RWP-2 (37); YPan-1 and YPan-2 against Capan 2 (38); DD9E7 against GER (39); F-30 against PK-1 (40); SPan-1 against SW-1990 (41); 1H1 against BxPC-3 (42); PA8-15 against SUIT-2 (43); 2C-8 against AsPC-1 (44); 5D-4 against AsPC-1 (45); Nd-2 against SW-1990 (46); and Ea6 against SW-1990 (47).

Acinar cells, which account for more than 80% of the pancreas, contain zymogen granules and secrete many of the enzymes required for digestion. Duct cells and centroacinar cells transport secretions of the acinar cells to the intestine and are the primary sites of water, electrolyte and mucin secretion. Ductal and acinar cells have interrelated functions; however, the full extent of biochemical interaction and communication between ductal cells and acinar cells is not well understood and this is an understudied subject area. Secretory functions in the pancreas are in part regulated by intestinal hormones including secretin, cholecystokinin (CCK), vasoactive intestinal peptide (VIP), and somatostatin. Several pancreatic tumor cell lines, including Capan-1, MIA-PaCa-2, and PANC-1 have been shown to respond to these intestinal hormones (48-55). Expression of mucin core proteins in the pancreas and alterations in mucin core protein expression in adenocarcinoma have been reported (56) and reviewed (57).

3. MOLECULAR GENETICS

Several genetic changes are associated with the development of pancreatic cancer. A summary of some specific genetic alterations reported for pancreatic tumor cell lines is presented in Table 3. One widely described abnormality is the presence of mutations in K-*ras* codons 12, 13 and 61,

Table 2 Pancreatic tumor cell lines: Differentiation, growth in nude mice as xenografts, unique features

Cell line	Differentiation status: *in vitro*/nude mouse xenograft	Unique Features
CaPa	MD-WD/n.r.	First reported pancreatic tumor cell line (2)
PANC-1	PD/PD	
Hs 700T	PD/n.r.	
Hs 766T	PD/n.r.	
HGC 25	PD/n.r.	
MIA PaCa-2	PD-MD/PD-MD	
Capan-1	WD/WD	
SW-850	n.r./n.r.	
SW-979	n.r./n.r.	
GER	WD	Used to produce MAb DDE9e7 (39)
QGP-1	PD – islet cell/n.r.	
COLO 357	WD/WD	
AsPC-1	PD-MD/n.r.	Used to produce MAbs 2c-8 (44) and 5D-4 (45)
RWP-1	MD-WD/n.r.	Used to produce MAbs AR2-20 and AR1-28 (37)
RWP-2	MD-WD/n.r.	Used to produce MAbs AR2-20 and AR1-28 (37)
HPAF	WD/MD-WD	Used to produce MAb DCU-PAN-2 (16)
HPC-Y1	WD/n.r.	
T3M-4	MD/WD	
SW-1990	WD/WD	Used to produce MAb Span-1 (41), Nd-2 (46), and Ea6 (47)
PK-1	n.r./n.r.	Used to produce MAb F-30 (40)
BXPC-3	MD/n.r.	Used to produce MAb 1H1 (42)
PSN-1	MD/n.r.	
HPC-YT	MD-WD/n.r.	
PK-8	MD/n.r.	
PK-9	MD/n.r.	
PK-12	MD/n.r.	
PK-14	PD/n.r.	
PK-16	MD/n.r.	
HPC-3	n.r.	
HPC-4	n.r.	
SUIT-2	MD-WD/MD-WD	Used to produce MAb PA8-15 (43)
SU.86	MD/n.r.	
KP-1N	MD/n.r.	
KP-2	MD/n.r.	
KP-3	adenosquamous/n.r.	
PC-1	PD/PD	
PC-2	MD/MD	
HPC-Y0	Acinar cell carcinoma/n.r.	
HPC-Y1	MD-WD/n.r.	
HPC-Y5	MD-WD/n.r.	
HPC-Y9	MD-WD/n.r.	
HPC-Y11	MD-WD/n.r.	
HPC-Y15	MD-WD/n.r.	

Continued on next page

Table 2 (continued)

Cell line	Differentiation Status: *in vitro*/nude mouse xenograft	Unique Features
HPC-Y19	MD-WD/n.r.	
HPC-YP	MD-WD/n.r.	
HPC-YS	MD-WD/n.r.	
HPC-Y25	MD-WD/n.r.	
A-818	WD/n.r.	
SU.86	MD-WD/n.r.	
CFPAC-1	WD/WD	Homozygous for CFTR mutation: ΔF508 (32)
Capan-2	WD/WD	Used to produce MAbs Ypan-1 and Ypan-2 (38)
JF305	n.r./n.r.	
HPAC	MD-WD/n.r.	
COLO 587	WD/WD	
MDAPanc-28	PD/PD	Expresses both ductal cell products and some acinar cell products (trypsin and ribonuclease) (36)

PD: Poorly differentiated; MD: moderately differentiated; WD: well differentiated; n.r.: not reported

which are found at a high frequency in primary pancreatic tumors (over 70% in most studies) and in most pancreatic tumor cell lines (58-61). Another common finding in pancreatic tumor cell lines is the presence of abnormalities in *p*53, including mutations, deletions, and loss of heterozygosity (LOH). Several studies have detected alterations in *p*53 in 40-70 % of pancreatic cancer, but not in benign pancreatic diseases (62-65); a similar incidence of *p*53 abnormalities is seen in pancreatic tumor cell lines (Table 3). The incidence of *BRCA2* mutations in pancreatic cancer appears to be approximately 10% of cases and cell lines (66-67); however, deletion of one allele of *BRCA*2 through LOH is a more common finding (30%-50% of tumors and tumor cell lines examined). Loss of heterozygosity affecting the *p*16 gene is reported in a high percentage of tumors and tumor cell lines examined (68-69). *p*16 gene alterations are also detected in intraductal precancerous pancreatic lesions (63). Approximately 10% of pancreatic tumors and tumor cell lines show complete loss of expression of the *RB* gene through deletion or mutation (68). Telomerase activity was detected in pancreatic cancer, but not in benign tumors and chronic pancreatitis (70). Northern blot analysis and immunohistochemical staining show overexpression of the c-myc gene and c-myc gene product in pancreatic cancer (22). Overexpression of the c-fos gene (59) and positive staining of c-erb B-2 in immunohistochemistry are also reported (71-72).

Table 3 Genetic alterations reported for human pancreatic tumor cell lines

Cell line	Gene				
	K-*ras*	p53	BRCA2	p16	Rb-1
CaPa					
PANC-1	Gly12Asp GGT>GAT	Arg273Cys CGT>TGT	LOH	LOH	Normal
Hs 700T					Normal
Hs 766T	Normal	Detected but uncharacterized	Normal	mut	3 bp deletion in intron 19
HGC 25					
MIA PaCa-2	Gly12Cys GGT>GAT	Arg248Trp CGG>TGG	Normal	LOH	Normal
Capan-1	Gly12Val GGT>GTT	Ala159Val GCC>GTC	6174DelT LOH	LOH	Normal
SW-850					
SW-979	Normal	Normal			Normal
GER					
QGP-1					Normal
COLO 357			Normal		Normal
AsPC-1	Gly12Asp GGT>GAT	Intron 4 splice donor ACG>ACT (200 bp deletion); 1 bp deletion at 135	LOH	mut	Normal
RWP-1	Gly12Asp GGT>GAT	Normal			
RWP-2	Gly12Asp GGT>GAT	Arg175His CGC>CAC			
HPAF	Gly12Asp GGT>GAT	Pro151Ser CCC>TCC			Normal
HPC-Y1		1 bp deletion 169 (frame shift)		LOH	Normal
T3M-4	His61Gln CAC>CAA				Normal
SW-1990					
PK-1					
BXPC-3	Normal	Tyr220Cys TAT>TGT	LOH	LOH	Normal
PSN-1	Gly12Arg GGT>CGT				
HPC-YT					
PK-8					
PK-9					
PK-12					
PK-14					
PK-16					
HPC-3		1 bp deletion 169 (frame shift)		LOH	Normal
HPC-4		1 bp deletion 169 (frame shift)		LOH	Normal

Continued on next page

Table 3 (continued)

Cell line	Gene				
	K-*ras*	p53	BRCA2	p16	Rb-1
SUIT-2	Gly12Asp GGT>GAT				
KP-1N	Gly12Asp GGT>GAT				
KP-2	Normal				
KP-3	Gly12Val GGT>GTT				
HPC-YP					
SU.86			LOH	LOH	Normal
CFPAC-1			LOH	Normal	Normal
Capan-2	Gly12Val GGT>GTT	Intron 4 splice donor ACG>ACT (200 bp deletion)			
JF305					
HPAC					
COLO 587					

LOH: loss of heterozygosity. mut: undefined mutation. Blank spaces indicate no published report. Data summarized in this table were taken from references 58–72

4. CELL LINES WITH UNIQUE CHARACTERISTICS

Most malignant tumors contain cells that are heterogeneous in morphology and biological properties. In the case of pancreatic tumors, individual cells often show differences in degree of differentiation, malignant transformation, and metastatic properties. Tumor cell lines derived from pancreatic tumors are often pleomorphic in the early stages of establishment, especially when these come from tumors that show mixed histological differentiation. Surprisingly, only two studies have been reported for pancreatic tumors in which clonal populations of cells were obtained with distinct morphologies; one report was the establishment of the CD11 (well differentiated) and CD18 (poorly differentiated) cell lines, which were derived from HPAF (73); another is the S2-series of cell lines derived from SUIT-2, which show a broad range of patterns of differentiation (74). It is likely that some of the different clones from these tumor cell lines represent distinct differentiated cell types from the pancreas that are fixed in different stages of differentiation. Thus, in addition to providing resources for studying the biology of tumors, these cell lines are unique resources that can be used to study the biological processes of different differentiated cell types of the pancreas. Another example of a pancreatic tumor cell line that is being used to study pancreatic cellular functions is CFPAC-1, a unique cell line that was derived from a ductal adenocarcinoma from a patient with cystic fibrosis (ΔF508 homozygous). CFPAC-1 cells exhibit defective ion transport activities consistent with cystic fibrosis and express the product of the CF gene – the cystic fibrosis transmembrane regulator (CFTR) (32).

One notable feature of pancreatic cancer, which is amenable to study through the use of unique cell lines, is invasiveness and metastasis. There are few reports of spontaneous invasion and metastasis by different pancreatic cancer cell lines, since most of the pancreatic cancer cell lines are tumorigenic but not invasive or metastatic in nude mice. Two cell lines, Capan 1 and SUIT-2, have been reported to demonstrate spontaneous metastasis and invasion in nude mice. Both subcutaneous and intraperitoneal tumors of these cell lines exhibit histological characteristics similar to the original tumors when transplanted to nude mice and they show spontaneous metastatic activity. Subcutaneous tumors of SUIT-2 metastasize to homolateral and contralateral lymph nodes and the lung. Among a series of clonal derivatives from SUIT-2, sublines S2-007, S2-013, S2-020, S2-028 show: different morphologies; differences in production of CEA, CA19-9, type 1 and type 4 collagenolytic activity; gland formation upon culture in matrigel; differences in histological grades of differentiation; and differences in metastatic potential in nude mouse xenografts (75-79).

5. CONCLUSION

The anatomic inaccessibility of the pancreas and the fact that it is an organ full of degradative enzymes has hindered our ability to study the biology of the normal pancreas and tumors that arise in this organ. Consequently, the diagnosis and treatment of pancreatic adenocarcinoma remains a difficult task. The development and use of pancreatic tumor cell lines has greatly enhanced our efforts to understand pancreatic cancer and yielded a great deal of information that would not have been obtained otherwise. We hope this short chapter is useful to both new and experienced investigators in advancing the assault on this terrible disease.

REFERENCES

1) Kobari, M., et al., *Jpn. J. Gastroenterol.*, 94, 1, 1997.
2) Dobrynin, Y. V. et al. *J. Natl. Cancer Inst.*, 31, 1173, 1963.
3) Gey, G., et al., *Cancer Res.*, 12, 264, 1952.

4) Yamaguchi, N., et al., *J. Natl. Cancer Inst.*, 75, 29, 1985.
5) Taniguchi, S. et al., *Hum. Cell*, 7, 207, 1994.
6) Lieber, M. et al. *Int. J. Cancer.* 15, 741, 1975.
7) Owens, R. B., et al. *J. Natl. Cancer Inst.*, 56, 843, 1976.
8) Akagi, T., et al., *Acta Path. Jpn.*, 27, 51, 1977.
9) Yunis, A. A., et al., *Int. J. Cancer,* 19, 128, 1977.
10) Fogh, J., et al., *J. Natl. Cancer Inst.*, 59, 221, 1977.
11) Grant, A. G., et al., *Br. J. Cancer*, 39, 143, 1979.
12) Kaku, M., et al., *Jpn. J. Cancer Res.*, 71, 596, 1980.
13) Morgan, R. T., et al., *Int. J. Cancer*, 25. 591, 1980.
14) Chen, W. H. et al., *In Vitro*, 18, 24, 1982.
15) Dexter, D. L., et al., *Cancer Res.*, 42, 2705, 1982.
16) Metzgar, R. S., et al., *Cancer Res.*, 42, 601, 1982.
17) Yamaguchi, N., et al., *Gastroenterol. Jpn.*, 18, 587, 1983.
18) Okabe, T., et al., *Cancer*, 51, 662, 1983.
19) Kyriazis, A. P., et al., *Cancer Res.*, 43, 4393, 1983.
20) Kobari, M., et al., *Tohoku J. Exp. Med.*, 143, 33, 1984.
21) Tan, M. H., et al., *Cancer Invest.*, 4, 15, 1986.
22) Yamada, H., et al., *Biochem. Biophys. Res. Comm.*, 140, 167, 1986.
23) Yamaguchi, N., et al., *Cancer Res.*, 46, 5353, 1986.
24) Kobari, M., et al., *Tohoku J. Exp. Med.*, 150, 231, 1986.
25) Sato, T., et al., *Tumor Res.*, 21, 9, 1986.
26) Iwamura, T., et al., *Jpn. J. Cancer Res.*, 78, 54, 1987.
27) Drucker, B. J., et al., *In Vitro Cell. Dev. Biol.*, 24, 1179, 1988.
28) Ikeda, Y., et al., *Jpn. J. Cancer Res.*, 81, 987, 1990.
29) Jie, C. et al, *Chinese Medical Journal*, 103, 369, 1990.
30) Yamaguchi, N., et al., *Cancer Res.*, 50, 658, 1990.
31) Schmigel, W. et al., *Digestion*, 46, 175, 1990.
32) Schoumacher, R. A., et al., *Proc. Natl. Acad. Sci. USA.*, 87, 4012, 1990.
33) Fogh, J. et al;., *Cancer Res.*, 53, 1437, 1993.
34) Li, X., et al., *Chung-Hua-Chung-Liu-Tsa-Chih.*, 16, 181, 1994.
35) Gower, W. R., et al., *In Vitro Cell. Dev. Biol.*, 30A, 151, 1994.
36) Frazier, Marsha L., et al., *Int J Pancreatology,* 18 31, 1996.
37) Chin, J., et al., *Cancer Res.*, 45,1723, 1985.
38) Yuan, S. Z., et al., *Cancer. Res.*, 45, 6179, 1985.
39) Grant, A. G., et al., *Br. J. Cancer,* 52, 543, 1985.
40) Kobari, M., et al., *Tohoku J Exp Med*, 148, 179, 1986.
41) Ho, J. J., et al., *Cancer Res*, 48, 3924, 1988.
42) Fujii, Y., et al., *Hum. Cell*, 1, 421, 1988.
43) Yoshimura, R., et al., *Jpn. J. Cancer Res.*, 79, 255, 1988.
44) Yanagie, H., et al., *Hum. Cell,* 2, 290, 1989.
45) Fujii, Y., et al., *Biomed. Pharmacother.*, 46, 405, 1992.
46) Chung, Y. S. et al., *Tohoku J. Exp. Med.*, 168, 397, 1992.
47) Ho, J. J., et al., *Cancer Res.*, 53, 884. 1993.
48) Ruellan, C. et al., *Peptides,* 1, 267, 1986.
49)Hierowski, M. T., et al., *FEBS-Lett.* 179, 252, 1986.
50) Liebow, C., et al., *Proc. Natl. Acad. Sci. USA,* 86, 2003, 1989.
51) Lee, M. T., et al. *Proc. Natl. Acad. Sci. USA,* 88, 1656, 1991.
52) Gillespie, J., et al., *Br. J. Cancer,* 66, 483, 1992.
53) Smith, J. P., et al., *Am. J. Physiol.*, 265, G149, 1993.
54) Radulovic. S., et al., *Pancreas,* 8, 88, 1993.

55) Sumi, S., et al., *Surg. Oncol.* 2, 267, 1993.
56) Hollingsworth, M. A., et al., *Int. J. Cancer,* 57, 198, 1994.
57) Hollingsworth, M. A. *Biliary and Pancreatic Ductal Epithelia* (eds. Sirica, A. E. & Longnecker D. S.), 409, 1997.
58) Almoguera, C., et al., *Cell,* 53, 549, 1988.
59) Wakita, K., et al., *Int. J. Pancreatol.* 11, 43, 1992.
60) Yasuda, D., et al., *Int. J. Oncol.*, 3, 641, 1993.
61) Christopher, A., et al, *Cancer Res.,* 57, 2140, 1997.
62) Barton, C. M., et al., *Br. J. Cancer,* 64, 1076, 1991.
63) Casey, G., et al., *Cancer lett*, 69, 151, 1991.
64) Scarpa, A., et al., *Am. J. Pathol*, 142,1534, 1993.
65) Redston, M. S., et al., *Cancer Res*., 54, 3025, 1994.
66) Teng, D., et al., *Nature Genetics,* 13, 241, 1996.
67) Goggins, M., et al., *Cancer Res,* 56, 5360, 1996.
68) Kaino, M., *Gastroenterol*, 32, 40, 1997.
69) Chen, Z.H., et al., *Cancer Res,* 56, 1083, 1996.
70) Hiyama, E., et al., *Cancer Res.*, 57, 326, 1997.
71) Kasahara, S., et al., *Kan-Tan-Sui,* 25, 1141, 1992.
72) Lemoine, N., et al., *J Pathol,* 168, 269, 1992.
73) Kim, Y. W., et al., *Pancreas,* 4, 353, 1989.
74) Iwamura, T., et al., J. *Gastroenterol. Hepatol.* 7, 512, 1992.
75) Taniguchi, S., et al., *Clinic. Exp. Met.*, 10, 259, 1992.
76) Yamanari, H., et al., *Exp. Cell Res*. 211, 175, 1994.
77) Taniguchi, S., et al., *Clinic. Exp. Met.,* 12, 238, 1994.
78) Kataoka, H., et al., *Int. J. Cancer,* 60, 123, 1995.
79) Iwamura, T., et al., *Cancer Res.* 57, 1206, 1997.

Chapter 6

Adrenal Cortex Tumors

William E. Rainey[1] and James J. Mrotek[2]
[1]*University of Texas Southwestern Medical Center, Department of Obstetrics & Gynecology, Division of Reproductive Endocrinology, Dallas, Texas, Tel: 001-214-648-4747; Fax:001-214-648-8066; E-mail: braine@mednet.swmed.edu and* [2]*Meharry Medical College, Department of Physiology, Nashville, Tennessee*

1. BACKGROUND

The adrenal gland is a compound endocrine gland composed of two developmentally unrelated tissues; an adrenal medulla and an adrenal cortex. The cells comprising the adrenal cortex are derived from mesoderm of the dorsal coelomic wall and thus have features in common with the steroidogenic cells within the gonads. In 1866, Arnold made the histological description and gave the associated nomenclature of the zona glomerulosa, zona fasciculata and zona reticularis to the three concentric zones of the mammalian adrenal cortex (1,2). While this description was based on the histological organization, it is now accepted that these zones have functionally distinct roles in steroid hormone production. Namely, the glomerulosa synthesizes mineralocorticoids, the fasciculata produces glucocorticoids and, in the human, the zona reticularis produces C_{19} steroids (including dehydroepiandrosterone [DHEA], DHEA-sulfate). The mechanisms regulating production of steroids from each zone are different, with angiotensin II and potassium regulating the glomerulosa and adrenocorticotropin (ACTH) regulating the fasciculata. Like all steroidogenic cells each adrenal zone synthesizes its steroid products from the same substrate, pregnenolone, which is formed from cholesterol. Within the human adrenal cortex, steroids can be metabolized by five forms of cytochrome P450 and the enzyme 3β-hydroxysteroid dehydrogenase (3βHSD) (illustrated in 1). These enzymes are distributed between the mitochondria and the endoplasmic reticulum (3). It is the differential expression of these enzymes within the

J.R.W. Masters and B. Palsson (eds.), Human Cell Culture Vol. 1, 123–135.

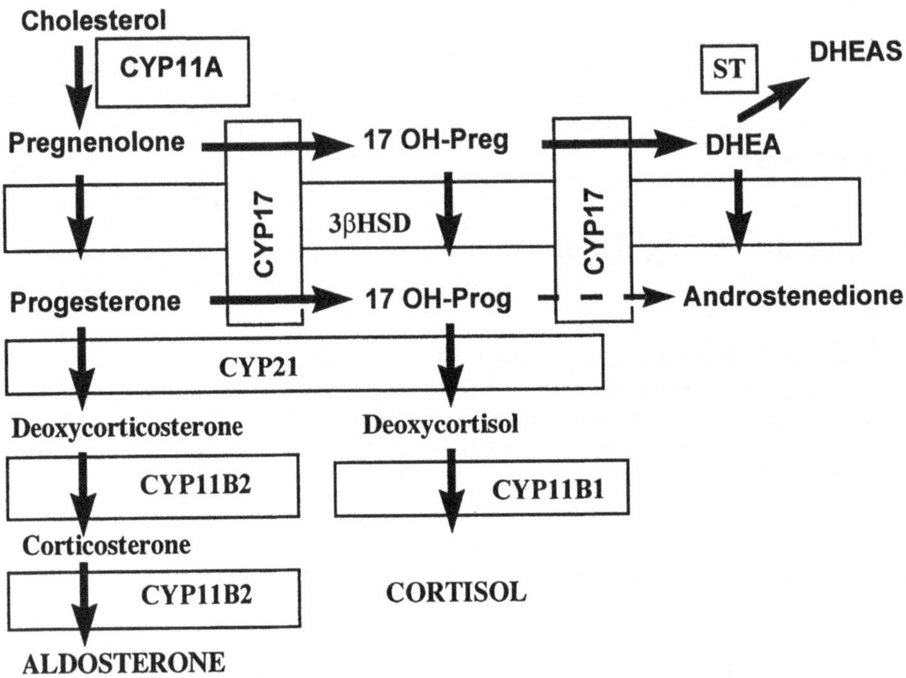

Figure 1. Human adrenal steroid biosynthetic pathways. CYP11A, cholesterol side chain cleavage; CYP17, 17α-hydroxylase; 3β-ST, 3β-hydroxysteroid sulfotransferase; 3βHSD, 3β-hydroxysteroid dehydrogenase; CYP21, 21-hydroxylase; CYP11B1, 11β-hydroxylase; CYP11B2, aldosterone synthase

three adrenocortical zones that causes the wide array of steroid hormones secreted by this gland. In addition, the production of each steroid group changes under normal development and during the development of pathological conditions including adrenocortical tumors (4).

Adrenal tumors can arise as functioning or nonfunctioning adenoma and carcinoma. The widespread application of abdominal computed tomography and magnetic resonance imaging scans has led to the incidental discovery of a large number of unsuspected adrenal adenomas. Estimates of the presence of adrenal "incidentalomas" in the general population range from 1–10 % (5). The number of these adrenal "incidentalomas" that result in clinical disease is not known. However, unilateral benign adrenal adenomas are the most common cause of low-renin aldosteronism, accounting for 60–70 % of these patients (6). These adenomas tend to be small (a few mm to 4 cm) and well circumscribed. The majority of bilateral adrenal lesions are metastases from another site and therefore do not produce steroids. Adrenocortical carcinoma (ACC) derived from steroid hormone-producing cells of the adrenal cortex is rare. The incidence of ACC is approximately one case per 1.7 million people and accounts for 0.02 % of cancers (7). These tumors usually reach a large size prior to diagnosis. Forty percent of ACC are functional with associated clinical endocrine features dependent on the steroid hormones produced by the tumor. Adrenal adenoma or carcinoma can synthesize any or all of the steroid hormones produced in the normal adrenal, but often demonstrate aberrant biosynthetic pathways. In addition, adrenal tumor steroidogenesis, as opposed to normal steroid hormone biosynthesis, can occur in a constitutive manner independent of hormonal regulation. The loss of feedback regulation is, in part, the reason many of these tumors lead to diseases associated with steroid excesses, i.e., hyperaldosteronism, Cushing's syndrome, or hirsutism.

2. CELL LINES

Because of the complex nature of human adrenocortical steroid hormone biosynthesis and regulation, an adrenocortical cell line would be extremely useful to help define the pharmacologic, biochemical, and molecular mechanisms that regulate adrenal function. *In vitro* investigations into adrenocortical physiology have been hampered by the difficulties associated with maintenance of long-term cultures of normal adrenocortical cells (8-10). Also, the constant requirement for freshly acquired tissue and the lack of readily available human tissue have made it difficult for laboratories to work in this area. Therefore, a human or animal cell line that retains the ability to secrete aldosterone, cortisol or C_{19} steroids could expedite research in this area. Until recently the *in vitro* study of adrenal steroidogenesis was conducted using primary steroidogenic cell cultures from rat, cow and sheep adrenals or the well-characterized Y-1 mouse adrenocortical tumor cell line. Primary cultures of adrenal cells appear to recapitulate physiologic responsiveness to hormones and production of steroid hormones. However, primary cultures of adrenal cells dedifferentiate with time in culture and therefore fresh tissue is required on a continuous basis. The Y-1 mouse adrenal cell line was originally developed by Sato and colleagues in 1966 (11). These cells offer rapid growth and responsiveness to adrenocorticotropin (ACTH) treatment. However, they have lost the ability to express steroid 21-hydroxylase, causing the cells to release a series of steroid precursors as their primary products (12). In addition these cells exhibit little response to either angiotensin II (13) or potassium (unpublished data), the primary physiologic regulators of aldosterone production.

Because of the diversity of physiologic regulators of adrenal steroid hormone synthesis, a cell line capable of synthesizing the spectrum of steroids in response to the appropriate physiological agonists (i.e., ACTH, angiotensin II [ANG II] and potassium ions) would be better suited to define the mechanisms regulating steroidogenesis. Such a cell line ideally would grow in culture as a monolayer for ease of handling and treatment, have a relatively short cell cycle duration to rapidly produce experimental material, and chronically maintain steroidogenic capabilities during prolonged culture to ensure continued comparability between experiments. Further, the ability of an adrenal cell line to act as an appropriate model system would require the expression of the various enzymes involved in adrenocortical steroid biosynthesis (Figure 1). Finally, this cell would have greatest model applicability if it were of human origin.

To develop a human adrenal cell line two experimental paradigms have been pursued. First, attempts have been made to immortalize human fetal adrenocortical cells using viral oncogenes. In these studies, cell lines were indeed developed, but the resulting cells no longer produced mineralocorticoids or glucocorticoids (14). Alternately, the isolation of cells from steroidogenic tumors has been used to develop steroid producing cell lines from animals (11,15-18). However, as mentioned above, human steroid hormone producing tumors are rare and even more rarely do they express a diversity of corticosteroid secretory products, including mineralocorticoids. Using this direction, however, two adrenal derived cell lines have been developed which are currently available from the American Type Culture Collection. One of these, the SW13 adrenal cell line is endothelial in appearance and has not been shown to have any steroidogenic capabilities and was most probably derived from a non-adrenocortical cell (19). Thus, the SW13 cell line does not appear to be an appropriate model for adrenocortical cell function and therefore will not be discussed further. More recently another cell line designated NCI-H295 was developed from an adrenocortical carcinoma (20). These cells have been shown to exhibit many of the characteristics of normal adrenocortical cells and have been the source for development of several additional cell strains used as *in vitro* models for studying many aspects of adrenal function.

3. THE NCI-H295 CELL LINE

In 1980, using computer-assisted tomography, an adrenocortical carcinoma was identified in a 48-year-old, black female who presented with weight loss, acne, facial hirsutism, edema, diarrhea and a recent cessation of menses (20) (see Table 1). During surgery, the malignancy was found to have invaded perinephric fat and vascular channels. The excised tumor was 14x13x11 cm and was a typical malignant adrenocortical carcinoma that contained necrotic foci and pleomorphic cells with abundant eosinophilic cytoplasm and large oval to round nuclei with prominent nucleoli. Subsequent metastases detected in lungs and liver were removed in February 1981, and again in September 1981. The patient died in 1982.

Table 1 Tumor origin of NCI-H295 cell lines

Patient age/sex	48/Female
Primary site	Right adnexal mass
Pathology stage and grade	Perinephric fat and vascular channel invasion
Specimen site	Non-necrotic adrenal tumor
Culture method	Minced and dissociated tissue
Authentication	HLA class II antigens were similar between the normal adrenal, the adenocarcinoma and the NCI-H295 cell (Marx, C., et al. *J Clin Endo and Metab* 81:4488, 1996)
Availability	NCI-H295 and NCI-H295R: American Type Culture Collection, Rockville, MD; NCI-295R: W.E. Rainey, UT Southwestern Medical Center, Dallas, TX
Primary reference	NCI-H295: Gazdar, A.F. *Cancer Res* 50:5488, 1990. NCI-H295R: Rainey, W.E. et al. *Mol Cell Endocr* 100:45, 1994

Steroidogenically, the patient demonstrated excesses of all adrenal steroid hormones (20). Serum cortisol was 11.9 μg/dl and 24-hour urinary cortisol, aldosterone and 17-ketosteroid excretion levels were greatly elevated. Thus, the tumor was responsible for causing the patient's virilization, Cushing's syndrome and mineralocorticoid-induced hypertension. Morphologically, cells in the adrenocortical tumor were relatively small and uniform with occasional multinucleated cells. Ultrastructurally, cells within the tumor were either multinucleated or the nuclei were highly convoluted since several nuclear fragments were typically observed in a single cell. The numerous mitochondria observed seemed to be sensitive to preparation and often were without limiting membranes and cleared cristae. The endoplasmic reticulum was not prominent. Epithelial-like intercellular desmosomal attachments and discontinuous basal lamina were observed.

Following development of viable cell cultures from finely minced tumor fragments, tumorigenicity, morphology, and steroid secretions were examined (20). Cultured cells were occasionally adherent, epithelioid and spindle-shaped or were floating cell aggregates without substrate attachment. Xenografts grown in athymic mice were used to confirm tumorigenicity. Xenograft cells were similar in appearance to the original adrenocortical carcinoma but with fewer multinucleated cells. Similar to cells within the tumor, cultured tumor cells maintained epithelioid desmosomal features with increased numbers of microvilli. Nuclei contained prominent nucleoli, while mitochondria were more apparent, more stable and contained lamellar and tubulovesicular cristae. Moderate amounts of vesicular smooth endoplasmic reticulum were observed. Few lipid droplets were observed. Between 7–10 years after the NCI-H295 cultures were established, 30 steroids characteristic of adrenocortical cells were detected in culture medium using gas chromatography and mass spectroscopy or radioimmunoassay. Of the 30 steroids detected, about 20 were identified. Based on secreted steroids, these cells appeared to contain all the adrenocortical enzyme systems which presumably were present in the original tumor, including cholesterol side-chain cleavage (CYP11A), 3β-hydroxysteroid dehydrogenase (3βHSD), 11β-hydroxylase (CYP11B1), 21-hydroxylase (CYP21), 17α-hydroxylase (CYP17), aldosterone synthase (CYP11B2), 3β-hydroxysulfotransferase, and low levels of aromatase (CYP19).

To establish the original NCI-H295 cell line, finely minced tumor tissue and the resulting suspension were maintained in various serum-containing and serum-free culture media for a one year period (20). At the end of one year floating, loosely aggregated cells lacking substrate attachment were selected. Initially, these cells were grown in serum-free medium containing selenium, insulin and transferrin. Subsequently these cells were grown in medium containing 2% bovine serum supplemented with selenium, insulin and transferrin. Cytogenetically the H295 cell is hypertriploid with 30% of the cells containing a modal chromosome number of 62. 3.8% of the cells had higher ploidies (Table 2). There are no Y or Y-like chromosomes. Karyotypically, 38% of the cells contained a total of 65 marker chromosomes, with 25 being common to most cells (Table 2). Chromosomes 1, 7, 9, 14, 16 and 20 contributed to the formation of these marker chromosomes. Most break-union sites occurred pericentrically. Karyotypes of most cells were similar. This cell line is presently available from the American Type Culture Collection as ATCC CRL-10296.

Two substrains were adapted from the NCI-H295 cell line using alternative growth conditions to encourage substrate attachment and shorter cell cycle times. The cell culture medium growth supplement, Ultroser G (2% – BioSepra SA, Villeneuve la Garenne Cedex, France), a relatively defined bovine derived serum substitute, was used to increase cell growth rate. This supplement was selected due to its ability to help retain steroidogenic cell function (21). NCI-H295 cells grown in Ultroser-supplemented medium decreased population doubling time significantly, but the cells still continued to grow as floating aggregates or loosely attached cells. Over a three month

Table 2

Main genetic changes in	NCI-H295
Chromosome translocations	38% of the cells contained marker chromosomes; chromosomes 1, 7, 9, 14, 16 and 20 contributed markers; break-union site were pericentric
Amplifications	Highly aneuploid, hypertriploid, 30% of the cells contained a modal chromosome number of 62

period cells which maintained attachment to plastic culture dishes were selected by changing culture medium every three days and discarding unattached cells. After three months, sufficient adherent cells existed in one dish to permit serial passage. After an additional 1.5 months, an actively growing NCI-H295 cell monolayer population was obtained. After characterization, it was subsequently designated as H295R to differentiate this strain from the original cells. In comparison to the parent H295 cell line, the H295R cells grow as a tightly adherent monolayer and the population doubling time was reduced from five to two days. This substrain is propagated in Dulbecco's modified Eagle's and Ham's F12 media (1:1) supplemented with pyridoxine HCl, L-glutamine, 15 mM HEPES, 1% ITS plus [Collaborative Biomedical Products, Bedford, MA, USA, containing insulin (6.25 mg/ml), transferrin (6.25 mg/ml), selenium (6.25 ng/ml), bovine serum albumin (1.25 mg/ml) and linoleic acid (5.35 mg/ml)], 1% penicillin/ streptomycin (Gibco), 0.01% gentamicin and 2% Ultroser G. Because of the difficulty in the importation of the Ultroser G serum substitute required for maintaining NCI-H295R, a population of these cells was selected to grow in a commercially available serum substitute (Nu-Serum type I – Collaborative Biomedical Products, Bedford, MA, USA). The strain which can be grown in Nu-Serum is available from the American Type Culture Collection as ATCC CRL-2128. Both H295R cell strains are grown and maintained in plastic cultureware at 37°C under a humidified atmosphere of 5% CO_2 in air. Medium is changed every three days. Every five days, on reaching 60–70% confluence, cells are subcultured using 0.25% (w/v) trypsin (split ratio, 1:3). Cells are trypsin sensitive and trypsin-containing cell suspensions must be inactivated with serum, centrifuged at 700 x g, and gently resuspended in standard culture medium before plating. The H295R cell strain has been characterized with regard to hormonal responsiveness, steroidogenesis and expression of steroid metabolizing enzymes (22–40). As detailed below, these cells appear to act as an appropriate model to study adrenocortical steroid hormone synthesis.

Subsequently, another strain of cells designated NCI-H295A have been described which grow as a monolayer (41). The method for isolation of the NCI-H295A is similar to that described above, relying on the removal of non-attached cells with medium changes and therefore selecting a population of cells that grow as a monolayer. The steroidogenic and hormonal responses of the NCI-H295A cells have not thus far been reported. However, as discussed later, the NCI-H295A cells have proven useful to define the mechanisms regulating the transcription of steroid metabolizing enzymes. Finally, the karyotypic changes in the H295R and the NCI-H295A cell strains have not been determined.

The NCI-H295 adrenocortical cell and its related substrains the NCI-H295R and NCI-H295A are not clonal cell populations since they were isolated from a tumor without the isolation of pure clones. Each population of human adrenal cells was selected either to grow as a suspension culture, in the case of the H295 cells, or as adherent cells in the case of the H295R and H295A cell strains. Altering culture conditions such as confluency, frequency of subculturing, serum and serum-substitutes or other conditions, may select for other cell strains with different physiological properties and responses.

The NCI-H295 cell line has been used to screen several anti-tumor and steroidogenic inhibitory drugs. Taxol was a potent cytotoxic, apoptotic, anti-proliferative and steroidogenic-regulating agent for the NCI-H295 human cell (42). This cell line was also used to compare effectiveness of the toxic adrenolytic drug, mitotane, and a methylated homologue, mitometh; mitotane was more effective in suppressing NCI-H295 cell growth (43). The adrenolytic effects of the antiparasitic drug, suramin, were examined using the NCI-H295 cell line. Suramin inhibited cell colony formation as well as glucocorticoid, mineralocorticoid and androgen secretion (44, 45).

4. THE H295 CELL AS A MODEL FOR ADRENOCORTICAL FUNCTION

4.1 Hormonal Response and Expression of Hormone Receptors

The primary physiologic regulators of steroid hormone production in the human adrenal gland are the peptide hormones ACTH and angiotensin II (ANG II), as well as circulating levels of potassium ions (K^+). The original description of NCI-H295 cells did not report on the hormonal responsiveness of these cells (20). One study was unable to induce levels of steroidogenic enzyme transcripts by ACTH or ANG II (46). However, the responses to ANG II, K^+ and ACTH treatment as well as expression of trophic hormone receptors have been characterized for the NCI-H295R cell strain developed in our laboratory. *In vivo* ANG II acts on the adrenal zona glomerulosa to increase production of aldosterone through type 1 ANG II (AT1) receptors. Studies of [^{125}I] radiolabeled Ang II binding to H295R cells in the presence of antagonists to the AT1 and AT2 receptors established that the H295R cells express AT1 receptors almost exclusively (23, 27). The AT1 receptor is coupled to phosphoinositidase C and increases the production of inositol phosphates in H295R cells (23). ANG II, through the AT1 receptor, also increases H295R cell production of aldosterone (23, 27,30). Treatment with Ang II resulted in a 4-fold increase in intracellular cytoplasmic free calcium in fura-2 loaded H295R cells (23). The responsiveness of the H295R cell strain is highly dependent on growth conditions. Because of the bovine components of Ultroser G, its importation into the USA requires a specific permit from the U.S. Department of Agriculture. For that reason we have also grown the H295R cells in NuSerum I (commercially available from Collaborative Research). These cells have maintained the ability to produce steroids in response to agonists of protein kinase A and to a lesser degree ANG II. However, K^+ responses have been lost. Thus when adapting the cells to alternate growth conditions, a characterization of responsiveness may be necessary. In addition to a model of ANG II action, the H295R cell line has been used to study the regulation of AT1 receptor expression. These cells exhibit readily detectable levels of AT1 receptor mRNA, as well as [^{125}I] ANG II binding (23, 27, 30). Expression of transcript and binding appears to be affected in parallel (27,30). Thus, the H295R cell line may be useful to define the mechanisms regulating adrenal cell responsiveness to ANG II.

The other major physiologic regulator of adrenal aldosterone production is K^+. Extracellular K^+ increases intracellular calcium levels in H295R cells, which appears to be the mechanism to increase aldosterone biosynthesis (26,34). Parathyroid hormone (PTH) and parathyroid hormone-related peptides (PTHRP), which have been shown to regulate aldosterone production by freshly isolated glomerulosa cells, also stimulate H295R cells to increase aldosterone synthesis (24). PTH and PTHRP activate steroidogenesis in a cAMP-dependent manner in normal cells and the H295R cell line. Considering these observations, the H295R cell strain appears to be an appropriate model to study the major physiologic regulators of aldosterone biosynthesis.

The primary hormonal regulator of adrenal cortisol production is ACTH. The H295R cell line is only mildly responsive to ACTH. While ACTH treatment did cause an acute increase in aldosterone synthesis, long-term stimulation could not be maintained (24). The low response to ACTH may reflect the low level of ACTH receptor expression in the H295R cell (33). Therefore, most experiments designed to examine the cAMP-dependent pathway require the addition of either forskolin (to activate adenylyl cyclase) or cAMP analogues (25). The low response to ACTH is therefore a drawback of this cell model. Studies directed toward ACTH action would need to be pursued using primary cultures of adrenal cells or the Y-1 mouse adrenal cell line which retains ACTH responsiveness (17). An alternate strategy would be to use transgenic technology to reinstate ACTH receptor expression in the H295 cell line.

4.2 Production of Adrenocortical Steroids

As previously noted, adrenocortical steroid hormone biosynthesis is complex and different in the glomerulosa, fasciculata, and reticularis zones. The clinical features of the patient with the adrenal tumor which gave rise to the H295 cell line indicated that steroids which normally arise from each of the adrenal zones were produced by the tumor. The steroids produced by the original H295 cell line as well as the H295R strain maintain this diversity. The ability of the cells to produce steroids which originate from multiple zones of the adrenal suggests that the H295R cell line remains pluripotent with regard to adrenocortical differentiation. In Gazdar's initial report (20) concerning these cells, a broad spectrum of basal steroid hormone synthesis was noted (30 steroids). The steroid profile was greatly influenced by serum conditions. Using the H295R cell line we have shown that these cells also produce an array of steroids even under basal conditions (23,25). However, treatment with agonists appears to selectively promote the synthesis of certain zone-specific steroid hormone groups. Specifically, treatment with ANG II or K^+ will promote the cells to produce aldosterone (23,26,29,34). While aldosterone does not constitute the major product, this is useful when one considers the difficulty in obtaining primary cultures of aldosterone producing cells. Treatment of the H295R cell strain with agonists working through the cAMP pathway produces a pattern of steroids approaching those of the zona fasciculata and reticularis. Steroids produced during treatment with forskolin include cortisol, 11β-hydroxyandrostenedione, DHEA, corticosterone, 11-deoxycortisol, and androstenedione (25). These data support the proposition that the H295R cell line can act as an appropriate model for adrenocortical steroid hormone biosynthesis and may be useful in defining the various mechanisms causing the synthesis of the "zone-specific" steroids.

4.3 Expression of Steroid-Metabolizing Enzymes

The original NCI-H295, as well as the H295R and H295A cell strains, have been used as genetic models for studying steroidogenic enzyme gene expression. The NCI-H295 adrenocortical cells express all of the enzymes participating in normal human adrenal steroidogenesis (46). In addition, the genes which encode these enzymes respond to the same second messengers controlling normal human adrenocortical function. Expression of genes encoding CYP11A1, CYP17, and CYP21 together with CYP11B1 and CYP11B2 were first studied in the NCI-H295 cells (42, 47). The enzymes encoded by these genes accumulated in response to agonists (8-bromo-cAMP, forskolin, cholera toxin, and 3-isobutyl-1-methylxanthine) which activate the protein kinase-A pathway. Consistent with normal adrenocortical tissue, stimulating the protein kinase-C pathway using phorbol esters resulted in decreased CYP11A1 and CYP17, but accumulation of CYP21. In a similar manner, 8-bromo-cAMP increased CYP11B1, CYP11B2 and unexpectedly aromatase (CYP19) gene activity. Recent studies have demonstrated that mRNAs encoding the five forms of cytochrome P450 known to be involved in adrenal steroidogenesis (CYP11A1, CYP17, CYP21, CYP11B2 and

CYP11B1) are also detectable in the NCI-H295R cell substrain (23-27,31, 35,38,39). The expression of CYP11B2 transcripts were also detectable and results indicate that this transcript could increase in a hormone-sensitive manner, confirming the potential use of this cell line for studying mechanisms which regulate aldosterone production.

In common with adrenal cells from other mammals (48, 49), the levels of mRNA encoding CYP17 and 3βHSD also appear to be differentially regulated in NCI-H295R cells (37,38). Ang II promoted a marked increase in mRNA level for 3βHSD, but only marginally increased the mRNA levels of CYP17. Potent activators of the protein kinase A pathway (forskolin and dbcAMP), while increasing the message for 3βHSD, have a greater effect on levels of CYP17 mRNA. Thus, activation of the protein kinase C and the calcium signaling pathways tended to support a glomerulosa-like steroidogenic enzyme expression, while activation of the protein kinase A pathway promoted fasciculata-like enzyme expression.

Another pair of steroidogenic enzymes which exhibit an adrenal zone-specific distribution are CYP11B2 and CYP11B1. CYP11B2 is expressed only in the zona glomerulosa and is essential for corticosterone conversion to aldosterone (Figure 1). On the other hand, CYP11B1 is expressed only in the zona fasciculata/reticularis and is essential for the production of cortisol from desoxycortisol (Figure 1). The transcripts encoding these two enzymes are increased by treatment of H295R cells with activators of the protein kinase-A pathway, although the effect on CYP11B1 mRNA levels is greater (35). In addition, levels of CYP11B2 mRNA are increased in H295R cells by treatment with ANG II (22,31,35,39) or K^+ (31,35,39). The effects of ANG II on the expression of CYP11B1 was not as pronounced as was observed for the CYP11B2, suggesting that the protein kinase-A and calcium signaling pathways differentially regulate the expression of CYP11B2 and CYP11B1.

The H295R and H295A have also been used to study mechanisms controlling transcription of steroid-metabolizing genes. Fusion genes containing the 5'-flanking DNA from CYP11B1, CYP11B2, CYP17 and 3βHSD have been studied using these cell lines (40,41,50,51). The H295R cell line is the first steroidogenic cell line to maintain the expression of CYP11B2 and CYP17 and therefore has been useful in defining the elements in the 5'-flanking region of these genes which regulate transcription.

ACKNOWLEDGEMENTS

Work on this review was supported by grants from the American Heart Association (95010570) and National Institute of Health (DK43140 and HL03720).

REFERENCES

1. Arnold J. *Arch Pathol Anat Physiol Klin Med.* 1866;35:64-107.
2. McNicol AM. Ed., James VHT *The Adrenal Gland.* 2nd ed. New York: Raven Press; 1992.
3. Simpson ER, Waterman MR. Ed., James VHT *The Adrenal Gland.* 2nd ed. New York: Raven Press; 1992.
4. Neville AM, O'Hare MJ, (Eds) *The Human Adrenal Cortex.* Berlin: Springer-Verlag; p. 35, 1982.
5. Cook DM, Loriaux DL. Eds., Mazzaferri EL, Samaan NA. *Endocrine Tumors.* p. 401, 1993.
6. White PC. *N Engl J Med.* 331:250, 1994.
7. Luton JP, Cerdas S, Billaud L, et al. *N Engl J Med.* 322:1195, 1990.
8. Hornsby PJ, O'Hare, Neville AM. *Endocrinology* 95:(5):1240, 1974.
9. Crivello JF, Hornsby PJ, Gill GN. *Endocrinology.* 113:235, 1983.
10. Crivello JR, Hornsby PJ, Gill GN. *Mol Endocrinol.* 5:1513, 1983.
11. Yasumura Y, Buonassisi V, Sato G. *Cancer Res.* 26:529, 1966.
12. Kowal J. *Biochemistry* 8(5):1821, 1969.
13. Langlois D, Saez JM, Begeot M. *Endocrine Research* 16:31, 1990.
14. Cheng CY, Flasch MV, Hornsby PJ. *Journal of Molecular Endocrinology* 9(1):7-17, 1992.
15. Buonassisi V, Sato G, Cohen AI. *Proc Natl Acad Sci USA*, 48:1148, 1962.
16. Ascoli M, Puett D. *Proc Natl Acad Sci USA*, 75:99, 1978.
17. Schimmer BP. *Methods Enzymol* 58:570, 1979.
18. Rebois RV. *Journal of Cell Biology* 94(1):70, 1982.
19. Leibovitz A, *McCombs WB III, Johnson D*, et al. In Vitro 8:433, 1973.
20. Gazdar AF, Oie HK, Shackleton CH, et al. *Cancer Res.* ;50:5488, 1990.
21. Hornsby PJ, McAllister JM. *Methods in Enzymology* 206:371, 1991.
22. Holland OB, Mathis JM, Bird IM, Rainey WE. *Molecular Cellular Endocrinology.* 94:R9, 1993.
23. Bird IM, Hanley NA, Word RA, et al. *Endocrinology.* 133:1555, 1993.
24. Hanley NA, Wester RM, Carr BR, Rainey WE. *Endocr J.* 1:447, 1993.
25. Rainey WE, Bird IM, Sawetawan C, et al. *J Clin Endocrinol Metab.* 77:731, 1993.
26. Rainey WE, Bird IM, Mason JI. *Mol Cell Endocrinol.* 100:45, 1994.
27. Bird IM, Mason JI, Rainey WE. *Endocrinology.* 134:2468, 1994.
28. Clyne CD, Nguyen A, Rainey WE. *Endocr Res.* 21:259, 1995.
29. Clark B, Pezzi V, Stocco D, Rainey WE. *Mol Cell Endocrinol.* 115:215, 1995.
30. Bird IM, Word RA, Clyne C, Mason JI, Rainey WE. *Hypertension.* 25:1129, 1995.
31. Bird IM, Mathis JM, Mason JI, Rainey WE. *Endocrinology.* 136:5677, 1995.
32. Watanabe G, Albanese C, Rainey WE, et al. *J Biol Chem.* 271:22570, 1996.
33. Mountjoy KG, Bird IM, Rainey WE, Cone RD. *Mol Cell Endocrinol.* 99:R17, 1994.
34. Pezzi V, Clark BJ, Ando S, Stocco DM, Rainey WE. *J Steroid Biochem Mol Biol.* 58:417, 1996.
35. Denner K, Rainey WE, Pezzi V, Bird IM, Bernhardt R, Mathis JM. *Mol Cell Endocrinol.* 121:87, 1996.
36. Bodart V, Rainey WE, Fournier A, Ong H, De Lean A. *Mol Cell Endocrinol* 118:137, 1996.
37. Bird IM, Imaishi K, Pasquarette MM, Rainey WE, Mason JI. *J Endocrinol.* 150:s165, 1996.
38. Bird IM, Pasquarette MM, Rainey WE, Mason JI. *J Clin Endocrinol Metab.* 81:2171, 1996.
39. Pezzi V, Clyne CD, Ando S, Mathis JM, Rainey WE. *Endocrinology.* 138:835, 1997.

40. Clyne CD, Zhang Y, Slutsker L, Mathis MM, White PC, Rainey WE. *Mol Endo.* 11:638, 1997.
41. Rogriquez H, Hum DW, Staels B, Miller WL. *J Clin Endocrinol* Metab. 82:365, 1997.
42. Fallo F, Pilon C, Barzon L, et al. *Endocr Res.* 22:709, 1996.
43. Schteingart DE, Sinsheimer JE, Counsell RE, et al. *Cancer Chemother Pharmacol.* 31:459, 1993.
44. La Rocca RV, Stein CA, Danesi R, Jamis-Dow CA, Weiss GH, Myers CE. *J Clin Endocrinol Metab.* 71:497, 1990.
45. Danesi R, La Rocca RV, Cooper MR, et al. *J Clin Endocrinol Metab.* 81:2238,1996.
46. Staels B, Hum DW, Miller WL. *Mol Endocrinol.* 7:423, 1993.
47. Winqvist O, Karlsson FA, Kampe O. *Lancet* 339:1559, 1992.
48. McAllister JM, Hornsby PJ. *Endocrinology.* 122:2012, 1991.
49. Rainey WE, Naville D, Mason JI. *Endocr Res.* 17:281, 1991.
50. Holland OB, Carr B, Brasier AR. *Endocr Res.* 21:455, 1995.
51. Leers-Sucheta S, Morohashi K, Mason JI, Melner MH. *J Biol Chem.* 272:7960, 1997.

Chapter 7

Thyroid Gland Tumors

Thomas Hoelting
Department of Surgery, University of Heidelberg, Im Neuenheimer Feld 100, 69120 Heidelberg, Germany. Tel: 0049-6221-56-6202; Fax: 0049-6221-56-5450

Tumors of the thyroid gland comprise a broad spectrum of neoplastic phenotypes which include benign and nonprogressive macrofollicular and microfollicular adenomas, well-differentiated adenomas, well-differentiated follicular and papillary carcinomas, and the invasive and always fatal anaplastic carcinomas. Both genetically inherited and environmentally acquired factors contribute to the pathogenesis of thyroid cancer. Thyroid follicle cell proliferation and function is controlled by a network of different regulatory mechanisms including autocrine and paracrine growth factors, oncogenes and growth suppressor genes. The thyroid follicular cell provides a model for studying the molecular genetics of epithelial cancer (66). In 1980 Ambesi-Impiombato described an immortalized thyroid cell line derived from a 5–6 week old rat, FRTL-5, that allowed extensive study of thyroid cell biology (2). The FRTL-5 cells maintained many differentiated characteristics of the thyroid gland, such as thyroglobulin biosynthesis, iodine uptake and thyrotropin (TSH) dependent cAMP accumulation (2).

Some human thyroid follicular cell lines, such as GEJ and HY2-15 are products of cell fusion between thyroid follicular cells and human lymphoid lines (47,52). HY2-15 cells are devoid of detectable post-receptor activity, lack TSH-dependent growth, radioiodine uptake and thyroglobulin secretion. In contrast, GEJ cells show enhanced cell growth, iodine organification (autoregulated transport), and thyroid hormone production in response to bovine TSH. These events are not paralleled by an increase in cyclic AMP production, thus bypassing the classical pathway of cAMP as a second messenger for TSH. The GEJ line has been described as unstable, reverting to a lymphoid phenotype, while the HY2-15 line retains only adenylate cyclase (AC) response to TSH.

J.R.W. Masters and B. Palsson (eds.), Human Cell Culture Vol. 1, 137–148.

Growth of differentiated thyroid carcinomas is thought to be influenced by TSH, as TSH binds to thyroid carcinomas and differentiated thyroid cancers express TSH receptor (10,11). However, TSH lacked effectiveness in differentiated thyroid tumors when thyroidal AC or thyrocyte cell proliferation was assessed *in vitro* or *in vivo* (31). Patients with differentiated thyroid cancer and/or metastases have thyroid tissue which remains responsive to endogenous TSH and this stimulation is manifested by a release of TG (61). Thus, an essential functional characteristic of thyroid tumor cells in culture is the increase of TG in response to (bovine) TSH. The SGHTL-34 human thyroid cell line has been shown to possess a specific TSH-sensitive AC-response, however it does not produce thyroglobulin (63).

Goretzki established a unique model of metastatic follicular thyroid cancer including a primary (FTC133) and two metastatic cell lines (FTC236: lymph node; FTC238: lung metastasis) from the same patient (14,33). FTC133 cells cultured as monolayers show an unorganized, fibroblast-like growth. In contrast, FTC236 and FTC238 cells show epithelial-like configurations and form dense monolayers. Growth of the FTC was similar in serum-free medium or in media containing FCS. The doubling time of FTC133 cells was 27 hours, compared to 31 hours of FTC238. These FTC cell lines have been demonstrated to lack TSH receptors and do not depend on TSH for growth. However, they contain thyroglobulin, have intact thyroid functions and respond to TSH. These cells accumulate cyclic AMP when stimulated by TSH, indicating an intact TSH-AC axis. Interestingly, when stably transfected with the human TSH receptor cDNA, FTC133 grew more slowly and bound more ^{125}I-TSH, suggesting a differentiating effect of TSH in these cells (16).

1. CELL CULTURE

The availability of defined media containing low serum concentrations enabled investigators to obtain clones of continously growing epithelial cells which maintain a differentiated phenotype in culture. The standard protocol for in vitro establishment of thyroid cancer cell lines is as follows. Tumor tissue obtained at surgery is minced into small pieces of less than 1mm^3 in size and washed three times. After centrifugation at 700 rpm for 5 minutes, the pellet is resuspended in RPMI-1640 medium containing 0.2 mg/mL collagenase I and IV (4:1) and incubated at 37°C overnight. The debris is macerated through a screen. The cells are harvested by centrifugation at 700 rpm for 5 minutes, transferred to RPMI-1640 containing 10% FCS, 100 U/mL penicillin and 100 µg/mL fungizone, and plated in culture flasks and the medium is changed every 24–48 hrs.

The pioneering work of Ambesi-Impiombato et al. provided the basis for selective culture conditions of follicular thyroid cells (2). A chemically defined culture medium was developed in which the serum component was partially replaced by six growth factors and hormones: insulin (10 μg/ml), hydrocortisone (10^{-8}M), transferrin (5 μg/ml), somatostatin (10 μg/ml), glycyl-L-histidyl-L-lysine acetate (10 μg/ml) and TSH (1nM). However, they showed that rat thyroid cells require the addition of a low serum concentration of 0.1-0.5% FCS, in the absence of which FRTL-5 failed to attach after trypsinization (2). This experience has been applied to cultures obtained from rat (2), porcine (44) and ovine glands (27).

During the first passages contaminating fibroblasts may be removed by selective detachment. Thereafter withdrawal or a low concentration of fetal calf serum (FCS) can inhibit fibroblast overgrowth in monolayer cultures (13). However, most cell lines do require serum supplement for growth in culture. The growth factors present in FCS may mask the effects of any growth factors tested by placing the cells in a state of enhanced growth stimulation. Cells stocked under liquid nitrogen conditions can be suspended in growth medium with 10% dimethyl sulfoxide (DMSO).

2. GROWTH FACTORS IN THYROID CANCER CELL LINES

Growth of normal and neoplastic thyroid cells is influenced by a network of hormones and growth factors. Increased expression of growth factors or growth factor receptors has been linked to tumor aggressiveness and metastatic disease in thyroid cancers. Tumor cells may achieve "growth autonomy" by producing autocrine or paracrine growth factors or by inactivating the effect of inhibitors. At the advanced stage of metastasis, cell growth may be completely independent of autocrine and paracrine growth factors or inhibitors (19,20,54). TSH is the classic stimulator of thyroid cell function. Clinically, treatment with thyroxin to suppress TSH decreased the risk of thyroid cancer recurrence and improved patient survival (51). Some thyroid cancers grow dramatically when patients are hypothyroid and have elevated serum TSH levels. Experimentally, TSH stimulates thyroid cell growth and differentiated functions. Most thyroid cells in culture require TSH for growth. TSH stimulates growth and invasion of some differentiated thyroid cancer cell lines (40,41,55). On the other hand, when these cells were transfected with and overexpressed the TSH receptor gene, TSH stimulation made them less aggressive (64).

Thyroid cells have EGF-receptors and differentiated thyroid carcinomas bind more EGF than normal thyroid tissue (18). In contrast to TSH, EGF stimulates thyroid cell growth and invasion, but inhibits its differentiated functions, such as iodine uptake and organification. Some papillary thyroid cancers express TGFα which binds to the EGF-receptor (35). Growth and invasion of some cell lines derived from metastases are increased by TGFα (41). IGF and IGF-binding protein are produced by normal and neoplastic thyroid tissues and stimulate their growth via autocrine and paracrine mechanisms (4). Transforming growth factor β (TGFβ) may be an important negative regulator of thyroid cell growth. TGFβ inhibits growth in rat and porcine thyroid cell lines and causes a reduction in TSH-stimulated AC-activity and a downregulation of TSH-receptors (12,53). TGFβ inhibits neoplastic thyroid cell growth and invasion. It probably acts by regulating IGF-1 production (39).

Other growth factors may also influence thyroid cell growth and function. Basic fibroblast growth factor is a strong angiogenesis factor and autocrine growth stimulator and is produced by thyroid cells (21). Upregulation of vascular endothelial growth factor and downregulation of placental growth factor were associated with increasing tumorigenic potential (62). Few investigations of other possible negative regulators of normal and neoplastic thyroid cells have been reported. Interferon-γ (IFN-γ) and tumor necrosis factor α (TNFα) are generally viewed as members of the family of proinflammatory cytokines. These factors upregulate cellular adhesion molecules in human thyroid cancer cell lines (5).

3. MOLECULAR GENETICS

Most benign and all malignant thyroid tumors are thought to arise from clones of thyrocytes with one or more mutations giving a growth advantage. Thyroid tumors of follicular origin appear to develop by means of multiple genetic changes in cells rather than by activation of a single oncogene. Early stages of thyroid tumorigenesis involve activation of at least five different oncogenes: *ras, ret, trk, gsp* and the *TSH-receptor* (*TSHr*).

The adenylate cyclase (AC) – protein kinase A (PKA) pathway stimulated by TSH is important for differentiated thyroid cell function and growth. Acting in synergy with other growth factors such as insulin growth factor I (IGF-I) and insulin, which trigger tyrosine kinase receptors, TSH induces thyroid cell growth (26). The protein kinase C (PKC) pathway induced by EGF and mimicked by phorbol esters is mediated independently and often antagonizes that of TSH (37,38,58). A primary human follicular thyroid cancer cell line transfected with the human TSH receptor cDNA grew more slowly and bound more ^{125}I – TSH than the wild type cells (17), indicating an important role for the TSH receptor in the control of thyroid carcinoma growth. *TSHr*-activating mutations result in the development of autonomously functioning thyroid adenomas. The *gsp*-oncogene codes for the a-subunit of the stimulating G-protein. Any stimulus or genetic alteration which constitutively increases GTP-binding protein activity may result in

Table 1 Differentiated Thyroid Cell Lines

Cell line	Origin	Reference	Patient sex
Animal			
FRTL-5	rat thyroid	Ambesi-Imp.	
PC-Cl-3	rat thyroid transformed with Kirsten murine sarcoma virus	Fusco	
WRT	rat thyroid	Brandi	
OVNIS 5H	ovine thyroid	Fayet	
Cloned cell lines			
GEJ	hybrid	Karsenty	
HY2-15	hybrid	Martin	
Ori-3	SV40	Lemoine	
Ori-5	SV40	Lemoine	
SGHTL-34	pSV 3neo	Whitley	
Benign human cell lines			
hAG	adenom. goiter	Hishinuma	female
KAK-1	foll adenoma	Ain	female
Human follicular cancer (FTC)			
FTC133	primary tumor	Goretzki	male
FTC236	metastasis (lymph node)	Goretzki	male
FTC238	metastasis (lung)	Goretzki	male
HTC-TSHr	primary	Derwahl	male
RO82-W1	metastasis (lymph node)	Estour	female
CGTH W-1	metastasis (sternum)	Lin	female
CGTH W-2	metastasis (skull)	Lin	female
WRO	primary	Fagin	female
XTC-UC1	hurthle cell (lymph node)	Zielke	male
Human papillary cancer (PTC)			
CGTH W-3	primary	Lin	female
PTC - UC3	primary	Hoelting	male
B-CPAP	primary	Fabien	female
PTC-1113A	local recurrence	Hermann	male
PTC-1113L	metastasis (lymph node)	Hermann	male
PTC-1113R	metastasis (lymph node)	Hermann	male
TPC-1	primary	Ishizaka	male
GLAG-66	primary	Antonini	female
K1	primary	Challeton	female
NPA	primary	Fagin	female
ONCO-DG1	primary	Grimm	male
hPTC	primary	Hishinuma	female
Human anaplastic cancer			
SW579	primary	Leibowitz	
ARO81	primary	Fagin	

Continued on next page

Table 1 (continued)

Cell line	Origin	Reference	Patient sex
Human anaplastic cancer			
HTh7	primary	Carlson	
DRO90	primary	Ain	
Human medullary cancer (MTC)			
TT	primary	Leong	
MTK-SK	primary	Pfragner	
DR081	primary	Van Herle	

thyroid cell growth. *Gsp*-activating mutations are found in functioning thyroid adenomas (59). Goretzki et al. showed *gsp*-mutations in 73% of thyroid tumors from Germany and in 20% of thyroid tumors from the United States (33). In these patients, the presence of *gsp* and *ras* mutations correlated with tumor metastasis and patient age, both predictors of poor prognosis.

Ras oncogene activation was associated with growth of rat thyroid derived FRTL-5 cells. TSH stimulated expression of *ras,* in FRTL-5 cells (15). After transfection with activated *K-ras* or *H-ras,* FRTL-5 cells became more aggressive and less differentiated (6,7). In these cells, TSH was no longer required for growth. In human thyroid tumors, *ras* oncogene activations were demonstrated in 20% of papillary and 53% of follicular thyroid cancers (49). However, 50% of microfollicular adenomas also had activated *ras* oncogenes. It is likely, therefore, that *ras* activation occurs early in thyroid tumorigenesis (25). The *ret* – proto – oncogene encodes a receptor tyrosine kinase. The *PTC* - oncogene (papillary thyroid cancer) is a rearranged version of the *ret* – tyrosine kinase – protooncogene, namely a product of the recombination of the tyrosine kinase domain of *ret* with an uncharacterized H4 – gene. The *ret* and the H4 – genes are both located on chromosome 10 (28). Large screening studies showed varying prevalences of *PTC* in papillary thyroid cancers from different geographic areas, ranging from 11% in French patients, 17% in American patients to 33% in Italian patients (29). Initially *ret/PTC* – mutations were considered specific for papillary thyroid cancers and may partly explain the differing clinical behaviour of the different thyroid tumor types. *ret* (*PTC*) – mutations, however, have also been detected in follicular adenomas, in adenomatous goiters and in medullary thyroid carcinomas.

Cellular differentiation may be inhibited by various oncogenes. Santoro et al. demonstrated that a rat thyroid epithelial cell line transfected with *PTC* no longer depends on six growth factors for growth. The *PTC* – oncogene blocked all thyroid differentiation markers: ability to concentrate iodide and expression of thyroglobulin, TSH – receptor and thyroperoxidase. When

Table 2 Features of representative thyroid cell lines in culture

Cell line	Morphology	Tumorigenicity
Animal		
FRTL-5	monolayer; doubling-time: 5-7 days; diploid; cells express thyroglobulin; concentrate iodide	non-tumorigenic
PC-Cl-3	monolayer; oncogene-transfected; malignant phenotype;	non-tumorigenic
WRT	monolayer; euploid; doubling-time: 80 hrs; require low serum concentration for growth; cells express thyroglobulin, concentrate iodide; TSH stimulates cAMP;	non-tumorigenic
OVNIS 5H	monolayer; grow in TSH-free serum; expression of thyroglobulin	non-tumorigenic
Cloned cell lines		
GEJ	hybrid of normal human thyroid cells + human lymphoblastoid cell line GM1500 6TG-A12 expression of TSH-receptor and thyroglobulin	non-tumorigenic
HY2-15	hybrid of human T-cell leukemia line Molt 4-8AG^{R} + human thyroid cells (Graves disease) monolayer, doubling time: 32 hrs; TSH stimulates cAMP and iodine uptake	non-tumorigenic
Ori-3	benign human thyroid cells immortalized after SV40 DNA transfection; monolayer; aneuploid low expression of thyroglobulin and iodide trapping; growth factor independent; positive staining for mutant p53	non-tumorigenic
Ori-5	benign human thyroid cells immortalized after SV40 DNA transfection; monolayer; aneuploid low expression of thyroglobulin and iodide trapping; growth factor independent; positive staining for mutant p53	non-tumorigenic
SGHTL-34	benign human thyroid cells immortalized after SV3 DNA transfection; monolayer; aneuploid thyrotropin-responsive	non-tumorigenic
Benign human cell lines		
hAG	monolayer, doubling time: 60 hrs; express functional TSH-receptors TSH stimulates cell growth	non-tumorigenic
KAK-1	monolayer, doubling time: 42 hrs; express functional TSH-receptor TSH stimulates cell growth	non-tumorigenic
Human follicular cancer (FTC)		
FTC133	monolayer; doubling time: 27 hrs; expression of vascular endothelial growth factor	tumorigenic in nude mice

Continued on next page

Table 2 (continued)

Cell line	Morphology	Tumorigenicity
FTC236	monolayer; doubling time: 34 hrs;	tumorigenic in nude mice
FTC238	monolayer; doubling time: 31 hrs;	tumorigenic in nude mice
HTC-TSHr	monolayer; TSHr-expression; cells secrete TG and concentrate iodine;	no data
RO82-W1	Monolayer culture without follicles; cell nuclei contain one or two nucleoli; cells secrete TG, but do not concentrate iodine;	slow tumor growth in nude mice
CGTH W-1	monolayer; doubling time: 18 hrs; thyroglobulin-negative	tumorigenic in nude mice spontaneous metastases
CGTH W-2	monolayer; doubling time: 20 hrs; thyroglobulin-negative	tumorigenic in nude mice spontaneous metastases
WRO	monolayer; doubling time: 48 hrs; thyroglobulin-expression;	no data
XTC-UC1	monolayer; expression of thyroglobulin and vascular endothelial growth factor;	tumorigenic in nude mice
Human papillary cancer (PTC)		
CGTH W-3	monolayer; doubling time: 20 hrs; thyroglobulin-negative;	tumorigenic in nude mice spontaneous metastases
PTC - UC3	monolayer; doubling time: 58 hrs; expression of epidermal growth factor, thyroglobulin;	tumorigenic in nude mice
B-CPAP	monolayer; positive staining for thyroglobulin, TGFβ1;	tumorigenic in nude mice
PTC-1113A	monolayer; doubling time: 2-5 days	no data
PTC-1113L	monolayer; doubling time: 2-5 days	no data
PTC-1113R	monolayer; doubling time: 2-5 days	no data
TPC-1	expression of thyroglobulin and vascular endothelial growth factor	no data
GLAG-66	monolayer; doubling time: 2 days	non-tumorigenic
K1	monolayer; doubling time: 38 hrs	non-tumorigenic
NPA	monolayer; doubling time: 40 hrs; p53 mutation	slow tumor growth in nude mice
ONCO-DG1	monolayer; positive staining for cytokeratin, vimentin, not for thyroglobulin	no data
hPTC	monolayer, doubling time: 18 hrs; express functional TSH-receptors; TSH inhibits cell growth	non-tumorigenic

Table 3 Molecular genetics

Cell line	TSH	p53	Other
FRTL-5	+	-	-
PC-C1-3	-	-	oncogene-transfected malignant phenotype (Kirsten murine sarcoma virus)
WRT	+	-	-
OVNIS 5H	+	-	-
GEJ	+	-	-
HY2-15	+	-	-
Ori-3	+	+	-
Ori-5	+	-	immortalized after SV40 DNA transfection
SGHTL-34	-	-	immortalized after SV3 DNA transfection
hAG	+	-	-
KAK-1	+	-	-
FTC133	-	+	-
FTC236	-	+	-
FTC238	-	+	-
HTC-TSHr	+	-	-
RO82-W1	+	+	p16
CGTH W-1	no data		
CGTH W-2	no data		
WRO	+	+	-
XTC-UC1	-	-	-
CGTH W-3	-	-	-
PTC-UC3	+	+	+
B-CPAP	+	+	NSE,S100,TGFβ1, no mutations of *PTC*,H-*ras,trk,ret*
PTC-1113A	-	-	*ret*;
PTC-1113L	+	-	-
PTC-1113R	+	-	-
TPC-1	+	-	*ret*
GLAG-66	-	+	-
K1	+	-	*p16*
NPA	+	+	-
ONCO-DG1	-	+	cytokeratin, vimentin
hPTC	+	-	-

additionally transfected with activated Ha – *ras* or Ki – *ras* oncogenes the cell line became highly undifferentiated and malignant (56,57).

Among the tumor suppressor genes that have been identified, *p53* is the most investigated (56,57). *p53* mutations are associated with undifferentiated carcinomas. Ito et al. reported *p53* mutations in 6 out of 7 undifferentiated carcinomas, compared with no mutations found in 10 differentiated papillary carcinomas (46). Fagin et al. showed mutations in 5 out of 6 anaplastic and in 1 out of 11 follicular carcinomas, but no mutations in normal thyroid tissues, follicular adenomas or papillary cancers (24). It is speculated that in human thyroid and other cancers one important biological effect of *p53* mutations may be escape from negative growth controls exerted by transforming growth factor beta. Wyllie et al. found a significant correlation between loss of TGFβ1 responsiveness and aberrant expression of *p53* in follicular thyroid cancer cell lines. TGFβ1 inhibited the growth of normal thyroid cells and thyroid adenoma, but not the growth of thyroid cancer cell lines transformed by SV40 (65). The role of altered expression of nucleoproteins, such as *myc, fos* and *jun* in the pathogenesis of metastatic tumors remains unclear. Three members of the *fos* (*c-fos, fos-B, fra-1*) and of the *myc* family (*c-myc*, L-*myc*, N-*myc*) have been identified. Terrier et al. found increased *c-myc* mRNA levels in 57% of thyroid carcinoma (n = 14/23) and increased *c-fos* mRNA levels in 61%. Poor prognosis was associated with an increased *c-myc* expression (60). Berlingieri et al. reported that the expression of the human *c-myc* oncogene enhanced the malignant phenotype of a rat thyroid epithelial cell line, with complete suppression of its differentiated functions. It has been proposed that transforming growth factor beta 1 mediates growth arrest by inhibition of the transcriptional initiation of *c-myc* (6,7).

4. PATHOLOGY OF XENOGRAFTS

Athymic mice have been used to grow functioning human thyroid xenografts, but few thyroid cancer cell lines have been tested. The metastatic follicular thyroid cancer lines CGTH W-1 and CGTH W-2 as well as the primary papillary thyroid cancer line CGTH W-3 grew well after subcutaneous injection of 10 million cells (50). After two months, subcutaneous tumors were 2–2.5 cm in size. Also, the mice developed spontaneous distant metastases to the infrahepatic area and the omentum. After subcutaneous inoculation of 10 million tumor cells of the follicular thyroid cancer cell line RO82-W1, a slower growth was observed. This line produced tumors that were visible after 4 months and had a size of only 1cm in diameter after 9 months. In contrast to the clinical behavior, the xenografts had a tendency to develop extensive necrosis instead of a follicular growth pattern (22).

FTC133, 236 and 238 were 100% tumorigenic in athymic nude mice, developing tumors of 2.5-3 cm^3 in 3–4 weeks after subcutaneous inoculation of unstimulated tumor cells. Post mortem dissections revealed no spontaneous metastases. After intraperitoneal injection of tumor cells, lymph node metastases were observed in all cases. Liver metastases (transsplenic inoculation) and lung metastases (tail vein injection) developed in 20% of animals (43). This metastatic FTC model has been used for therapeutic studies with antitumor agents (42).

REFERENCES

1. Ain KB, Taylor KD, Banks ER et al. *Thyroid* 2:49, 1992
2. Ambesi-Impiombato FS, Parks LAM, *Coon HG Proc Natl Acad Sci USA* 77:3455, 1980
3. Antonini P, Linares G, Gaillard N et al. *Cancer Genet Cytogenet* 67:117, 1993
4. Bachrach LK, Eggo MC, Mak WW et al. *Biochem Biophys Res Comm* 154:861, 1988
5. Bassi V, Maio M, Altomonte M et al. *Thyroid* 4:33, 1994
6. Berlingieri MT, Potella G, Grieco M et al. *Mol Cell Biol* 8:2261, 1988
7. Berlingieri MT, Akamizu T, Fusco A et al. *Biochem Biophys Res Comm* 173:172, 1990
8. Brandi ML, Rotella CM, Mavilia C et al. *Mol Cell Endocrinol* 54:91, 1987
9. Challeton C, Schlumberger M, Gaillard N et al. *J Endocrinol Invest* 15:32, 1992
10. Clark OH. World J Surg 5:39, 1981
11. Clark OH, Gerend P, Goretzki PE et al. *J Clin Endocrinol Metab* 57:140, 1983
12. Coletta G, Cirafici AM, Vecchio G. *Science* 233:458, 1986
13. Davies TF, Platzer M, Schwartz AE et al. *Clin Endocrinol* 23:469, 1985
14. Demeure MJ, Damsky CH, Elfman F et al. *World J Surg* 16:770, 1992
15. Dere WH, Hirayu H, Rapoport B. FEBS Lett 196:305, 1986
16. Derwahl M, Broecker M, Meyer K. *Acta Endocrinol* 126:50, 1992
17. Derwahl M, Kuemmel M, Goretzki PE et al. *Biochem Biophys Res Comm* 191:1131, 1993
18. Duh QY, Gum ET, Gerend PL et al. *Surgery* 98:1000, 1985
19. Dumont JE, Jauniaux JC, Roger PP. *Trends Biochem Sci* 14:67, 1989
20. Eggo MC, Bachrach LK, Burrow GN. *Growth Factors* 2:99, 1990
21. Emoto N, Isozaki O, Arai M et al. *Endocrinology* 128:58, 1991
22. Estour B, Van Herle AJ, Juillard GJF et al. *Virchows Arch (Cell Pathol)* 57:167, 1989
23. Fabien N, Fusco A, Santoro M et al. *Cancer* 73:2206, 1994
24. Fagin JA. *J Clin Endocrinol Metab* 75:1398, 1992
25. Fagin JA, Matsuo K, Karmakar et al. *J Clin Invest* 91:179, 1993
26. Farid NR, Shi Y, Zou M. *Endocrinol Rev* 15:202, 1994
27. Fayet G, Aouani A, *Hovsepian S FEBS Lett* 194:287, 1986
28. Fusco A, Berlingieri MT, Di Fiore PP et al. *Mol Cell Biol* 7:3365, 1987
29. Fusco A, Grieco M, Santoro M et al. *Nature* 328:170, 1987
30. Gagel RF, Zextinoglu F, Voelkel EF et al. *Endocrinology* 107:516, 1980
31. Goretzki PE, Koob R, Koller C et al. *Acta Endocrinol Suppl* 281:273, 1987
32. Goretzki PE, Grussendorf M, Frilling A et al. *Ann Endocrinol* 50:145, 1989
33. Goretzki PE, Lons J, Stacy-Phipps S et al. *World J Surg* 16: 576, 1992
34. Grimm D, Hofstaedter F, Hoehne HM et al. *Verh Dtsch Ges Pathol* 75:347, 1992
35. Haugen DR, Akslen LA, Varhaug JE et al. *Int J Cancer* 55:37, 1993
36. Hermann ME *Cancer Genet Cytogenet* 89: 14, 1996
37. Hishinuma A, Yamanaka T, Kasai K et al. *Thyroid* 5:41, 1995
38. Hoelting Th, Tezelman S, Siperstein AE et al. *Biochem Biophys Res Comm* 195:1230,1993
39. Hoelting Th, Siperstein AE, Clark OH et al. *J Clin End Metab* 79:401, 1994
40. Hoelting Th, Zielke A, Siperstein AE et al. *Thyroid* 5: 35, 1995
41. Hoelting Th, Siperstein AE, Duh QY et al. *Eur J Endocrinol* 132:229, 1995
42. Hoelting Th, Duh QY, Clark OH et al. *J Clin Endocrinol Met* 81:2638, 1996
43. Hoelting Th, unpublished data
44. Hovsepian S, Aouani A, Fayet G. *Mol Cell Endocrinol* 45:119, 1986
45. Ishizaka Y, Itoh F, Ikeda I et al. *Jpn J Cancer res* 80:1149, 1989
46. Ito T, Seyama T, Mizuno T et al.. *Cancer Res* 52:1369, 1992
47. Karsenty G, Michel-Bechet M, *Charreire J Proc Natl Acad Sci USA* 82:2120, 1985
48. Lemoine NR, Mayall ES, Wyllie FS *Cancer Res* 48:4459, 1988
49. Lemoine NR, Mayall ES, Jones T et al. *Br. J Cancer* 60:897, 1989

50. Lin JD, Chao TC, Weng HF et al. *J Surg Oncol* 63:112, 1996
51. Mazzaferri EL, Young RL, Oertel JE et al. *Medicine* 56:171, 1977
52. Martin A, Platzer M, Davies TF *Mol Cell Endocrinol* 60:233, 1988
53. Morris JC III, Ranganathan G, Hay ID et al. *Endocrinology* 123:1385, 1988
54. Nicolson GL. *Cancer Met Rev* 12: 325, 1993
55. Roger PP, Dumont JE. *Biochem Biophys Res Comm* 149:707, 1987
56. Santoro M, Carlomagno F, Hay ID et al. *J Clin Invest* 89:1517, 1992
57. Santoro M, Melillo RM, Grieco M et al. *Cell Growth Diff* 4: 77, 1993
58. Smith P, Wynford-Thomas D, Stringer BMJ et al. *Endocrinol* 119:1439, 1986
59. Suarez HG, Du Villard JA, Caillou B. *Oncogene* 2:403, 1988
60. Terrier P, Sheng ZM, Schlumberger M et al. *Br J Cancer* 57:43, 1988
61. Van Herle AJ, Uller RP. *J Clin Invest* 56:272, 1975
62. Viglietto G, Maglione D, Rambaldi M et al. *Oncogene* 11:1569, 1995
63. Whitley G ST J, Nussey SS, Johnstone AP *Mol Cell Endocrinol* 52:279, 1987
64. Wright PA, Lemoine NR, Goretzki PE. *Oncogene* 6:1693, 1991
65. Wyllie FS, Dawson T, Bond JA et al. *Mol Cell Endocrinol* 76:13, 1991
66. Wynford-Thomas D. *Trends Endocrinol Metab* 4:224, 1993
67. Zielke A, Tezelman S, Jossart G et al. *Thyroid* 9, 1994

Chapter 8

Pituitary Gland Tumors

Leo J. Hofland and Steven W.J. Lamberts
Department of Internal Medicine III, University Hospital Dijkzigt, Dr Molewaterplein 40, 3015 GD Rotterdam, The Netherlands. Tel: 0031-10-463-4633; Fax: 0031-10-463-5430; E-mail: hofland@inw3.azr.nl

Six types of human pituitary gland adenomas can be distinguished using morphological, ultrastructural and immunocytochemical criteria. These include somatotrophic adenomas (growth hormone (GH) secreting adenomas), prolactinomas (prolactin (PRL) secreting adenomas), corticotrophic adenomas (proopiomelanocortin derivate secreting adenomas), thyrotrophic and plurihormonal adenomas (thyrotrophin (TSH) and multiple hormone secreting adenomas), gonadotrophic adenomas (luteinizing hormone (LH) and/or follicle stimulating hormone (FSH) or α-subunit secreting adenomas) and non-functioning adenomas (α-subunit secreting adenomas and adenomas unable to synthesize or secrete hormones). Within these groups of pituitary gland adenomas there is considerable cellular heterogeneity. For example, in GH-secreting pituitary adenomas the percentage of GH-secreting cells varies from 5–100% (1), whereas in approximately 50% of these adenomas the cells produce more than one hormone (PRL and/or α-subunit) (2–5). Cellular heterogeneity has been demonstrated also in the other types of pituitary adenomas (1,6). Taking into account this heterogeneity between, as well as within, these tumors it is clear that the use of pituitary cell lines as a model to study the mechanisms governing the regulation of hormone secretion and synthesis of human pituitary adenomas must be interpreted with care. A number of murine pituitary cell lines have been established (rat GH cells and mouse AtT20 cells (7)) and extensively studied with respect to their responsiveness to hypothalamic regulatory peptides and the mechanisms involved in the regulation of hormone secretion. However, few human pituitary cell lines have been established, perhaps related to the difficulty of maintaining the capacity of the tumor cells to produce and synthesize hormones during long-

J.R.W. Masters and B. Palsson (eds.), Human Cell Culture Vol. 1, 149–159.

term culture. In the majority of cases, hormone production rapidly declines to undetectable levels. Most studies on the regulation of hormone release and synthesis in human pituitary adenomas have been performed with primary cultures. In this chapter we describe the methods and culture conditions for primary cultures of human pituitary adenomas and cell lines. In addition, an overview is presented of the studies that have been described using human pituitary tumor xenografts.

1. PRIMARY CULTURE

Tissue dissociation: The texture of pituitary adenoma tissue resected by transsphenoidal surgery varies considerably. While most adenomas have a tight texture, some tumors are so loose that a dispersed tumor cell suspension can be obtained by simple mechanical dispersion with an all-glass Dounce tissue grinder (Wheaton Scientific, Millville, USA) (8). Other adenomas have to be minced with scissors, while tissue fragments are subsequently enzymatically dissociated into single cell suspensions using dispase, a gentle, neutral protease from Bacillus polymyxa (8). Some investigators have used trypsin or collagenase to prepare dissociated cell suspensions, while others used a nonenzymatic procedure only (9). We have obtained a high cell yield when the adenoma tissue is collected by aspiration during transsphenoidal surgery. A minor disadvantage is that the cell suspensions are contaminated with some leukocytes and large numbers of erythrocytes. The erythrocytes, as well as non-viable cells, are easily removed by density centrifugation with Ficoll-Isopaque (d=1.077 g/ml) (8).

Culture requirements: Due to the considerable variability between pituitary adenomas with respect to the amount of hormone(s) secreted and their sensitivity to stimulatory and inhibitory regulatory neuropeptides and drugs, it is difficult to establish an optimum culture medium. In our hands, good results have been obtained by culturing the cells in Minimal Essential Medium (MEM) supplemented with non-essential amino acids (1%), sodium pyruvate (1 mM), L-glutamine (2mM), penicillin (100 U/ml) and 10% fetal calf serum (FCS) (8,10). MEM containing D-valine may be used instead of MEM, because of its fibroblast growth inhibitory property. In our experience, the characteristics of hormone secretion, as well as the sensitivity of hormone secretion to drugs like the somatostatin analog octreotide and the dopamine agonist bromocriptine, closely resemble that of the pituitary adenoma in vivo. Other investigators have used different media with good results. In Ham's F10 medium supplemented with 20% FCS or 15% horse serum (HS)/2.5 % FCS, functional pituitary adenoma cell cultures responsive to thyrotrophin-releasing hormone (TRH) and dopamine could be maintained and propagated in monolayer or suspension culture for up to 9 months (9).

Serum-free defined media supplemented with TRH, thyroid hormone (T_3), parathyroid hormone (PTH), insulin and transferrin have been used (9,11). In a GH_3 rat pituitary tumor cell line these substances, together with IGF-I and FGF were shown to be essential serum-replacing compounds (11). Wilfinger et al. (12) compared various media and supplements and showed in rat anterior pituitary cell cultures that the quantity of PRL production by the cells in α-MEM with 5% HS was consistently higher than that in all other combinations of media and sera examined (12). Finally, because serum (FCS and/or HS) contains many undefined factors, it is advisable to test each batch of serum which is used for primary cultures (and for cell lines) on primary cultures of rat anterior pituitary cells or on established rat pituitary tumor cell lines.

Glucocorticoids and T_3 increase GH synthesis by human GH-secreting pituitary adenomas in vitro and may therefore be used to prevent the decline in GH production by cultured human GH-secreting pituitary adenoma cells (9,13,14). However, hormonal additives should be used with caution, since they may alter the responsiveness of the tumor cells to stimulatory and inhibitory drugs and neuropeptides (14).

Characteristics of primary cultures: A major difficulty of the use of primary cultures of human pituitary adenomas is the rapid decline of hormone secretion and synthesis within weeks to months of culture (9,14–16). This decline may be related to the absence of (stimulatory) hypothalamic influences on the tumor cells in vitro.

Plating efficiency of human pituitary tumor cells varies considerably. We have found that after three days of attachment plating efficiency was greater than 95% for GH-secreting adenoma cells, but lower for PRL-secreting adenoma cells (60–80%) and ACTH-secreting pituitary adenoma cells (about 80%) (8). The morphological appearance of the cells derived from different types of pituitary adenomas is quite different. While the majority of cultured human GH-secreting pituitary adenoma cells show spindle-shaped cells (8, 15, figure 1A), PRL-secreting pituitary adenoma cells are often rounded and form strains. Clinically non-functioning pituitary adenoma cells appear either as attached epithelial cobble-stone like cells or they form strains of rounded cells (our experience). The majority of studies concerning human pituitary adenoma cell cultures have used regular multiwell culture plates. For long-term studies we obtained excellent results by culturing the adenoma cells in collagen-coated tissue culture inserts with microporous membranes (Transwell®-COL, Costar, Badhoevedorp, The Netherlands) (17–19). These have the advantage of preventing selective cell loss during medium replenishment because the Transwells® containing the cells can be easily transferred to wells containing fresh medium. Our experience is that there is a less rapid decline of hormone production by the tumor cells, compared with cells cultured in regular multiwell culture plates. Other investigators have published similar findings (20). Figure 2 shows hormone production by human pituitary

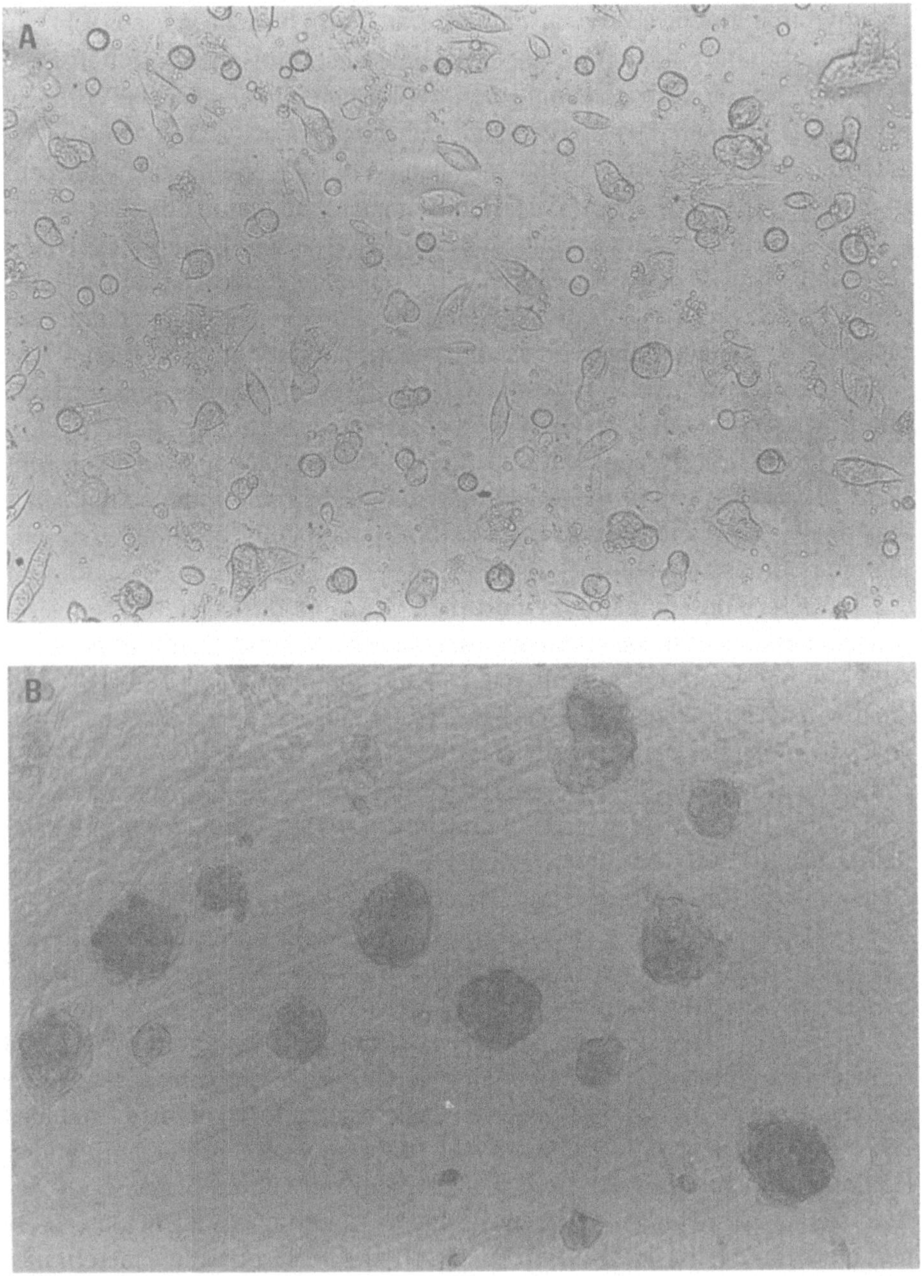

Figure 1 continued on next page

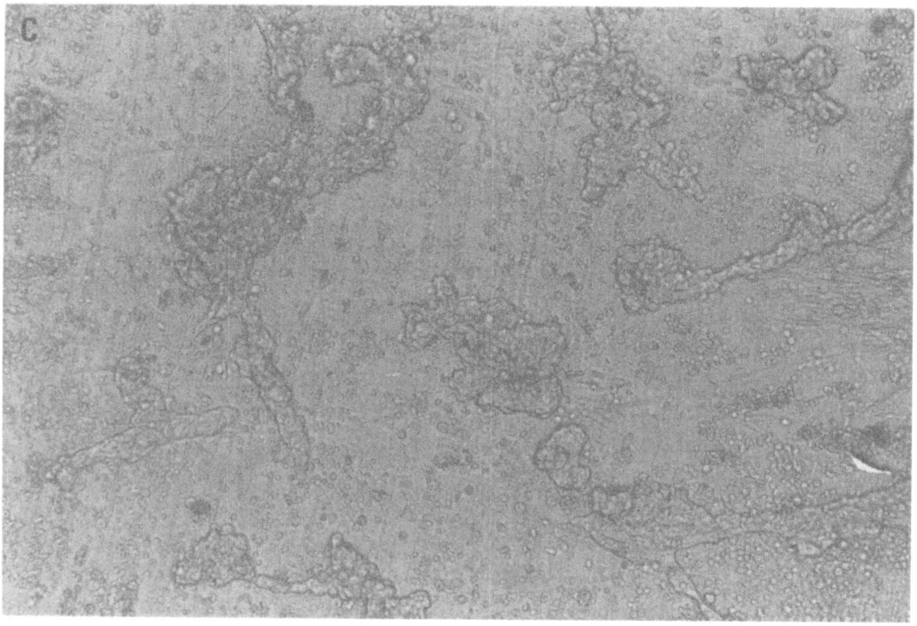

Figure 1. Morphological appearance of cultured cells from three human GH-secreting pituitary adenomas. **A**: cells cultured in a regular multiwell culture plate, **B** and **C**: cells cultured in Transwell®-COL microporous tissue culture inserts. Note the difference in appearance between adenoma cells shown in **A** and that of adenoma cells shown in **B** and **C**, as well as the difference between the two adenoma cell cultures shown in **B** and **C**

adenoma cells cultured for a period up to 3 weeks in Transwell®-COL inserts.

When cultured in Transwells®, the cells usually have a different morphology compared with cells cultured in regular multiwell culture plates. In the majority of the cultures, the tumor cells appear as cell clusters (figure 1B and 1C). Responsiveness to inhibitory drugs like bromocriptine and octreotide is also retained in Transwell®-cultured pituitary adenoma cells. Figure 3 shows a typical example of the effect of somatostatin (SS-28) and the somatostatin analog octreotide on α-subunit secretion by cells of a clinically nonfunctioning pituitary adenoma cultured in Transwells®. After drug withdrawal on day 14 of culture, α-subunit secretion recovered, suggesting that the increasing inhibitory action of both compounds represented inhibition of hormone release and/or production and not inhibition of cell proliferation.

2. ESTABLISHED CELL LINES

The availability of established cell lines of human pituitary adenomas would help in the study of the mechanisms involved in the regulation of hormone production by human pituitary adenomas, especially because of the

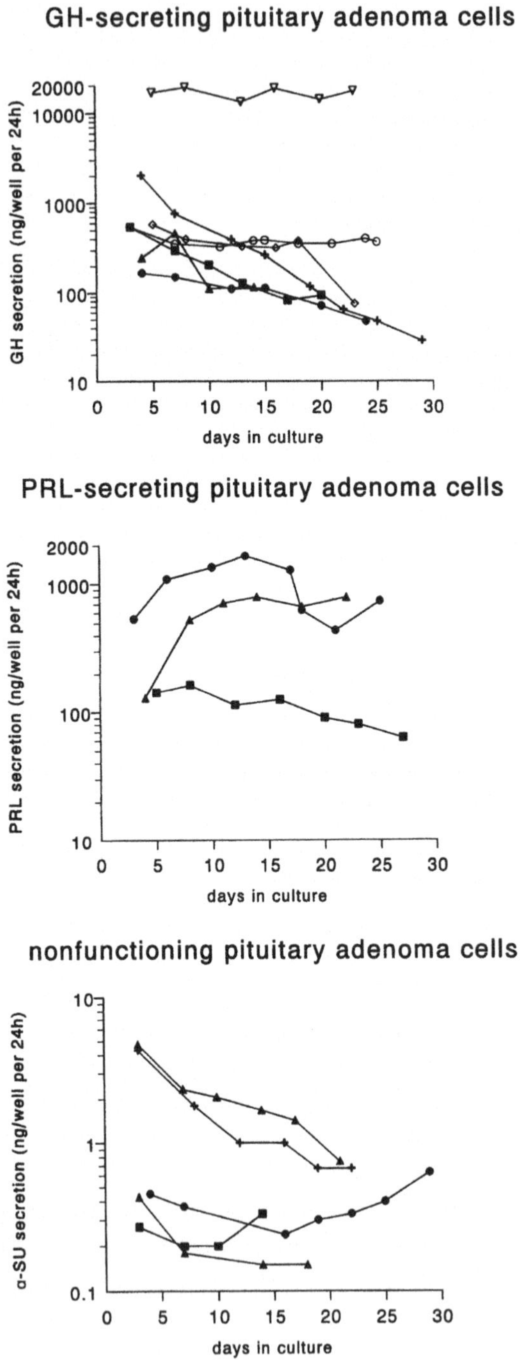

Figure 2. Hormone secretion by primary cultures of human GH-secreting pituitary adenomas (seven cultures, upper panel), prolactinomas (three cultures, middle panel) and clinically nonfunctioning pituitary adenomas (five cultures, lower panel), on various days of culture, expressed in ng/well per 24h. Values are the mean of four wells and SEM's were less than 10%. The cells were cultured in Transwell®-COL tissue culture inserts.

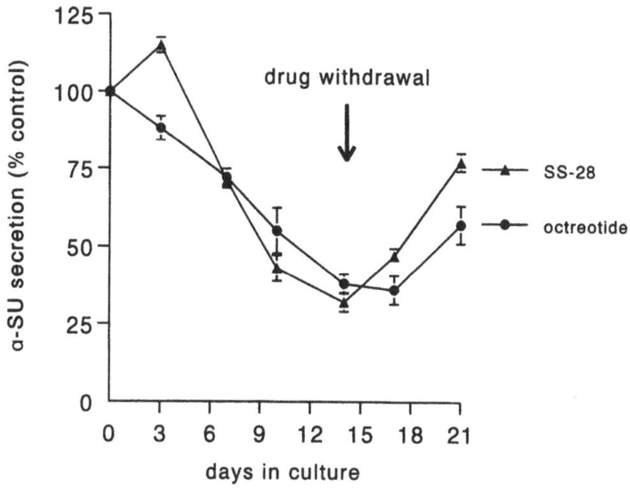

Figure 3. The effect of long-term incubation with somatostatin (SS-28) and the SS-analog octreotide on α-subunit secretion by a primary culture of human clinically nonfunctioning pituitary adenoma cells. The arrow at day 14 of culture represents the time-point of drug withdrawal. The cells were cultured in Transwell®-COL tissue culture inserts.

often limited availability of adenoma tissue obtained during transsphenoidal surgery. However, no well-characterized cell lines derived from human pituitary adenomas exist. This seems to be related to the fact that tumor cell survival during prolonged culture is limited and that hormone production often declines to undetectable levels after several weeks to months in culture. Loss of differentiated characteristics is a frequent feature of long-term culture of most cell types. In addition, proliferative indices of human pituitary adenomas are usually very low. The number of proliferating cells in the majority of pituitary adenomas or primary cultures amounts to less than 0.5% of the total tumor cell population (21). The three human pituitary lines so far established are derived from a PRL-secreting (HPA, ref. 22), a GH-secreting pituitary adenoma (GX, ref. 23) and a PRL-secreting cell from a human pituitary adenoma of unknown origin (24). A fourth 'human' pituitary tumor cell line (designated as "18-54 and 18-54, SF cells", refs. 25-27), was originally reported to be derived from an acromegalic patient, but was later found to be of rat origin (28). These cell lines are not available from cell banks. Finally, a human embryonic pituitary cell line has been described (Flow 9000, refs. 29–32). To our knowledge, no established cell lines derived from other types of pituitary adenomas exist. The few pituitary cell lines descibed are listed in tables 1 and 2, and some of their characteristics are described below.

HPA: This cell line was established from a human PRL-secreting pituitary adenoma (22). The HPA cell line was prepared from enzymatically dispersed

Table 1 Characteristics of human pituitary cell lines

Cell line	Patient age/sex	Specimen site	Culture method	Authentication	Primary reference
HPA	?	pituitary	D	not investigated	22
GX	6yr/M	pituitary	D	not investigated	23
Not named	?	pituitary	D	not investigated	24
Flow 9000	?	pituitary	D	not investigated	29,30

D = dissociated cells

Table 2 Pathology of human pituitary cell lines

Cell line	Description of tumor pathology	In vitro features	Xenograft pathology
HPA	prolactinoma	PRL secretion, PRL secretion inhibited by bromocriptine, 9% colony forming efficiency	slowly growing, non-metastasizing, cells smaller than original tumor, high number of mitoses (original tumor mitoses virtually absent)
GX	somatotroph hyperplasia	GH and PRL mRNA expression, no GH and PRL production, GH mRNA expression stimulated by T_3	not investigated
not named	prolactinoma	PRL secretion, PRL secretion stimulated by TRH, VIP and EGF	not investigated
Flow 9000	–	no pituitary hormone secretion, cells express functional CCK-8 and muscarinic receptors	not investigated

cells cultured in RPMI 1640 supplemented with 10% FCS. The culture medium was changed every 3 days and the cells passaged when a confluent monolayer was achieved. After several passages (more than 2 months after initiation of the primary culture), pure cultures of morphologically transformed cells were maintained and cultured in medium containing 10% FCS. After 30 passages in vitro, the cells demonstrated a short doubling time (14 h) and a low plating efficiency (9%). PRL secretion by the cells rapidly declined from the primary culture at an initial mean level of 756 ng/ml to 148 ng/ml on day 22. After day 90, PRL production continued to decline and amounted to only 5–10 ng/ml after day 150 of maintenance. Interestingly, the HPA cells were tumorigenic in vivo in nude mice and remained responsive to the dopamine-agonist bromocriptine in vivo and in vitro.

GX: The human somatomammotroph GX pituitary cell line (23) was originally isolated from pituitary tissue of a 6-year old boy with gigantism.

This tissue was histologically characterized as somatotroph hyperplasia with focal adenomatous transformation (33). Enzymatically dispersed cells were plated in α-MEM supplemented with T_3 (5 nmol/L), corticosterone (2 nmol/L), human growth hormone releasing hormone (GHRH, 0.1 nmol/L), and antibiotics. At intervals of 2–3 weeks the cells were passaged by trypsinization and the GX cell line has been maintained in culture for over 2 years in the plating medium, modified by the omission of GHRH. For experimental studies the cells were cultured in medium containing 10% charcoal-absorbed horse serum. Initially, the GX cultures secreted a large amount of GH. However, after 5 months in culture GH secretion had declined to undetectable levels, although there was persistence of transcription of the GH and PRL genes. T_3 stimulated GH mRNA, but not PRL mRNA, accumulation in these cells.

PRL-secreting cell line (not named): This cell line was established from material from a patient with a pituitary adenoma (24). The type of pituitary adenoma was not described. Enzymatically dissociated cells were plated in Ham's F10 medium containing 10% FCS. After about 6 weeks in culture dividing cells were observed. Aliquots of these cells were stored frozen in liquid nitrogen. The remaining cells were subcultured every 1–2 weeks and continued to secrete PRL for at least 3 months. In vitro, PRL secretion by these cells was stimulated by Vasoactive Intestinal Peptide (VIP), thyrotrophin releasing hormone (TRH) and epidermal growth factor (EGF).

Flow 9000 embryonic pituitary cells: The human Flow 9000 cell line was originally derived from human pituitary. So far, this cell line has not been extensively used or investigated. Flow 9000 cells (passage 17–26) are routinely cultured in Ham's F10 medium containing glutamine (4 mM), non-essential amino acids (1%), 10% HS, 2.5% FCS, and antibiotics. The cells have been shown to express functional CCK-8 and muscarinic receptors (29,30).

3. XENOGRAFTS

Probably due to the benign character of the majority of human pituitary adenomas, the results of xenografting these adenomas in athymic nude mice have been poor. While many malignant human tumors have been shown to grow rapidly in athymic nude mice (34–36), no significant "take" was observed when human pituitary adenoma tissue was transplanted (37–44). In the majority of the studies, the transplanted adenoma tissue showed no macro- or microscopic signs of growth and eventually decreased in size. It has been proposed that this lack of xenograft growth in nude mice was due to the absence of stimulating hypothalamic influences (38,45,46). Some evidence for this hypothesis has been presented by Landolt et al. (38). These investigators transplanted fragments from four pituitary adenomas (3 prolactinomas, 1 ACTH-producing pituitary adenoma) into the pituitary fossa of total-body irradiated, hypophysectomized rats. After two weeks they demonstrated vascularization of the grafts from the pituitary stalk, survival of some adenomas, and numerous mitoses in an ACTH-secreting tissue specimen. However, these experiments resulted in a high rate of mortality of the experimental animals, which hampers the practical use of this model.

In one study, a higher success rate was reported. Puchner et al. (47) were able to maintain 37% of transplanted adenoma fragments up to 46 days without alterations in morphological characteristics of GH-secreting pituitary adenomas. However, they did not observe tumor growth. The response rate of the transplanted GH-secreting pituitary adenomas to GHRH was as high as that in acromegalic patients. The model was used to demonstrate receptor binding in vivo of [^{125}I]-GHRH to tumor tissue (48).

4. SUMMARY

Few human pituitary tumor cell lines have been established. This scarcity may be related to the fact that hormone production rapidly declines within weeks to months after initiation of the culture. In addition, the number of proliferating cells in pituitary adenomas is very low. The use of primary cultures of human pituitary adenomas to study the mechanisms underlying the regulation of hormone synthesis and release is limited by the small quantity of tumor tissue which can be obtained by transsphenoidal surgery. The use of primary cultures yields considerable information regarding the sensitivity of hormone secretion by pituitary adenoma cells to drugs like somatostatin analogs and dopamine agonists as well as to various stimulatory and inhibitory neuropeptides. The establishment of growing tumors in nude mice transplanted with human pituitary adenomas has also been shown to be difficult and of limited success. Therefore, the study of human pituitary gland tumor cell function is limited mainly to primary cultures of human pituitary adenomas.

REFERENCES

1. Trouillas J, et al. In: A.M. Landolt, P.U. Heitz, J. Zapf, E. Giraud, E. del Pozo (eds) *Advances in pituitary adenoma research*, vol. 69 pp. 11-20, 1987.
2. Kanie N, et al. *J Clin Endocrinol Metab* 57: 1093, 1983.
3. Riedel M, et al. *Virch Arch (Pathol Anat)* 407: 83, 1985.
4. Landolt AM, et al. *Virch Arch (Pathol Anat)* 409, 417, 1986.
5. Bassetti M, et al. In: A.M. Landolt, P.U. Heitz, J. Zapf, J. Giraud and E. del Pozo (eds). *Advances in pituitary adenoma research vol 69*. pp. 23-26, 1987.

6. Heshmati HM, et al. *Acta Endocrinol (Copenh)* 118: 533, 1988.
7. Tashjian AH. *Methods Enzymol* 58: 527, 1979.
8. Oosterom R, et al. *J Endocrinol Invest* 7: 307, 1984.
9. Melmed S, et al. *In Vitro* 18: 35, 1982.
10. Oosterom R, et al. *Mol Cell Endocrinol* 29: 197, 1983.
11. Hayashi I, et al. *In Vitro* 14: 23, 1978.
12. Wilfinger WW, et al. *Endocrinology* 105: 530, 1979.
13. Oosterom R, et al. *Endocrinology* 113: 735, 1983.
14. Oosterom R, et al. *J Endocrinol* 100: 353, 1984.
15. Adams EF, et al. *J Clin Endocrinol Metab* 49: 120, 1979.
16. Kohler PO, et al. *Metabolism* 18: 782, 1969.
17. Kwekkeboom DJ, et al. *J Clin Endocrinol Metab* 71: 718, 1990.
18. Hofland LJ, et al. *Clin Endocrinol* 37: 240, 1992.
19. Hofland LJ, et al. *J Clin Endocrinol Metab* 82: 3011, 1997.
20. Kabuto M, et al. *Neurol Res* 15: 304, 1993.
21. Atkin SL, et al. *J Neurosurg* 87: 85, 1997.
22. Kikuchi Y, et al. *Cancer Res* 45: 5722, 1985.
23. Chomczynski P, et al. *J Clin Endocrinol Metab* 77: 381, 1993.
24. Prysor-Jones RA, et al. *J Endocrinol* 114: 119, 1986.
25. Noteboom WD, et al. *J Steroid Biochem* 16: 633, 1982.
26. Hosojima H, et al. *Horm Res* 21: 240, 1985.
27. Wyche JH, et al. *Endocrinology* 104: 1765, 1979.
28. Rosenfeld RG, et al. *Biochem Biophys Res Commun* 138: 304, 1986.
29. Lo WWY, et al. *FEBS Lett* 220: 327, 1987.
30. Lo WWY, et al. *FEBS Lett* 226: 67, 1987.
31. Lo WWY, et al. *Biochem J* 251: 625, 1988.
32. Sharif NA, et al. *Neurosci Lett* 86: 279, 1988.
33. Zimmerman D, et al. *J Clin Endocrinol Metab* 76: 216, 1993.
34. Rae-Venter B, et al. *Cancer Res* 40: 95, 1980.
35. McManus MJ, et al. *Cancer Res* 38: 2343, 1978.
36. Sharkey FE, et al. In: Fogh J, Giovanelli BC (eds), *The nude mouse experimental and clinical research,* pp. 187-214, Academic Press, New York, 1978.
37. O'Sullivan JP, et al. *J Endocrinol* 79: 139, 1978.
38. Landolt AM, et al. In: Drome PJ, Jedynak CP, Peillon F (eds), Pituitary adenomas. Biology, physiopathology and treatment, pp. 49-57, Asclepios Publishers, Paris, France, 1980.
39. Ueyama Y, et al. In: Nomura T, Ohsawa N, Tamaoki N, Fujiwara K (eds) *Proceedings of 2nd International Workshop on Nude Mice,* pp. 365-374, Fisher, Stuttgart, 1977.
40. Shin S, et al. In: Nomura T, Ohsawa N, Tamaoki N, Fujiwara K (eds) *Proceedings of 2nd International Workshop on Nude Mice,* pp. 375-393, Fisher, Stuttgart, 1977.
41. Oosterom R, et al. *Horm Res* 17: 57, 1983.
42. Bullard DE, et al. *J Neuropathol Exp Neurol (abstr)* 38: 306, 1979.
43. Kameya T, et al. In: Nomura T, Ohsawa N, Tamaoki N, Fujiwara K (eds) *Proceedings of 2nd International Workshop on Nude Mice*, pp. 406-416, Fisher, Stuttgart, 1977.
44. Ohsawa N, et al. In: Nomura T, Ohsawa N, Tamaoki N, Fujiwara K (eds) *Proceedings of 2nd International Workshop on Nude Mice*, pp. 396-404, Fisher, Stuttgart, 1977.
45. Landolt AM, et al. *Cell Tissue Res* 221: 269, 1981.
46. Reid LCM, et al. In: Fogh J, Giovanelli BC (eds), *The nude mouse experimental and clinical research,* pp. 313-351, Academic Press, New York, 1978.
47. Puchner MJA, et al. *Horm Res* 35: 198, 1991.
48. Puchner MJ, et al. *Mol Cell Endocrinol* 85: 157, 1992.

Chapter 9

Salivary Gland Tumors

Mitsunobu Sato
Second Department of Oral and Maxillofacial Surgery, Tokushima University School of Denistry, 3 Kuramoto-cho, Tokushima 770, Japan. Tel: 0081-886-33-7353; Fax: 0081-886-33-7462; E-mail: satomitu@dent.tokushima-u.ac.jp

It has been proposed that tumors of the salivary gland arise from the intercalated duct cells, which can differentiate into both myoepithelial cells and acinar cells (1-3). In order to examine this hypothesis, a neoplastic intercalated duct cell line HSG, established from a human submandibular gland, has been used as a model for studying mechanisms regulating cytodifferentiation. In 1977, HSG cells were established from the submandibular gland resected from a patient who had received therapeutic irradiation for carcinoma of the floor of the mouth (4). Irradiation to the salivary gland causes acute lethal damage to various types of epithelial cell, including acinar, striated, and intercalated duct cells, and disturbs the glandular structure. Although this degenerative change persists without repair for a long period, intercalated duct cells can proliferate excessively among the damaged epithelia (5,6). This proliferation of surviving intercalated duct cells may be associated with the appearance of salivary gland tumors.

In 1987, two sublines (HSG-AZA1 and HSG-AZA3) were cloned after treatment of HSG with 5-azacytidine (7). HSG-AZA1 cells exhibit a phenotype similar to that of myoepithelial cells, including microfibrils and myosin expression, and form a myoepithelioma on transplantation into athymic nude mice. HSG-AZA3 cells have a phenotype similar to that of acinar cells, including cytoplasmic expression of secretory granules containing salivary amylase, and xenotransplantation of HSG-AZA3 cells results in the production of acinar cell carcinoma. This differentiation indicates that the neoplastic human salivary intercalated duct cell line HSG can differentiate into both myoepithelial cells and acinar cells. In addition, HSG-AZA11 cells,

J.R.W. Masters and B. Palsson (eds.), Human Cell Culture Vol. 1, 161–178.

which have a neuron-like cell phenotype, were cloned from 5-azacytidine-treated HSG cells in a similar manner (8).

In 1987, the parotid gland adenocarcinoma cell line HSY was established from an explant culture of a xenograft in a nude mouse (9). HSY cells have an ultrastructure similar to that of human salivary intercalated duct cells and express both amylase and vasoactive intestinal polypeptide (VIP).

In 1986, a neoplastic epithelial cell line, TYS, was isolated from an explant culture of a well-differentiated squamous cell carcinoma expressing carcinoembryonic antigen (CEA) that arose in human oral mucosa (10). Expression of CEA and amylase as well as ample tonofilaments were detected in cultured TYS cells. Transplantation of the cells into athymic nude mice resulted in the development of adenosquamous cell carcinoma containing CEA and amylase. Cultivation of TYS cells in the presence of sodium butyrate resulted in suppression of cell growth and production of secretory granules with human salivary amylase in the cytoplasm of the cells. When the sodium butyrate-treated cells were transplanted into athymic nude mice, a small mass developed transiently at the inoculation site and then disappeared. A biopsy of this mass was histopathologically interpreted as acinic cell carcinoma with squamoid features. These findings suggest that we have established a human adenosquamous cell line, presumably derived from a minor salivary gland present in oral mucosa.

In 1993, immortalized normal human salivary gland cells were established (11). Primary cultures of human submandibular gland cells were transfected with origin-defective mutant DNA of SV40. Using limiting dilution, 4 cell clones (NS-SV-DC, NS-SV-MC, NS-SV-SC, NS-SV-AC) with distinct morphologies were isolated. NS-SV-DC, NS-SV-MC, NS-SV-AC and NS-SV-SC have characteristics of duct, myoepithelial, acinar and squamous cells, respectively. Integration and expression of SV40 DNA were confirmed by Southern blot and indirect immunofluorescence staining. The DNA fingerprint analysis showed that the banding patterns of the 4 cell clones are identical, indicating that these cell clones are derived from a single donor. Anchorage-independent growth in semisolid agar and tumorigenicity in athymic nude mice were not observed and therefore the clones are considered to be immortal, but non-neoplastic.

Three cell lines (ACCS, ACCY and ACCAY) were isolated from three individuals with salivary adenoid cystic carcinoma (12,13). These cell lines form pseudocysts, which are a specific architectural feature of adenoid cystic carcinoma, and can produce abundant extracellular matrix including basal lamina components such as fibronectin, laminin, type IV collagen and glycosaminoglycans.

A salivary myoepithelial cell line UNC4 was isolated from a human salivary pleomorphic adenoma (14). UNC4 cells exhibit unique chromosome rearrangement and mucin production.

1. CULTURE CONDITIONS

HSG cells were established from explant culture of irradiated submandibular gland, grown in Eagle's minimal essential medium (MEM) supplemented with 10% calf serum and 2mM L-glutamine. Epithelioid cells were harvested by treatment with 0.08% trypsin and 1.4% ethylenediamine tetraacetic acid (EDTA) in calcium- and magnesium-free phosphate-buffered saline (PBS(-); pH7.2). The single cell suspension was seeded into a 100-mm plastic Petri dish at a density of $3x10^6$ cells in 15 ml of growth medium. The cells were passaged at 5–6 day intervals. At the third passage the colony forming ability in semisolid agar was 9%. Suspensions of 100 or 1000 cells in growth medium containing 0.3% Special Agar Noble (Difco Laboratories, Detroit, Mich) were poured into agar medium thickened by the addition of 0.6% agar. After about 20 days' incubation, colonies were isolated with Pasteur pipets, and the cells were cultured in 3 ml of growth medium in 30-mm plastic Petri dishes until confluent cell monolayers were formed. A total of six clones were isolated from the colonies and examined using electron microscopy. These cells showed characteristics of intercalated duct-type cells of human salivary gland. The clone which grew best was designated HSG. Currently, HSG cells are cultured in MEM supplemented with 10% newborn calf serum and 2mM L-glutamine at 37°C in a 5% CO_2 incubator. HSG cells continue to grow logarithmically in the serum-free medium for three days after plating, although the number of HSG cells gradually decreases after four days of incubation.

HSG-AZA1 and HSG-AZA3 cells were cloned as follows. Briefly, HSG cells (10^6 cells/dish) were seeded in 60-mm plastic Petri dishes in MEM supplemented with 10% newborn calf serum and 2mM L-glutamine. After 24h, 5-azacytidine was added at a concentration of 5 μM. The HSG cells were cultured in the continued presence of 5-azacytidine for 5 days, changing the growth medium daily. Then the treated cells were subcultured once in the growth medium without 5-azacytidine and cells were cloned using the colony-forming technique in semisolid agar medium as described above. The 7 subclones isolated were classified morphologically into 2 groups. Two subclones were composed of spindle-shaped or stellate cells and 5 were composed of polygonal cells with numerous secretory granules in the cytoplasm, whereas the parental HSG clonal cells were cuboidal and conical, grew with occasional tubular arrangement, and eventually formed multilayered foci. The subclones which showed the most stable growth in each of the 2 groups were designated HSG-AZA1 and HSG-AZA3, respectively. In another experiment, HSG-AZA11 cells were isolated from 5-azacytidine-treated HSG cells (8). HSG-AZA1 cells express a myoepithelial cell phenotype, HSG-AZA3 cells an acinar cell phenotype and HSG-AZA11 cells a neuron-like cell phenotype. The cells are cultured in MEM supplemented with 10% newborn calf serum

Table 1 Characterization of human salivary cell lines

Cell line	Patient age/sex	TNM category	Path stage and grade	Primary site	Specimen site	Culture method	Description	Differentiation	Primary reference
HSG	54/male			submandibular gland	irradiated salivary gland	explant culture of operation material	salivary intercalated duct cell[4,12] expression of carcinoembryonic antigen(CEA), secretory component and lactoferrin[12]	Differentiation into myoepithelial, acinar, keratinizing squamous, neuronal and smooth muscle cells	(4)
HSG-AZA1 (HSG-AZA1, HSG-AZA3 and HSG-AZA11 cells were cloned in vitro from the 5-azacytidine-treated HSG cells.)							DNA methylation level of HSG-AZA1, HSG-AZA3 and HSG-AZA11 cells significantly decreased when compared with HSG cells HSG, HSG-AZA1 and HSG-AZA11 possesses the same restriction fragment length polymorphisms (RFLP)[8] salivary myoepithelial cell[7] expression of myosin, microfibrils with focal condensation and pinocytic vesicles[7]	Differentiation into neuronal cells	(7)

Continued on next page

Table 1 (continued)

Cell line	Patient age/sex	TNM category	Path stage and grade	Primary site	Specimen site	Culture method	Description	Differentiation	Primary reference
HSG-AZA3							expression of amylase[7]	Differentiation into chondrocyte, osteoblast, and keratinizing squamous cells	(7)
HSG-AZA11							induction of synaptophysin and neuron specific enolase[8]	Formation of neurite-like structures	(8)
HSY	51/female	T2N0M0	Stage 2 grade 2	parotid gland	primary tumor	explant culture of nude mouse-grown tumor	Amylase isozyme analysis salivary type expression of amylase and VIP[9]	Differentiation into neuronal cells	(9)
TYS	81/female	T2N1M0	stage 3 grade 1	minor salivary gland	primary tumor	explant culture of biopsy material	Amylase isozyme analysis salivary type[10] expression of CEA and amylase	Differentiation into acinar cells	(10)

Continued on next page

Table 1 (continued)

Cell line	Patient age/sex	TNM category	Path stage and grade	Primary site	Specimen site	Culture method	Description	Differentiation	Primary reference
NS-SV-DC	35/male			submandibular gland	normal salivary gland	explant culture of operation material	NS-SV-DC, NC-SV-MC, NS-SV-SC and NC-SV-AC possess the same RFLP and integrate SV40 DNA[11]	Morphogenesis of human salivary gland[13)]	(11)
NS-SV-DC, NS-SV-MC, NS-SV-SC NS -SV-AC were cloned from the primary culture of human salivary gland which was transfected with origin-defective mutant DNA of SV40.							salivary intercalated duct cell[11]		
							expression of secretory component[11]		
NS-SV-MC							salivary myoepithelial expression of myosin, myofilaments and pinocytic vesicles[11]		
NS-SV-AC							salivary acinar cell[11]		
							expression of amylase[11]		
NS/SV-SC							squamous cell[11]		
							expression of tonofilaments, desmosomes and 68kD cytokeratin[11]		

Continued on next page

Table 1 (continued)

Cell line	Patient age/sex	TNM category	Path stage and grade	Primary site	Specimen site	Culture method	Description	Differentiation	Primary reference
ACCS	56/female			maxillary sinus	primary sinus	explant culture of biopsy material	expression of cytokeratin, vimentin, S-100 protein and lactoferrin synthesis of type IV collagen and glycosaminoglycans	production of extracellular matrix including basal lamina components	(12)
ACCY	54/male			tongue	primary tumor	explant culture of biopsy material	expression of cytokeratin, vimentin, S-100 protein and lysozyme synthesis of type IV collagen and glycosaminoglycans	production of extracellular matrix including basal lamina components	(12)
ACCAY	63/female			submandibular gland	primary tumor	explant culture of operation material	expression of type IV collagen, fibronection, laminin, chondroitin 6-sulfate proteoglycan and heparin sulfate	production of basement membrane components	(13)
UNC4	51/female			parapharynx	primary tumor	in vitro culture of operation material	salivary myoepithelial cell expression of cytokeratin desmin, and epithelial membrane antigen	genetic rearrangement and salivary mucin production	(14)

Table 2 Tumor pathology and in vitro features of neoplastic and immortalized human salivary gland cell lines

Cell line	Original tumor pathology	In vitro features of cell line	Xenograft pathology
HSG	Irradiated submandibular gland. Parenchyma damage, but no neoplastic appearance, is visible.	Cultured HSG cells form multilayed foci with occasional tubular arrangement and indicate anchorage-independent growth.	Adenocarcinoma with solid and trabecular pattern
		HSG cells have ultrastructure of salivary intercalated duct cells; presence of desmosomes, tight junctions and intercellular digitations formed by papillary infoldings of the cytoplasmic processes.	Ultrastructurally, junctional complexes between neighboring cells represent desmosomes, tight junctions and intercellular digitations.
		The cytoplasmic organelles are often situated in one side of the cytoplasm.	
		HSG cells proliferated in the sponge matrix forming scattered cords with tubular structure.	
HSG-AZA1		HSG-AZA1 cells are spindle or stellate in shape and exhibit phenotypes similar to human salivary myoepithelial cells, such as microfibrils and myosin.	Myoepithelioma HSG-AZA1 nude mouse tumor cells have ultrastructure of salivary myoepithelial cells; e.g. presence of ample microfibrils with some areas of focal condensation in parallel fashion and pinocytic vesicles as well as microfilament systems which react positively with anti-myosin serum.

Continued on next page

Table 2 (continued)

Cell line	Original tumor pathology	In vitro features of cell line	Xenograft pathology
HSG-AZA3		Cultured HSG-AZA3 cells are composed of polygonal cells with numerous secretory granules in the cytoplasm and contain salivary amylase that is specific to salivary acinar cells.	Acinic cell carcinoma HSG-AZA3 nude mouse tumor cells have ultrastructure of human salivary acinar cells; e.g. presence of abundant secretory granules which react positively to anti-amylase serum.
HSG-AZA11		Cultivation of HSG-AZA11 cells in the presence of dibutyryl cAMP results in formation of neurite-like structures and induction of synaptophysin and neuron-specific enolase.	
HSY	Adenocarcinoma with solid and trabecular pattern. The tumor is composed of polygonal cells with a pale-staining eosinophilic cytoplasm, which are arranged to form ducts or tubules surrounded by hyalin-like substances or fibrous tissues.	Cultured HSY cells are characterized by large epithelial cells with large oval nucleus and reticulated cytoplasm as well as by formation of multilayered foci. Ultrastructurally, the profiles of cytoplasmic organelles in HSY cells resemble those in tumor cells in the original tumor or xenograft	Adenocarcinoma with solid and trabecular pattern Ultrastructure of xenograft is similar to that of the original tumor.

Continued on next page

Table 2 (continued)

Cell line	Original tumor pathology	In vitro features of cell line	Xenograft pathology
	Ultrastructurally, intercalated duct-type cells are found mainly in the tumor tissue; e.g. presence of desmosomes, tight junctions and intercellular digitations formed by papillary infoldings of the cytoplasmic processes.	HSY cells express VIP and amylase.	
TYS	Well-differentiated squamous cell carcinoma expressing CEA.	Cultured TYS cells are polygonal in shape, and grow in a pavement-like formation with occasional multilayered foci.	Adenosquamous cell carcinoma Keratinizing squamous cells and adenoid structure
		Almost all of the TYS cell population show strongly positive staining for CEA in the cytoplasm.	
		Ultrastructurally, TYS cells show desmosomes, tight junctions, ample microvilli as well as well-developed tonofilaments and secretory granules.	When TYS cells treated with sodium butyrate are transplanted into athymic nude mice, the tumor formed is interpreted as acinic cell carcinoma containing a squamous cell area
		Ultrastructure of the cultured TYS cells treated with sodium butyrate shows ample tonofilament, numerous secretory granules or vacuoles reactive to anti amylase serum and presence of the substance reactive to anti-CEA serum in the rims of vacuoles	

Continued on next page

Table 2 (continued)

Cell line	Original tumor pathology	In vitro features of cell line	Xenograft pathology
NS-SV-DC	Normal submandibular gland	NS-SV-DC cells show round- or polygonal-shaped morphology and duct-like structure, and express secretory component.	Co-injection of NS-SV-DC cells with Matrigel into athymic nude mouse results in formation of cord similar to a duct-like structure.
		NS-SV-DC cells seeded on Matrigel form round clusters consisting of refractile cells, and gradually slough off the Matrigel surface.	
		Ultrastructurally, NS-SV-DC cells have intercellular digitations formed by papillary infoldings of the cytoplasmic processes.	
NS-SV-MC		NS-SV-MC cells exhibit spindle-shaped morphology, and express myosin.	Co-injection of NS-SV-MC cells with Matrigel into athymic nude mouse results in formation of amorphous structure with area of pericellular lysis.
		NS-SV-MC cells seeded on Matrigel penetrate into the Matrigel.	
		Ultrastructurally, NS-SV-MC cells have myofilaments and pinocytic vesicles.	
NS-SV-AC		NS-SV-AC cells show polygonal-shaped morphology with numerous secretory granules, and express amylase.	

Continued on next page

Table 2 (continued)

Cell line	Original tumor pathology	In vitro features of cell line	Xenograft pathology
NS-SV-SC		NS-SV-SC cells exhibit flattened-shaped morphology, and have tonofilaments and desmosomal junctions as well as 68kD cytokeratin.	
ACCS	Adenoid cystic carcinoma with a cribriform and solid pattern	formation of cyst-like acellular space presence of ample extracellular matrix and microfibrils with occasional aggregation	no tumor formation
ACCY	Adenoid cystic carcinoma with a cribriform and trabecular pattern	formation of cyst-like acellular space presence of ample extracellular matrix and microfibrils with occasional aggregation	no tumor formation
ACCAY	Adenoid cystic carcinoma with a cribriform pattern	formation of cyst-like acellular space	no tumor formation
UNC4	pleomorphic adenoma	Butyrate or DMSO stimulation of UNC4 cultures results in upregulation of mucin production. When UNC4 cells are cultured on Matrigel, the cells grow in organized clumps and ductules within the substance of the matrix.	not determined

Table 3 Genetic changes of human salivary tumor cell lines

Cell Line	Chromosome	p53	oncogene
HSG	range from 49 to 72 (average: 66.3) including XY: triploid range (49-69)	wild type	expression of H-ras gene
HSG-AZA1	range from 60 to 79 (average: 66.7) including XY; triploid range (60-72)	wild type	expression of H-ras gene
HSG-AZA3	range from 63 to 76 (average: 67.8) including XY; triploid range (63-72)	wild type	expression of H-ras gene
HSY	range from 56 to 128 (average 81.3) including XX; triploid range (58-63)	not determined	not determined
TYS	not determined	mutant type[39] (codon 281$^{Asp\text{-}His}$)	not determined
ACCS	range from 54 to 63 (average:59)	not determined	not determined
ACCY	range from 47 to 56 (average:50)	not determined	not determined
ACCAY	range from 43 to 46 + marker chromosome	not determined	not determined
UNC4	rearrangement between one chromosome 8 and the two chromosome 9s involving the 8q12 locus	not determined	not determined

and 2mM L-glutamine. These cells can grow logarithmically in the serum-free medium for at least 4 days.

HSY cells were established from explants of xenografts of a human parotid gland adenocarcinoma. TYS cells were established from explants of biopsy material obtained from primary tumor. Both HSY and TYS cells were cloned in a similar manner as follows. The fragments of tumor tissue were cultured in MEM supplemented with 10% newborn calf serum and 2mM L-glutamine. Explants with outgrowths of epithelioid cells alone were selected. When a monolayer of epithelioid cells was formed, the cells were harvested by treatment with 0.08% trypsin and 1.4% EDTA in PBS (-). These cells, dispersed in a single-cell suspension, were seeded into a 100-mm plastic Petri dish at a density of 3 million cells in 15ml of growth medium. Of colonies isolated at random, those that showed the fastest growth were designated HSY and TYS. Subculturing is done at 4-5 day intervals.

Immortalized human salivary gland cells (NS-SV-DC, NS-SV-MC, NS-SV-SC, NS-SV-AC) were established. Culture of normal human submandibular gland showing no pathological change, obtained during radical surgery for the treatment of submandibular sialolithiasis under general anaesthesia, was performed as follows: salivary gland specimens obtained at the operation were immediately cut into approximately 1mm^3 explants and placed (approximately 10 explants/60-mm plastic Petri dish) onto type 1 pig tendon collagen gel-coated plastic dishes. Collagen gels were prepared according to the recommendation of the manufacturer (Nitta Gelatin, Osaka, Japan) and pre-equilibrated in growth medium for at least 24 hours before use. The growth medium employed in this study was SFM (GIBCO BRL, New York, NY) supplemented with 5ng/ml of recombinant epidermal growth factor (GIBCO) (SFM+). Cultures were grown in gassed, humidified incubators and the medium was changed every 2 to 3 days. When cells reached confluence, they were subcultured using 0.1% trypsin and 1.4% EDTA in PBS(-). Normal human salivary gland cells cultured in SFM+ were transfected using a liposome-mediated method. The SV40 ori-mutant DNA (obtained from the Japanese Cancer Research Resources Gene Bank, Tokyo, Japan) is a hybrid DNA consisting of plasmid (pMK16) and the full genome of SV40 DNA less 6 nucleotides at the BglI site. Transfection procedures were performed according to the recommendation of the manufacturer (GIBCO) with a minor modification. In brief, cells plated on collagen gel-coated dishes at 80 to 85% confluence were incubated for 24 h with lipofectin reagent-DNA complex, containing 10µg of SV40 ori-mutant DNA and 40 µg of lipofectin reagent. Approximately 2-3 weeks later, colonies were picked using a cloning syringe and reseeded on collagen gel-coated dishes in SFM+. Cloning of cells was by the limiting dilution method. Four cell clones showing cuboidal (NS-SV-DC), spindle (NS-SV-MC), flattened (NS-SV-SC) and polygonal (NS-SV-AC) morphology were established. Characterization of cell clones by ultrastructural examination and expression of specific antigens showed the similarity of NS-SV-DC, NS-SV-MC, NS-SV-AC and NS-SV-SC to duct, myoepithelial, acinar, and squamous cells, respectively.

ACCS, ACCY and ACCAY cells were established from explants of biopsy or resected primary tumor. These cell lines were isolated from outgrowths consisting only of epithelioid cells, grown in Dulbecco's modified MEM or MEM supplemented with 10% fetal bovine serum and 2mM L-glutamine.

UNC4 cells were established by culturing the cell suspension prepared from primary tumor. The cells were cultured in a base keratinocyte medium (GIBCO) supplemented with nonessential amino acids (0.1mmol/L), glutamine (0.2mmol/L), insulin, hydrocortisone and transferrin (10mg/ml each), and epidermal growth factor (5ng/ml) without serum supplementation.

2. PATHOLOGY

HSG and its derivatives (HSG-AZA1 and HSG-AZA3) can differentiate into myoepithelial, acinar or neuronal cells as well as keratinizing squamous cells, chondrocytes, osteoblasts and smooth muscle cells, with a concomitant decrease of anchorage-independent and anchorage-dependent growth and decreased tumorigenicity in athymic nude mice. HSG cells can differentiate into myoepithelial cells (HSG-AZA1), acinar cells (HSG-AZA3) or neuronal cells (HSG-AZA11) in response to 5-azacytidine (7,8), into myoepithelial cells in response to dibutyryl cAMP (15) or sodium butyrate (16), into keratinizing squamous cells in response to retinoic acid (17), into smooth muscle cells in response to etoposide (18), and into neuronal cells in response to epidermal growth factor (19). In addition, treatment of HSG-AZA1 cells with nerve growth factor (20), protein kinase inhibitor H-7 (21), or dibutyryl cAMP (22) results in their differentiation into neuronal cells. HSG-AZA3 cells can differentiate into chondrocytes in response to 22-oxa-1α,25-dihydroxyvitamin D3 (23) and into keratinizing squamous cells in response to dibutyryl cAMP (24). When HSG-AZA3 cells are treated with 22-oxa-1α, 25-dihydroxyvitamin D3 in the presence of β-glycerophosphate, the treated cells exhibit enhanced expression of osteopontin and osteonectin mRNA and formation of bone nodules (25). When the tumors produced by transplantation into athymic nude mice of HSG-AZA3 cells are treated with 22-oxa-1α,25-dihydroxyvitamin D3, growth is significantly suppressed and bone formation is induced in the treated tumor. The tumor cells around bone express human osteopontin and osteonectin mRNA, detected by in situ hybridization. These findings indicate that the emergence of osteoblasts in the HSG-AZA3 cells occurs in the presence of 22-oxa-1α,25-dihydroxyvitamin D3 and β-glycerophosphate.

It has been found that the HSG cell line and its derivatives HSG-AZA1, HSG-AZA3 and HSG-AZA11 express neurofilaments (Mr 200,000, 160,000, and 68,000) and specific antigens such as tubulin α and β chain, HNK-1 antigen and laminin, and stain with Bodian impregnation (7,8), the binding site for which is located in the b domain in the extended tail segment of neurofilament polypeptides (26). In addition, the findings that the HSG-AZA11 cells treated with dibutyryl cAMP form neurite-like structures and express synaptophysin indicate that the parental HSG cells may be neuroectodermal in origin (8). Moreover, the human parotid adenocarcinoma cell line HSY, cultured in the presence of dibutyryl cAMP, expresses some antigens specific to neuronal cells, such as a triplet of neurofilament polypeptides, neuron-specific enolase, synaptophysin, α or β-chains of tubulin, and HNK-1 antigen, whereas these antigens are not detected in the untreated cells (27). Furthermore, it has been found that HSG cells express intermediate-sized filaments such as cytokeratin, vimentin and desmin (28) as well as epidermal growth factor and transforming growth factor-β (29).

HSG cells form a glandular morphology and differentiate into acinar cells when cultured on the reconstituted basement membrane, Matrigel, which contains primary laminin as well as type IV collagen, heparin sulfate proteoglycan, and a number of growth factors such as epidermal growth factor, fibroblast growth factor, and transforming growth factor-β. Laminin, one of the main components of Matrigel, is the major initiation factor in the differentiation of these cells (30). When SV40-immortalized cell clones with duct (NS-SV-DC) or myoepithelial phenotype (NS-SV-MC) are seeded on Matrigel in serum-free culture conditions, they fail to develop a morphogenesis consistent with salivary gland. On the other hand, three-dimensional morphogenesis of cells co-injected with Matrigel into the backs of athymic nude mice results in the characteristic features of each cell clone (31). NS-SV-DC cells align themselves into a cord, which is similar to the duct-like structure observed in normal salivary gland in vivo (32), whereas NS-SV-MC cells form an amorphous structure consisting of cuboidal and spindle-shaped cells, similar to the histologic appearance of a neoplastic myoepithelial cell line grown in the backs of athymic nude mice (33). The net balance of proteolytic activity is important in morphogenesis (34,35).

An in vitro system has been established in which conversion from non-metastasizing to metastasizing human adenocarcinoma cells can be induced (36). A human salivary adenocarcinoma cell clone HSGc with no metastatic ability, which was cloned from HSG cells, was exposed to N-methyl-N-nitrosourea (MNU). Following exposure to MNU, cells with altered morphology were cloned, which exhibited a flattened morphology with cytoplasmic processes, whereas HSGc cells were cuboidal or round. Following subcutaneous inoculation into athymic nude mice, MNU-treated HSGc clones formed metastatic foci in various organs, including lung, liver, spleen, pancreas and lymph node. Five metastasizing clones were isolated. The tumors growing at the inoculation site were diagnosed as adenocarcinoma with a solid and trabecular pattern, while the metastasizing clones produced tumors at the inoculation site which were classified as highly invasive clear-cell trabecular adenocarcinoma. Emergence of a clear-cell variant has been documented in a benign mixed tumor which recurred as a trabecular adenocarcinoma with clear-cell pattern (37). MNU-treated HSGc and metastasizing cell clones all metastasized to axillary and/or inguinal lymph nodes at a frequency of 100%. DNA fingerprint analysis confirmed that the metastasizing cell clones were of HSGc origin (36).

MNU-treated HSGc and metastasizing clones were found to secrete high levels of tissue-type plasminogen activator (tPA), while HSGc produced undetectable levels of this enzyme. Expression of urokinase-type plasminogen activator (uPA) was not observed in any of the cell clones. Metastasizing clones produced higher levels of 57-and 32-kD, but not of 92- or 72-kD gelatinases, as compared to HSGc cells. Although tissue inhibitor of metalloproteinases-1 (TIMP-1) was detected in all cell clones, metastatic clones secreted less TIMP-1 than HSGc cells. These findings suggest that the acquisition of metastatic ability by HSGc cells is closely associated with increased secretion of several metalloproteinases as well as decreased or altered TIMP-1 expression. In addition, Northern blot analysis showed that a small amount of transforming growth factor-β receptor-II mRNA was detectable in HSGc cells, while no significant bands were detected in metastatic cell clones (38).

When xenografts of TYS were treated with a differentiation-inducing agent, vesnarinone [(3,4-dihydro-6-[4-(3,4)-dimethoxybenzoyl)-1-piperazinyl]-2(1H)-quinolinone], given per os, a significant suppression of tumor growth was observed with tumor nests indicating keratinocyte and acinar cell differentiation (39). The tissue sections from vesnarinone-treated TYS tumors showed positive reaction with 3'-OH nick-end labelling and were stained strongly by monoclonal antibody (MAb) directed to carbohydrate antigen Le^Y, whereas the untreated tumors showed negative reaction with nick-end labelling and were infrequently stained by anti-Le^Y MAb. Within Le^Y-positive areas of the vesnarinone-treated tumors, keratinocyte and acinar cell differentiation as well as DNA fragmentation were frequently observed. It has been found that TYS cells have a mutant p53 gene at codon $281^{Asp\text{-}His}$ (40), and that expression of $p21^{WAF1}$ and transforming growth factor-β (TGF-β1) mRNA is up-regulated by treating TYS cells with vesnarinone. In addition, treatment of TYS cells with vesnarinone resulted in the enhanced generation of latent TGF-β1 and the expression of TGF-β receptor (TβR), including TβR-I, TβR-II and TβR-III, was detected on TYS cells. These findings suggest that vesnarinone might directly induce expression of the $p21^{WAF1}$ gene in TYS, the product of which may be associated with the inhibition of cell growth and the induction of differentiation. Thus, TYS cells might be a model suitable for studying differentiation therapy of salivary gland cancer.

Detailed banding pattern analysis of UNC4 cells revealed rearrangements of chromosomes 8 and 9. The long arm of chromosome 8 is translocated to the long arm of chromosome 9. A distal segment of the long arm of chromosome 9 is translocated to the short arm of chromosome 9. A segment of the short arm of chromosome 9 is translocated to the truncated long arm of chromosome 8. It has recently been reported that salivary pleomorphic adenomas have structural chromosomal abnormalities including the following: translocations between chromosomes 3 and 8, chromosomes 6 and 16, chromosomes 8 and 9, chromosomes 8 and 12, chromosomes 8 and 14, and chromosomes 8 and 21 (41).

REFERENCES

1. Eversole, I.L. *Arch. Pathol.*, 92: 443, 1971.
2. Batsakis, J. G. *Oral Surg.*, 49: 229, 1980.
3. Pierce, G. B. *Am. J. Pathol.*, 77: 103, 1974.
4. Shirasuna, K., et al. *Cancer*, 48: 745, 1981.
5. Cherry, C. P., et al. *Br. J. Radiol.*, 32: 596, 1959.

6. Kashima, H. H., et al. *Am. J. Roentgenol.*, 94: 271, 1965.
7. Sato, M., et al. *Cancer Res.*, 47: 4453, 1987.
8. Sato, M., et al. *Cancer J.*, 6: 26, 1993.
9. Hayashi, Y., et al. *J. Nat. Cancer Inst.*, 79: 1025, 1987.
10. Yanagawa, T., et al. *Am. J. Pathol.*, 124: 496, 1986.
11. Azuma, M., et al. *Lab. Invest.*, 69: 24, 1993.
12. Shirasuna, K., et al. *Cancer Res.*, 50: 4139, 1990.
13. Sobue, M., et al. *Virchows Archiv B Cell Pathol.*, 57: 203, 1989.
14. Witsell, D. L., et al. *Arch Otolaryngol Head Neck Surg.*, 119: 1151, 1993.
15. Yoshida, Y., et al. *Cancer*, 57: 1011, 1986.
16. Azuma, M., et al. *Cancer Res.*, 46: 770, 1986.
17. Azuma, M., et al. *Cancer Res.*, 48: 7219, 1988.
18. Yoshida, H., et al. *Cancer J.*, 6: 220, 1988.
19. Aladib, W., et al. *Cancer Res.*, 50: 7650, 1990.
20. Iga, H., et al. *Cancer Res.*, 49: 6708, 1989.
21. Yoshida, H., et al. *Cancer J.*, 4: 267, 1991.
22. Kawamata, H., et al. *Cancer Invest.*, 10: 111, 1992.
23. Azuma, M., et al. *Cancer Res.*, 49: 5435, 1990.
24. Kawamata, H., et al. *Cancer J.*, 3: 274, 1990.
25. Sato, M., et al. *Cancer Lett.*, 115: 149, 1997.
26. Weber, K., et al. *Cold Spring Harbor Symp. Quant. Biol.*, 47: 717, 1983.
27. Nagamine, S., et al. *Cancer Res.*, 50: 6396, 1990.
28. Sato, M., et al. *Cancer Res.*, 45: 3878, 1985.
29. Sato, M., et al. *Cancer Res.*, 45: 6160, 1985.
30. Royce, L. S., et al. *Differentiation*, 52: 247, 1993.
31. Azuma, M., et al. *Lab. Invest.*, 70: 217, 1994.
32. Martinez-Madrigal, F., et al. *Am. J. Surg. Pathol.*, 13, 879, 1989.
33. Shirasuna, K., et al. *Cancer*, 45: 297, 1980.
34. Bacharach, E et al. *Proc Natl Acad Sci USA*, 89: 10686, 1992.
35. Montesano, R., et al. *Cell*, 62: 435, 1990.
36. Azuma, M., et al. *Int. J. Cancer*, 54: 669, 1993.
37. Evance, R. W., et al. In: Epithelial tumors of the salivary glands, pp276-277, Saunders, Philadelphia, 1970.
38. Azuma, M., et al. *Int. J. Cancer*, 68: 802, 1996.
39. Sato, M., et al. *Cancer Lett.*, 91: 1, 1995.
40. Sato, M., et al. *Cancer Lett.*, 112: 181, 1997.
41. Mark, H. F. L., et al. *Oral Surg Oral Med Oral Pathol Oral Radiol Endod.*, 82: 187, 1996.

Chapter 10

Esophageal Cancers

Yutaka Shimada
Department of Surgical Basic Science, Graduate School of Medicine, Kyoto University, 54 Shogoin Kawara-Cho, Sakyo-Ku, Kyoto 606-8507, Japan. Tel: 0081-75-751-3626; Fax: 0081-75-751-4390; E-mail: shimada@kuhp.kyoto-u.ac.jp

In 1976, Bey et al described the SN cell line derived from human esophageal cancer. In the same year, the EC-109 and CaEs-17 cell lines were developed in China (Department of Cell Biology of CICAMS, Laboratory section of Peking Medical College). Since then, another 3 cell lines, listed in Table 1, have been reported from China. In general, esophageal cancer cell lines have been established in countries with a high incidence of the disease; namely China, South Africa, and Japan.

The key references to the cell lines that have been established are shown in Table 1. Nishihira et al started to establish the TE-series in 1979 and then went on to establish 15 esophageal cancer cell lines. In 1980, Robinson et al started to establish the HCU series (9 cell lines) from resected specimens and the B series (3 cell lines) from biopsied specimens. In 1981, Shinbo et al started the SGF series of 12 cell lines. Iizuka et al reported the TH and TS cell lines in 1981. In 1984 Hu et al reported the CE series (3 cell lines) and in 1986 Banks-Schlegel et al reported the HCE series (8 cell lines). In 1986, Shimada et al started to establish the KYSE series of 40 cell lines. In 1987 Mok et al reported the EC/CUHK1 cell line and same institute reported another cell line, EC/CUHK2, derived in 1989 from a biopsy. In 1987, Sato et al reported the EC-GI cell line which produced PTHrp (parathormone related polypeptide) and Matsuoka et al started the KSE series (2 cell lines). In 1991, Nakamura et al began to establish the YES series of 6 cell lines. In 1993, Altorki et al established the SK-GT-4 cell line. In 1994, Tomita et al began to establish the KE series of 5 cell lines. Recently in 1997, Rockett et al reported the JROECL series (5 cell lines). Although more than 100 esophageal cancer cell lines have been reported, the HCU series has cross contamination (Helden et al 1988) and many cell lines have been lost for various reasons.

J.R.W. Masters and B. Palsson (eds.), Human Cell Culture Vol. 1, 179–212.

Table 1 List of esophageal cancer cell lines

No	Cell line	Age/ Sex	TNM stage	Histology	Primary site	Specimen site	Culture method	Authentication	Primary reference
1	KYSE-30	64/M	T4N1MO stage 3	WD SCC	Esophagus	Esophagus	Xenograft	DNA,HLA	Shimada (Cancer 1992)
2	KYSE-50	58/M	T2N1MO stage 2b	PD SCC	Esophagus	Esophagus	Xenograft	DNA,HLA	Shimada (Cancer 1992)
3	KYSE-70	77/M	T2N1MO stage 2b	PD SCC	Esophagus	Esophagus	Xenograft	DNA,HLA	Shimada (Cancer 1992)
4	KYSE-110	63/M	T4N1MO stage 3	PD SCC	Esophagus	Esophagus	Explant	DNA,HLA	Shimada (Cancer 1992)
5	KYSE-140	54/M	T3N1MO stage 3	MD SCC	Esophagus	Esophagus	Explant	DNA,HLA	Shimada (Cancer 1992)
6	KYSE-150	49/F	T4N1MO stage 3	PD SCC	Esophagus	Esophagus	Explant	DNA,HLA	Shimada (Cancer 1992)
7	KYSE-170	53/F	T3N0MO stage 2a	MD SCC	Esophagus	Esophagus	Explant	DNA,HLA	Shimada (Cancer 1992)
8	KYSE-180	53/M	T4N1MO stage 3	WD SCC	Esophagus	Esophagus	Explant	DNA,HLA	Shimada (Cancer 1992)
9	KYSE-190	69/F	T2N1MO stage 2b	WD SCC	Esophagus	Esophagus	Xenograft	DNA,HLA	Shimada (Cancer 1992)
10	KYSE-200	73/M	T3N1MO stage 3	PD SCC	Esophagus	Esophagus	Explant	DNA,HLA	Shimada (Cancer 1992)
11	KYSE-220	78/M	T3N1MO stage 3	MD SCC	Esophagus	Esophagus	Xenograft	HLA	Shimada (Cancer 1992)
12	KYSE-240	53/M	T2N1MO stage 2b	PD SCC	Esophagus	Esophagus	Explant	HLA	Shimada (Cancer 1992)
13	KYSE-270	79/M	T2N0MO stage 2a	WD SCC	Esophagus	Esophagus	Explant	HLA	Shimada (Cancer 1992)
14	KYSE-280	75/F	T3N1MO stage 3	MD SCC	Esophagus	Esophagus	Xenograft	-	Shimada (Cancer 1992)
15	KYSE-330	53/M	T3N1M1 stage 4	MD SCC	Esophagus	Esophagus	Explant	-	Shimada (Br J Surg 1993)
16	KYSE-350	66/M	T4N1MO stage 3	MD SCC	Esophagus	Esophagus	Explant	DNA,HLA	Shimada (Cancer 1992)
17	KYSE-360	58/M	T3N1MO stage 3	MD SCC	Esophagus	Esophagus	Explant	DNA,HLA	Shimada (Cancer 1992)
18	KYSE-390	57/F	T3N1MO stage 3	PD SCC	Esophagus	Esophagus	Explant	DNA,HLA	Shimada (Cancer 1992)
19	KYSE-410	51/M	T3N1MO stage 3	PD SCC	Esophagus	Esophagus	Explant	DNA,HLA	Shimada (Cancer 1992)
20	KYSE-450	59/M	T1N1MO stage 2b	WD SCC	Esophagus	Esophagus	Explant	DNA,HLA	Shimada (Cancer 1992)
21	KYSE-510	67/F	T4N1MO stage 3	WD SCC	Esophagus	Esophagus	Explant	DNA,HLA	Shimada (Cancer 1992)
22	KYSE-520	58/F	T3NOMO stage 2a	MD SCC	Esophagus	Esophagus	Explant	DNA,HLA	Shimada (Cancer 1992)
23	KYSE-590	39/M	T3N1MO stage 3	MD SCC	Esophagus	Esophagus	Explant	DNA,HLA	Kanda (Int J Cancer 1994)
24	KYSE-770	46/M	T4N1MO stage 3	MD SCC	Esophagus	Esophagus	Explant	DNA,HLA	Kanda (Int J Cancer 1994)
25	KYSE-790	68/M	T3NOMO stage 2a	WD SCC	Esophagus	Esophagus	Explant	DNA,HLA	Kanda (Int J Cancer 1994)
26	KYSE-850	54/M	T4NOMO stage 3	MD SCC	Esophagus	Esophagus	Explant	DNA,HLA	Kanda (Int J Cancer 1994)

Continued on next page

Table 1 (continued)

No	Cell line	Age/ Sex	TNM stage	Histology	Primary site	Specimen site	Culture method	Authentication	Primary reference
27	KYSE-890	61/M	T3N1MO stage 3	MD SCC	Esophagus	Esophagus	Explant	DNA,HLA	Kanda (Int J Cancer 1994)
28	KYSE-960	58/M	T4N1MO stage 3	MD SCC	Esophagus	Esophagus	Explant	DNA,HLA	Kanda (Int J Cancer 1994)
29	KYSE-1040	85/M	T4N1M1 stage 4	PD SCC	Esophagus	ascites	Dissociated tissue	-	-
30	KYSE-1140	74/M	T1NOMO stage 1	PD SCC	Esophagus	Esophagus	Explant	DNA,SCC	Tanaka (Int J Cancer 1996)
31	KYSE-1170	56/M	T3N1M1 stage 4	PD SCC	Esophagus	Esophagus	Explant	DNA,HLA	Tanaka (Int J Cancer 1996)
32	KYSE-1190	67/M	T3N1M1 stage 4	PD SCC	Esophagus	Esophagus	Explant	DNA,HLA	Tanaka (Int J Cancer 1996)
33	KYSE-1230	59/M	T3N1MO stage 3	Spindle cell	Esophagus	Esophagus	Explant	-	Tanaka (Int J Cancer 1996)
34	KYSE-1240	52/M	T3N1MO stage 3	MD SCC	Esophagus	Esophagus	Explant	DNA,HLA	Tanaka (Int J Cancer 1996)
35	KYSE-1250	81/M	T1NOMO stage 1	MD SCC	Esophagus	Esophagus	Explant	DNA,HLA	Tanaka (Int J Cancer 1996)
36	KYSE-1260	52/M	T4N1MO stage 3	MD SCC	Esophagus	Esophagus	Explant	DNA,HLA	Tanaka (Int J Cancer 1996)
37	KYSE-1320	68/M	T4NOMO stage 3	MD SCC	Esophagus	Esophagus	Explant	-	-
38	KYSE-1440	71/M	T2N1MO stage 2b	PD SCC	Esophagus	Esophagus	Explant	DNA,HLA	Tanaka (Int J Cancer 1997)
39	KYSE-1830	76/M	T1NOMO stage 1	WD SCC	Esophagus	Esophagus	Explant	DNA,HLA	
40	KYSE-1860	77/M	T2N1M1 stage 4	MD SCC	Esophagus	Esophagus (Biopsy)	Explant	DNA,HLA	Tanaka (Int J Cancer 1997)
41	SUM/c	-	T3N1MO stage 3	WD SCC	Esophagus	thoracic duct	Xenograft	-	
42	HSA/c	-	T3N1MO stage 3	PD SCC	Esophagus	thoracic duct	Xenograft	-	
43	TE-1	58/M	T3NOMO stage 2a	WD SCC	Esophagus	Esophagus	Explant	-	Nishihira (Jpn J Cancer Res 1979)
44	TE-2	56/M	T4N1MO stage 3	PD SCC	Esophagus	Esophagus	Explant	-	Nishihira (Jpn J Cancer Res 1979)
45	TE-3	48/M	T4N1M1 stage 4	WD SCC	Esophagus	muscle metastasis	Explant	-	Kuriya (Tohoku J exp Med. 1983)
46	TE-4	48/F	T3N1MO stage 3	WD SCC	Esophagus	LN metastasis	Xenograft	-	Akaishi (Excerpta Medica 1986)
47	TE-5	73/F	T4N1MO stage 3	PD SCC	Esophagus	Esophagus	Explant	-	Akaishi (Excerpta Medica 1986)
48	TE-6	71/M	T4N1MO stage 3	WD SCC	Esophagus	Esophagus	Dissociated tissue	-	Akaishi (Excerpta Medica 1986)
49	TE-7	72/M	T2N1MO stage 2b	AD	Esophagus	Esophagus	Dissociated tissue	-	Akaishi (Excerpt Medica 1986)
50	TE-8	63/M	T3N1MO stage 3	MD SCC	Esophagus	Esophagus	Dissociated tissue	-	Akaishi (Excerpta Medica 1986)

Continued on next page

Table 1 (continued)

No	Cell line	Age/ Sex	TNM stage	Histology	Primary site	Specimen site	Culture method	Authentication	Primary reference
51	TE-9	48/M	T2N1M1 stage 4	PD SCC	Esophagus	pleural effusion	Dissociated tissue	-	Akaishi (Excerpta Medica 1986)
52	TE-10	50/M	JSEC stage 4	WD SCC	Esophagus	Esophagus	Dissociated tissue	-	Akaishi (Excerpta Medica 1986)
53	TE-11	58/M	JSEC stage 4	MD SCC	Esophagus	Esophagus	Dissociated tissue	-	Akaishi (Excerpta Medica 1986)
54	TE-12	54/M	JSEC stage 3	MD SCC	Esophagus	Esophagus	Dissociated tissue	-	Akaishi (Excerpta Medica 1986)
55	TE-13	65/F	JSEC stage 4	PD SCC	Esophagus	Esophagus	Dissociated tissue	-	Akaishi (Excerpta Medica 1986)
56	TE-14	57/M	JSEC stage 4	MD SCC	Esophagus	Esophagus	Dissociated tissue	-	Akaishi (Excerpta Medica 1986)
57	TE-15	58/F	JSEC stage 3	WD SCC	Esophagus	Esophagus	Dissociated tissue	-	Akaishi (Excerpta Medica 1986)
58	HCE-1	50/F	-	MD-WD SCC	Esophagus	Esophagus	Xenograft	-	Banks-Schlegel (Cancer Res 1986)
59	HCE-3	48/M	-	MD SCC	Esophagus	Esophagus	Explant	-	Banks-Schlegel (Cancer Res 1986)
60	HCE-4	50/F	-	MD SCC	Esophagus	Esophagus	Explant	-	Banks-Schlegel (Cancer Res 1986)
61	HCE-5	57/M	-	PD SCC	Esophagus	Esophagus	Explant	-	Banks-Schlegel (Cancer Res 1986)
62	HCE-6	42/M	-	MD SCC	Esophagus	Esophagus	Explant	-	Banks-Schlegel (Cancer Res 1986)
(63	HCE-7 NM)	50/F	-	MD SCC	Esophagus	Esophagus	Xenograft	-	Banks-Schlegel (Cancer Res 1986)
64	HCE-8	67/M	-	PD SCC, AD	Esophagus	Esophagus	Explant	-	Banks-Schlegel (Cancer Res 1986)
65	HCE-9	61/M	-	SCC	Esophagus	Esophagus	Explant	-	Banks-Schlegel (Cancer Res 1986)
(66	HCU 10)	40/F	T3N1MO stage 3	PD SCC	Esophagus	Esophagus	Explant	cross-contamination	Robinson (Clinical Oncology 1980)

Continued on next page

Table 1 (continued)

No	Cell line	Age/ Sex	TNM stage	Histology	Primary site	Specimen site	Culture method	Authentication	Primary reference
67	HCU 13	50/M	T4N1MO stage 3	WD SCC	Esophagus	Esophagus	Explant	DNA	Robinson (Clinical oncology 1980)
(68	HCU 18)	40/M	TxN1M1 stage 4	PD SCC	Esophagus	Esophagus	Explant	cross-contamination	Robinson (Clinical oncology 1980)
(69	HCU 33)	45/M	T4N1MO stage 3	WD SCC	Esophagus	Esophagus	Explant	cross-contamination	Robinson (Clinical oncology 1980)
70	HCU 35)	66/M	T4N1M1 stage 4	WD SCC	Esophagus	Esophagus	Explant	-	Robinson (Clinical oncology 1980)
(71	HCU 37)	60/F	T4N1MO stage 3	PD SCC	Esophagus	Esophagus	Explant	cross-contamination	Robinson (Clinical oncology 1980)
(72	HCU 39)	70/M	T4N1MO stage 3	WD SCC	Esophagus	Esophagus	Explant	cross-contamination	Robinson (Clinical oncology 1980)
73	HCU 50	59/M	-	MD SCC	Esophagus	Esophagus	-	-	Rabin (Cancer of the esophagus 1982)
74	HCU 57	-	-	MD SCC	Esophagus	Esophagus	Explant	-	Robinson (JNCI 1983)
75	B5	65/M	T4NxMO	PD SCC	Esophagus	Esophagus (biopsy)	Explant	-	Robinson (Clinical oncology 1980)
76	B17	44/F	T4NxMO	MD SCC	Esophagus	Esophagus (biopsy)	Explant	-	Robinson (Clinical oncology 1982)
77	B29	41/M	T1NxMO	MD SCC	Esophagus	Esophagus (biopsy)	Explant	-	Robinson (Clinical oncology 1980)
78	SN	62/M	TxN1MO	WD SCC	Esophagus	Esophagus	Dissociated tissue	DNA, Isozyme	Bey(In vitro 1976)
79	CaEs-17/EC1	65/M	T3N1MO stage 3	SCC	Esophagus	LN metastasis	Explant	-	Surgical Lab (CMJ 1976)
80	EC8712	45/M	-	WD SCC	Esophagus	not described	Explant	-	Feng (Science in China 1992)

Continued on next page

Table 1 (continued)

No	Cell line	Age/ Sex	TNM stage	Histology	Primary site	Specimen site	Culture method	Authentication	Primary reference
81	EC8501	55/M	-	MD SCC	Esophagus	not described	Explant	-	Feng (Science in China 1992)
82	EC-56	35/M	-	SCC	Esophagus	Esophagus	-	-	Li (ACTA Zool 1979)
83	ECa109	37/F	-	MD SCC	Esophagus	Esophagus	-	-	Pan (CAMS 1980)
84	CE-48T/VG1	58/M	-	epidermoid Ca.	Esophagus	Lung metastasis	Explant	-	Hu (JNCI 1984), Wang (JNCI 1984)
85	CE-69T/VG1	51/F	-	WD SCC	Esophagus	Esophagus	Explant	-	Hu (JNCI 1984), Wang (JNCI 1984)
86	CE-81T/VG1	57/M	-	WD SCC	Esophagus	Esophagus	Explant	-	Hu (JNCI 1984), Wang (JNCI 1984)
87	EC/CUHK 1	72/M	-	WD SCC	Esophagus	-	Explant	-	Mok (Anticancer Res 1987)
88	EC/CUHK 2	-	-	PD SCC	Esophagus	Esophagus	Explant	-	Chew (Recent Advances in Cancer 1989)
89	JROECL 21	74/M	stage 2a	MD SCC	Esophagus	Esophagus	Explant	HLA	Rockett (Br J Cancer 1997)
90	JROECL 24	68/M	stage 2b	PD AD	Esophagus	Esophagus	Explant	HLA	Rockett (Br J Cancer 1997)
91	JROECL 33	73/F	stage 2a	PD AD	Esophagus	Esophagus	Explant	HLA	Rockett (Br J Cancer 1997)
92	JROECL 47	76/M	stage 2a	PD SCC, AD	Esophagus	Esophagus	Explant	HLA	Rockett (Br J Cancer 1997)
93	JROECL 50	71/F	stage 2a	MD PD AD	Esophagus	Esophagus	Explant	HLA	Rockett (Br J Cancer 1997)
94	SK-GT-4	89/M	T2N1MO stage 2b	WD AD	(Esophagus	Esophagus	Explant	-	Altorki (Cancer 1993)
95	TT	67/M	TxNxM1 stage 4	SCC	Esophagus	Oral cavity metastasis	Explant	Isozyme	Naitoh (Am J Path 1995)
(96	TTn)	67/M	TxNxM1 stage 4	SCC	Esophagus	Oral cavity metastasis	Xenograft	Isozyme	
97	TH	56/F	TxN1MO	PD SCC	Esophagus	LN metastasis	Dissociated tissue	-	Iizuka (Jpn J Oncol 1981)
98	TS	68/M	-	MD SCC	Esophagus	Esophagus	Dissociated tissue	-	Iizuka (Jpn J Oncol 1981)
99	SH-1	50/M	-	PD SCC	Esophagus	Esophagus	-	-	Saito (Cancer 1992)

Continued on next page

Table 1 (continued)

No	Cell line	Age/ Sex	TNM stage	Histology	Primary site	Specimen site	Culture method	Authentication	Primary reference
100	KSE-1	68/M	T2NOMO stage 2a	PD SCC	Esophagus	LN recurrence	Dissociated tissue	-	Matsuoka (Cancer Res 1987)
101	KSE-2	72/F	T2NOMO stage 2a	PD SCC	Esophagus	LN recurrence	Dissociated tissue	-	Ueo (Cancer Res 1990)
102	EC-Gi	65/M	TxN1M1	SCC	Esophagus	LN metastasis	Xenograft	-	Sato (Jpn J Cancer Res 1987)
103	YES 1	50/M	T2N1MO stage 2b	PD SCC	Esophagus	Esophagus	Xenograft	-	Nakamura (Arch Jpn Chir 1991)
104	YES 2	81/M	T4N1M1 stage 4	MD SCC	Esophagus	Esophagus	Xenograft	-	Murakami (Nippon Gekagakkai Zasshi 1991)
105	YES 3	66/M	T3N1MO stage 3	MD SCC	Esophagus	pleural effusion	Dissociated tissue	-	Murakami (Nippon Gekagakkai Zasshi 1991)
106	YES 4	57/M	T3N1MO stage 3	MD SCC	Esophagus	Esophagus	Explant	-	
107	YES 5	66/M	T3N1MO stage 3	MD SCC	Esophagus	Esophagus	Explant	-	
108	YES 6	71/M	T3N1MO stage 3	MD SCC	Esophagus	Esophagus	Explant	-	
109	HPL-EsC-1	50/M	-	SCC	Esophagus	pleural effusion	Dissociated tissue	not done	Morikawa (Acta Histochem Cytochem 1986)
110	HPL-EsC-2	59/M	-	epidermoid Ca.	Esophagus	pleural effusion	Dissociated tissue	not done	Morikawa (Acta Histochem Cytochem 1986)
111	EC1NU	-	-	-	Esophagus	-	-	-	Murase (J Surg Oncology 1996)
112	EC2NU	-	-	-	Esophagus	-	-	-	Murase (J Surg Oncology 1996)
113	WSSC	-	-	-	Esophagus	-	Xenograft	-	Murase (J Surg Oncology 1996)
114	SGF3	44/M	T1N1MO stage 1	MD SCC	Esophagus	Esophagus	Explant	-	Saito (In Vitro Cell Dev Biol 1990)
115	SGF 4	64/M	T4N1MO stage 3	MD SCC	Esophagus	Esophagus	Explant	not done	Saito (Human Cell 1994)
116	SGF 5	78/M	T4N1MO stage 3	WD SCC	Esophagus	LN metastasis	Explant	-	Saito (In Vitro Cell Dev Biol 1990)
117	SGF 7	71/M	T4N1MO stage 3	MD SCC	Esophagus	Esophagus	Explant	-	Saito (Human Cell 1994)

Continued on next page

Table 1 (continued)

No	Cell line	Age/ Sex	TNM stage	Histology	Primary site	Specimen site	Culture method	Authentication	Primary reference
118	SGF 8a	52/M	T3N1MO stage 3	PD SCC	Esophagus	LN metastasis	Xenograft	-	Saito (Human Cell 1994)
119	SGF 8b	52/M	T4N1M1 stage 4	PD SCC	Esophagus	LN metastasis	Xenograft	-	Saito (Human Cell 1994)
120	SGF 9	58/M	T4N1MO stage 3	WD SCC	Esophagus	Esophagus	Dissociated tissue	-	
121	SGF11	61/M	T3N1MO stage 3	MD SCC	Esophagus	Esophagus	Dissociated tissue	-	
122	SGF12	79/M	T3N1MO stage 3	MD SCC	Esophagus	Esophagus	Dissociated tissue	-	
123	E-oh	70/M	T3N1MO stage 3	MD SCC	Esophagus	Esophagus	Xenograft	-	
124	E-Um	73/M	T4N1MO stage 3	SCC	Esophagus	Esophagus	Xenograft	-	
125	E-Th	65/M	T3N1MO stage 3	MD SCC	Esophagus	Esophagus	Xenograft	-	
126	E-1	-	-	-	Esophagus	LN metastasis	Xenograft	-	
127	KE-1	70/M	T3N1M1 stage 4	WD SCC	Esophagus	LN metastasis	Xenograft	not done	Jpn J Surg Soc 1993
128	KE-2	54/M	T3N1MO stage 3	PD SCC	Esophagus	LN metastasis	Xenograft	not done	Recent advances in diseases of the esophagus 1993
129	KE-3	62/M	T3N1MO stage 3	MC SCC	Esophagus	-	Dissociated tissue	HLA	Jpn J Cancer Chem 1994
130	KE-4	50/M	T4N1MO stage 3	PD SCC	Esophagus	Esophagus	Dissociated tissue	HLA	Nakao (Cancer Res 1995)
131	KE-5	69/M	T3NOMO stage 2a	WD SCC	Esophagus	Esophagus	Dissociated tissue	-	
132	AEN-1	71/M	-	WD SCC	Esophagus	LN metastasis	-	-	Akita J Med 1985
133	AEN-2	75/M	-	MD SCC	Esophagus	LN metastasis	-	-	

Abbreviations used are as follows: SCC: squamous cell carcinoma, AD: adenocarcinoma; PD: poorly differentiated, MD: moderately differentiated; WD: well differentiated; DNA: DNA fingerprint; JSEC: Japanese staging of esophageal cancer. ( ) indicate cross-contaminated cell lines or sublines derived from xenografts

1. CULTURE CONDITIONS

Our success rate for cell line establishment was 38/137, ie 27.8% (Shimada et al. 1996), and currently is 21.1% (40/190). Robinson et al (1982) reported that 8 cell lines (17.4%) were established from 46 resected tumor samples and two cell lines (5.5%) were established from 36 tumor biopsy samples. Rockett et al (1997) reported that cell attachment and growth did not improve using feeder cells or collagen-coated tissue culture flasks, with a success rate of 12.5% (5/40). Akaishi et al (1984) described two dissociation methods (Dispase® and collagenase method). The success rate of the collagenase method was 2/9 (22.2%), whereas that of the Dispase® method was 0/5.

The average time to first passage of the KYSE series ranged from 2 weeks to several months. Other reports describe similar time periods. A wide variety of culture media have been used, including MEM, DMEM, Medium 199, Ham's F12, RPMI 1640, and Ham's F12/RPMI1640. The fetal calf serum (FCS) concentration ranged from 2% to 20% (Shimada et al. 1992), but is usually 5% to 10%. The antibiotics used included penicillin, streptomycin, gentamycin and amphotericin B. The use of these drugs enabled the culture of esophageal cancer cells from the lumen of the esophagus without bacterial and fungal contamination.

A major problem for the establishment of esophageal cell lines is overgrowth of fibroblasts. Fibroblasts can be removed by differential trypsinization or mechanical scraping. Fibroblasts have a tendency to detach differentially from the surface of the flask during trypsinization. When the fibroblasts start to round up and detach from the flask, most of the tumor cells are still attached, and therefore the fibroblasts can be removed from the flask by washing. However, excessive removal of fibroblasts can suppress tumor cell growth, because in the early stages of culture the tumor cells need additional support. In order to scrape fibroblasts off easily, Shimada et al (1992) used 60mm culture dishes, and to avoid contamination they covered the 60mm dish with a 90mm Petri dish. Another method to prevent fibroblast overgrowth is serum-free culture. Some tumor cells can grow in a serum-free environment, although it usually takes a long time and the tumor cells tend to senesce. KYSE 110 cells were started in a serum-free environment and then at a later stage a low concentration of FCS was added. This method can result in the successful establishment of cell lines without overgrowth of fibroblasts.

Most of the cell lines grow as an adherent monolayer, although a few cell lines grow as floating aggregates. Although the samples were obtained from solid tumors, TS cells (Iizuka et al 1981) initially grew as a floating ball on the surface of the medium for 2 to 3 weeks until they attached to the glass substrate. KYSE-50 and 360 cells also showed less attachment ability (Shimada et al. 1992). On the other hand, samples obtained from pleural or abdominal effusions, such as TE-9, KYSE-1040, YES-3 and HPL-EsC-1 grew as adherent monolayers. HPL-EsC-2, which was derived from a pleural effusion, is the one cell line that grows mainly as floating cell aggregates. Care should be taken to preserve tumor cells which have a tendency to float or have little attachment capability, particularly during the first weeks of culture.

Some cell lines can grow in a serum-free environment (Shimada et al. 1991a,b, 1992; Oku et al. 1991, Iihara et al. 1993, Murakami et al. 1991). The KYSE series was cultured in Ham's F12/RPMI 1640 (1:1, v/v), TE-2 and 3 were cultured in Ham's F12/MEM (1:1, v/v). Thirteen of 21 cell lines of the KYSE-series can grow in a protein-free environment, and there was no relation between the number of EGF receptors and the growth effect of EGF on these cells. One third of the tested KYSE series were inhibited by EGF, which demonstrated that EGF is not always conducive to the growth of cells in a protein-free culture. The growth of these cells was not affected by conditioned medium, but the growth of NIH3T3 cells and human fibroblasts was stimulated by conditioned media (Shimada et al 1991a). Iihara et al (1993) and Oku et al (1991) reported that cloned cell lines derived from TE-2 and TE-3 cells (TE-2-NS, TE-3-OS) can grow in protein-free culture and possess IGF and TGFα binding sites. They also reported that these cell lines were stimulated by IGFs, EGF and TGFα. In summary, the cells which can grow in a protein-free environment may either grow in an autocrine manner, or may not require additional growth factors, or may have a loss of growth inhibition mechanisms such as tumor suppressor genes.

2. DERIVATION

Esophageal cancer cell lines have been derived mainly from primary cancers, but some are from xenografts, lymph node metastases or pleural effusions. There is no correlation between establishment and the histological grade of the tumors. Tumors were cut into small fragments and used for explant cultures (Nishihira et al 1979, Robinson et al. 1980, Shimada et al. 1992), although some cell lines (for example, the TE-series) were established following enzymatic dissociation.

Although many cell lines have been derived from the cardia of the stomach (Altorki et al 1993), few esophageal adenocarcinoma cell lines have been reported (Nishihira et al. 1985, Altorki et al. 1993). Recently Rockett et al. (1997) reported five newly established esophageal carcinoma cell lines, which included 3 from esophageal adenocarcinoma. Two were derived from Barrett's adenocarcinoma and one from a squamous cell carcinoma with focal adenocarcinoma.

The KYSE series and the TE series are now widely used and are available (Shimada et al., 1992; Nishihira, 1993). The YES series, the KE, JROECL the SGF series are also available (Saito et al., 1990; Nakamura et al., 1991; Nakao et al., 1995; Rockett et al., 1997). The HCE cell lines are probably available (Gemma et al. 1996, Zhou et al 1995a). The availability of the cell lines which were originally established in China, Taiwan or South Africa is unknown, although some of these cell lines may be maintained in the USA (Jiang et al. 1992, 1993).

DNA fingerprint analysis revealed that a number of the cell lines in the HCU series have become cross-contaminated. HCU 10 and HCU 13 are genetically distinct, but HCU 10, 18, 33, 37 and 39 are genetically identical. Low passage number samples of these lines also showed the same results, so van Helden et al (1988) concluded that these cell lines are in fact subcultures of the same cells. Karyotypic analysis of lines HCU 18, 33 and 39 showed a large degree of commonality, but differences were observed in marker chromosomes. Within the KYSE-series 19 cell lines were reported to be genetically distinct in 1992 and recently we rechecked 30 cell lines out of 40 in the KYSE-series by DNA fingerprint analysis. This time we detected one cross contamination between KYSE-110 and 200, but this cross-contamination occurred during DNA extraction. Re-analysis revealed that the cell lines KYSE-110 and 200 are genetically distinct. Therefore the KYSE-series has been free of cross-contamination for more than 5 years.

The JROECL series and YES series were not infected by mycoplasma. TT, TH, TS, KSE-1, KE-2, KE-3 and KE-4 were also free from mycoplasma infection. The other cell lines were not tested for mycoplasma infection, or there is no information.

3. PATHOLOGY

Some cell lines grow in tightly adherent clusters, whereas others are not densely packed. Most cell lines are pleomorphic, varying greatly in size and shape, and demonstrate the tendency to pile up in an organized fashion. Cells that are less epithelioid and more spindle-shaped in appearance have been observed in some cell lines (Banks-Schlegel et al. 1986a). The KYSE-50 cell line has a tendency to bunch together in a grape-like manner (Shimada et al. 1992). The HCU series grow as multilayers with the superficial cells being more differentiated than the cells located basally (Robinson et al. 1983a). The presence of varying numbers of tonofilaments and desmosomes provide evidence of the epithelial nature of each line.

The xenograft take rate of 21.9% recorded in Robinson's study is low compared with other reports. Tumors were classified as mobile (noninvasive) or infiltrating, depending on the extent of infiltration into surrounding tissues. Infiltrating tumors developed from 23 xenografts, with metastasis being observed in a single mouse. Thirty mice were observed for 6 months

Table 2 Tumor pathology

No.	Cell line	Tumor pathology	Xenograft pathology	Doubling time	Modal chromosome number	Serum free culture	Tumor marker	Receptor	Mycoplasma, HPV, EB
1	KYSE-30	WD SCC	MD SCC	20.8h	93	possible	Cytokeratin19		
2	KYSE-50	PD SCC	-	28.2h	110	possible	RTHrP		
3	KYSE-70	PD SCC	MD SCC	17.1h	66	possible			
4	KYSE-110	PD SCC	MD SCC	19.1h	65	possible	SCC		
5	KYSE-140	MD SCC		27.4h	118	possible	Cytokeratin19		
6	KYSE-150	PD SCC	PD SCC	13.7h	84	impossible	SCC		
7	KYSE-170	MD SCC	MD SCC	32.5h	65	possible			
8	KYSE-180	WD SCC	MD SCC	15.2h	63	possible	SCC, Cytokeratin19		
9	KYSE-190	WD SCC	WD SCC	23.5h	62	impossible	SCC, Cytokeratin19		
10	KYSE-200	PD SCC	PD SCC	20.5h	65	possible			
11	KYSE-220	MD SCC	WD SCC	29.9h	55	possible	Cytokeratin19		
12	KYSE-240	PD SCC		17.7h	-	NT			
13	KYSE-270	WD SCC	PD SCC	37.7h	77	possible	SCC, Cytokeratin19		
14	KYSE-280	MD SCC		61.1h	-	NT			
15	KYSE-330	MD SCC		24.2h	-	NT			
16	KYSE-350	MD SCC	WD SCC	50.7h	95	impossible			
17	KYSE-360	MD SCC		26.4h	46	impossible			
18	KYSE-390	PD SCC		-	-	NT			
19	KYSE-410	PD SCC	MD SCC	24.2h	52	possible	CEA, Cytokeratin19		
20	KYSE-450	WD SCC	WD SCC	24.4h	64	possible			
21	KYSE-510	WD SCC	WD SCC	26.4h	78	possible	SCC, Cytokeratin19		
22	KYSE-520	MD SCC		30.2h	67	impossible	SCC		
23	KYSE-590	MD SCC		62.5h	79	NT	SCC		
24	KYSE-770	MD SCC		46.2h	50	possible	SCC, Cytokeratin19		
25	KYSE-790	WD SCC	WD SCC	29.7h	92	impossible	Cytokeratin19		

Continued on next page

Table 2 (continued)

No.	Cell line	Tumor pathology	Xenograft pathology	Doubling time	Modal chromosome number	Serum free culture	Tumor marker	Receptor	Mycoplasma, HPV, EB
26	KYSE-850	MD SCC	MD SCC	42.2h	62	impossible	SCC, Cytokeratin19		
27	KYSE-890	MD SCC	PD SCC	36.8h	64	impossible	SCC,CEA,Cytokeratin19		
28	KYSE-960	MD SCC		23.1h	73	possible	SCC,Cytokeratin19		
29	KYSE-1040	PD SCC		-	-	NT			
30	KYSE-1140	PD SCC		27.1h	92	NT	Cytokeratin19		
31	KYSE-1170	PD SCC		35.1h	81	impossible	SCC		
32	KYSE-1190	PD SCC		22.4h	58	possible	SCC, Cytokeratin19		
33	KYSE-1230	Spindle cell		-	-	NT			
34	KYSE-1240	MD SCC		26.9h	57	NT	SCC, Cytokeratin19		
35	KYSE-1250	MD SCC		39.7h	70	NT	SCC, Cytokeratin19		
36	KYSE-1260	MD SCC		31.1h	73	NT	CEA, Cytokeratin19		
37	KYSE-1320	MD SCC		12.9h	not tested	NT			
38	KYSE-1440	PD SCC		35.1h	138	NT			
39	KYSE-1830	WD SCC				NT			
40	KYSE-1860	MD SCC				NT			
41	SUM/c	WD SCC		52.9h	58	NT			
42	HSA/c	PD SCC		40.8h	60	NT			
43	TE-1	WD SCC	MD SCC	60h	55		CEA(+/-)		EBNA(-)
44	TE-2	PD SCC	PD SCC	72h	72	Possible			EBNA(-)
45	TE-3	WD SCC	WD SCC	48h	72	Possible			EBNA(-)
46	TE-4	WD SCC	WD SCC	94h	55				EBNA(-)
47	TE-5	PD SCC	PD SCC	48h					
48	TE-6	WD SCC	-						
49	TE-7	AD	-						
50	TE-8	MD SCC							

Continued on next page

Table 2 (continued)

No.	Cell line	Tumor pathology	Xenograft pathology	Doubling time	Modal chromosome number	Serum free culture	Tumor marker	Receptor	Mycoplasma, HPV, EB
51	TE-9	PD SCC							
52	TE-10	WD SCC							
53	TE-11	MD SCC							
54	TE-12	MD SCC							
55	TE-13	PD SCC							
56	TE-14	MD SCC							
57	TE-15	WD SCC							
58	HCE-1	MD-WD SCC			63/65				
59	HCE-3	MD SCC	PD-MD SCC	33h	55				
60	HCE-4	MD SCC	PD-MD SCC	33h	56				
61	HCE-5	PD SCC	-	60h	98				
62	HCE-6	MD SCC	MD-WD SCC	27h	70				
63	HCE-7 (NM)	MD SCC	PD-MD SCC	53h	54				
64	HCE-8	PD SCC, AD		66h	42				
65	HCE-9	SCC			52				
66	HCU 10	PD SCC	PD or WD SCC						
67	HCU 13	WD SCC	WD,MD or PD SCC		53				
(68	HCU 18)	PD SCC	WD SCC		87		prostaglandin		
(69	HCU 33)	WD SCC	WD SCC or MD SCC		87/94				
70	HCU 35	WD SCC	WD SCC						
(71	HCU 37)	PD SCC							
(72	HCU 39)	WD SCC	WD SCC						
73	HCU 50	MD SCC							
74	HCU 57	MD SCC							

Continued on next page

Table 2 (continued)

No.	Cell line	Tumor pathology	Xenograft pathology	Doubling time	Modal chromosome number	Serum free culture	Tumor marker	Receptor	Mycoplasma, HPV, EB
75	B5	PD SCC	WD SCC						
76	B17	MD SCC	MD SCC						
77	B29	MD SCC							
78	SN	WD SCC		36h	60.5				EB virus negative
79	CaEs-17/EC	SCC		65h	73				
80	EC8712	WD SCC	+		100-110				
81	EC8501	MD SCC	+		60-70				
82	EC-56	SCC			48-88				
83	ECa109	MD SCC	invasive characteristics	54h	63				
84	CE-48T/VGH	epidermoid Ca.	-	55h	78	CEA			
85	CE-69T/VGH	WD SCC	+	48h	76		CEA		
86	CE-81T/VGH	WD SCC	+	72h	32-137		CEA		
87	EC/CUHK 1	WD SCC	WD SCC	12h	85				EB virus negative
88	EC/CUHK 2	PD SCC			43	impossible			HPV-16 DNA (+)
89	JROECL 21	MD SCC	+		17 LOH				free from Mycoplasma
90	JROECL 24	PD AD (Barrett)							free from Mycoplasma
91	JROECL 33	PD AD (Barrett)	+						free from Mycoplasma
92	JROECL 47	PD SCC AD	+						free from Mycoplasma
93	JROECL 50	MD PD AD	NT						free from Mycoplasma

Continued on next page

Table 2 (continued)

No.	Cell line	Tumor pathology	Xenograft pathology	Doubling time	Modal chromosome number	Serum free culture	Tumor marker	Receptor	Mycoplasma, HPV, EB
94	SK-GT-4	WD AD (Barrett)		39h	59				
95	TT	SCC		27h	85-87				
96	TTn	SCC	+	27h	82-84				
97	TH	PD SCC	similar to original tumor	17h	46				free from Mycoplasma
98	TS	MD SCC	similar to original tumor	47h	59				free from Mycoplasma
99	SH-1	PD SCC		20h					
100	KSE-1	PD SCC	PD SCC (spindle shape)	19.5h	67			ER(+), TR(+)	Mycoplasma(-), EBV(-)
101	KSE-2	PD SCC	+	24.9h	62				
102	EC-G1	SCC	+	25h	59-101		IL-1, PTHrP		negative
103	YES 1	PD SCC	+	31.4h	51	possible	SCC, CEA, TPA		negative
104	YES 2	MD SCC	+	23.7h	60	possible	SCC, SLX, CA19-9,CA-125		negative
105	YES 3	MD SCC		22.3h	50	possible	TPA		negative
106	YES 4	MD SCC		35.0h	64	impossible	cytokeratin 19		negative
107	YES 5	MD SCC		30.7h	62	impossible	cytokeratin 19		negative
108	YES 6	MD SCC		29.5h	67	impossible	cytokeratin 19		negative
109	HPL-EsC-1	SCC	M:WD SCC,S: PD SCC		50h				
110	HPL-EsC-2	epidermoid	Cakeratin(-)	15day					
111	EC1NU	-							
112	EC2NU	-							
113	WSSC	-							

Continued on next page

Table 2 (continued)

No.	Cell line	Tumor pathology	Xenograft pathology	Doubling time	Modal chromosome number	Serum free culture	Tumor marker	Receptor	Mycoplasma, HPV, EB
114	SGF 3	MD SCC	+	33h	52	impossible	TNF receptor		
115	SGF 4	MD SCC	+	26h	71	impossible	CEA	TNF receptor	
116	SGF 5	WD SCC	+	30h	105	impossible	SCC	TNF receptor	
117	SGF 7	MD SCC	+	35h	48	no information		SCC	TNF receptor
118	SGF 8a	PD SCC			63	no information		TNF receptor	
119	SGF 8b	PD SCC		38.5h	64	no information		TNF receptor	
120	SGF 9	WD SCC		28h		no information			
121	SGF11	MD SCC		30.1h	4	no information			
122	SGF 12	MD SCC		25.1h	50	no information			
123	E-oh	MD SCC				no information			
124	E-Um	SCC				no information			
125	E-Th	MD SCC				no information			
126	E-1	-				no information			
127	KE-1	WD SCC	similar to original tumor	60h		impossible			-
128	KE-2	PD SCC	similar to original tumor		48h	no information			no
129	KE-3	MD SCC	48h			possible			no
130	KE-4	PD SCC	-	48h		impossible			no
131	KE-5	WD SCC				impossible			-
132	AEN-1	WD SCC							
133	AEN-2	MD SCC							

Abbreviations used are as follows:
NT: not tested. ND: not done. Others as Table 1

following removal of the primary tumor, and a single instance of recurrent carcinoma was noted (Robinson et al. 1983). Botha et al (1986) reported that HCU 18 cells produced metastatic tumors (2/50) with high production of prostaglandin. With the KYSE series, tumors developed from 17 of 23 (73.9%) cell lines inoculated (Kanda et al. 1994). Comparing the histology of the original tumors and the xenografts, 8 cases were completely matched, one case (KYSE-270) was completely different and the other 8 cases were thought probably to be consistent.

Robinson et al. (1986) reported that tumors induced by the inoculation of different cell lines showed similarities. Cell lines derived from well differentiated tumors are characterized by stratified growth, an abundance of desmosomes and tonofilaments and relatively sparse microvilli, while lines derived from poorly differentiated tumors typically comprise loosely packed angular cells with few desmosomes and microvilli.

Banks-Schlegel et al (1986a) reported that the histological appearance of the xenografts were similar to the original tumor, although one cell line also exhibited areas of squamous pearl formation. The major *Mr* 52,000 and 61,000 keratins, characteristic of normal human esophageal epithelium, were missing in some esophageal cancer cell lines. HCE-4 and HCE-7 (derived from a xenograft of the same sample as HCE-4) have the same keratin expression. Interestingly, when cell lines from poor to moderately differentiated squamous cell cancer (SCC) were injected into athymic nude mice, the xenografts showed a reduction or loss of the *Mr* 44,000 and *Mr* 52,000 keratins, concomitant with the appearance of a *Mr* 67,000 keratin, characteristic of keratinizing squamous epithelium. Banks-Schlegel et al. also reported that some cell lines exhibit high levels of envelope-forming capabilities. In summary, similar to the results of Robinson and Maistry, Banks-Schlegel found no correlation between tumorigenicity, latency periods, and the extent of histological differentiation.

4. MOLECULAR AND CYTOGENETICS

Whang-Peng et al (1990) reported detailed chromosomal analysis of the HCE, HCU and TE series. The presence of extensive numerical and structural abnormalities involving every chromosome, including the sex chromosomes, in the primary explant indicates that these abnormalities develop early in the malignant process. They reported that the chromosomes most frequently involved in structural abnormalities are chromosomes 1, 3, 9, and 11 (Whang-Peng et al 1990). Representative karyotypic aberrations of the TE series were also reported (Nishihira et al. 1994).

CyclinD1 has been reported to be amplified and overexpressed in esophageal cancer (Naitoh et al. 1995, Jiang et al. 1992). There are some

Table 3a Oncogene amplification, other genetic changes

No	Cell line	Tumor pathology	erbB	overexpression	CyclinD1	int2	hst1	myc	Ras	MDM2	bFGF	Mage	HLA,MHC
1	KYSE-30	WD SCC	x12, amplification		x4	amplification		x2		amplification		Mage3	MHC class1(+), class2(-)
2	KYSE-50	PD SCC	x2		x1	amplification		x2				Mage1.3	MHC class1(+), class2(-)
3	KYSE-70	PD SCC	x2		x1	no amplification		x1		over expression			MHC class1(+), class2(-)
4	KYSE-110	PD SCC	not detectable		amplification	amplification	amplification	x1		amplification		Mage1	MHC class1(+), class2(-)
5	KYSE-140	MD SCC	x2		amplification	amplification	amplification	x1				Mage1.3	MHC class1(+), class2(-)
6	KYSE-150	PD SCC	x8		x4	no amplification		x1			over expression	Mage1	MHC class1(+), class2(-)
7	KYSE-170	MD SCC	not detectable		x1	no amplification		x6		amplification	over expression	Mage1.3	MHC class1(+), class2(-)
8	KYSE-180	WD SCC	x4		x4	amplification	amplification	x2			over expression	Mage1	MHC class1(+), class2(-)
9	KYSE-190	WD SCC	not detectable		x0.2	not tested		x4					MHC class1(+), class2(-)
10	KYSE-200	PD SCC	x2		x2	amplification		x4				Mage 1	MHC class1(+), class2(-)
11	KYSE-220	MD SCC	x2		x2	amplification	amplification	x2, amplification					MHC class1(+), class2(-)
12	KYSE-240	PD SCC	not checked		NT	NT		NT					
13	KYSE-270	WD SCC	x4		x1	NT		x4 expression			over expression	Mage 3	MHC class1(+) class2(+),
14	KYSE-280	MD SCC	not checked		NT	NT		NT					
15	KYSE-330	MD SCC	not checked		NT	NT		NT					
16	KYSE-350	MD SCC	x2		x1	NT		x2		no amplification			MHC class1(+), class2(-)
17	KYSE-360	MD SCC	not detectable		not detectable	NT		x0.2					
18	KYSE-390	PD SCC	not checked		NT	NT		NT					
19	KYSE-410	PD SCC	x1		x4	amplification	amplification	x1				Mage1.3	MHC class1(+), class2(-)
20	KYSE-450	WD SCC	x1		x1	NT		x6,amplification		amplification	over expression		MHC class1(+), class2(-)
21	KYSE-510	WD SCC	x1		amplification		amplification	amplification	amplification	no amplification	over expression	Mage 3	MHC class1(+), class2(-)

Continued on next page

Table 3a (continued)

No	Cell line	Tumor pathology	erbB	overexpression	CyclinD1	int2	hst1	myc	Ras	MDM2	bFGF	Mage	HLA,MHC
22	KYSE-520	MD SCC	x8		x4			x2		no amplification		Mage1.3	MHC class1(+) class2(-)
23	KYSE-590	MD SCC	x2, amplification		x2			x3, amplification		no amplification			MHC class1(+), class2(-)
24	KYSE-770	MD SCC	x2		x4, amplification		amplification	x1				Mage3	MHC class1(+), class2(-)
25	KYSE-790	WD SCC	x2		x12, amplification		amplification	x3, amplification					MHC class1(+), class2(-)
26	KYSE-850	MD SCC	x1		x1			x0.5			over expression		MHC class1(+), class2(-)
27	KYSE-890	MD SCC	x2		x10, amplification		amplification	x2				Mage3	MHC class1(+), class2(-)
28	KYSE-960	MD SCC										Mage1,3	MHC class1(+), class2(-)
29	KYSE-1040	PD SCC											MHC class1(+), class2(-)
30	KYSE-1140	PD SCC										Mage 3	MHC class1(+), class2(-)
31	KYSE-1170	PD SCC											MHC class1(+), class2(-)
32	KYSE-1190	PD SCC											MHC class1(+), class2(-)
33	KYSE-1230	Spindle cell											
34	KYSE-1240	MD SCC										Mage 1	MHC class1(+), class2(-)
35	KYSE-1250	MD SCC											MHC class1(+), class2(-)
36	KYSE-1260	MD SCC										Mage 3	MHC class1(+), class2(-)
37	KYSE-1320	MD SCC											MHC class1(+), class2(-)
38	KYSE-1440	PD SCC										Mage1.3	MHC class1(+), class2(-)
39	KYSE-1830	WD SCC											
40	KYSE-1860	MD SCC										Mage1.3	MHC class1(+), class2(-)

Continued on next page

Table 3a (continued)

No	Cell line	Tumor pathology	erbB	overexpression	CyclinD1	int2	hst1	myc	Ras	MDM2	bFGF	Mage	HLA,MHC
41	SUM/c	WD SCC											
42	HSA/c	PD SCC											
43	TE-1	WD SCC	amplification	EGFR,EGF, TFGα				K-ras mutation	(codon 12)	expression			
44	TE-2	PD SCC		EGFR,TGFα					K-ras mutation) (codon 12		expression		
45	TE-3	WD SCC	amplification	EGFR,TGFα							expression		
46	TE-4	WD SCC									expression		
47	TE-5	PD SCC	TGFα								expression		
48	TE-6	WD SCC				amplification	amplification				expression		
49	TE-7	AD	amplification	EGFR,TGFα							expression		
50	TE-8	MD SCC	amplification	EGFR,TGFα, erbB-2		amplification	amplification		H-ras mutation (codon 12)		expression		
51	TE-9	PD SCC			amplification	amplification	amplification				expression		
52	TE-10	WD SCC			amplification	amplification	amplification				expression		
53	TE-11	MD SCC			amplification	amplification	amplification				expression		
54	TE-12	MD SCC		EGFR,TGFα						expression			
55	TE-13	PD SCC								expression			
56	TE-14	MD SC											
57	TE-15	WD SCC											
58	HCE-1	MD WD SCC											
59	HCE-3	MD SCC											
60	HCE-4	MD SCC			amplification								
61	HCE-5	PD SCC											
62	HCE-6	MD SCC											
63	(63 HCE-7) (NM)	MD SCC			amplification								
64	HCE-8	PD SCC, AD											
65	HCE-9	SCC											
66	HCU-10	PD SCC											
67	HCU-13	WD SCC											
68	HCU-18	PD SCC											
69	HCU-33	WD SCC											
70	HCU-35	WD SCC											
71	HCU-37	PD SCC											
72	HCU-39	WD SCC											

Continued on next page

Table 3a (continued)

No	Cell line	Tumor pathology	erbB	overexpression	CyclinD1	int2	hst1	myc	Ras	MDM2	bFGF	Mage	HLA,MHC
73	HCU-50	MD SCC											
74	HCU-57	MD SCC											
75	B5	PD SCC											
76	B17	MD SCC											
77	B29	MD SCC											
78	SN	WD SCC											
79	CaEs-17/EC	SCC			no amplification		amplification						
80	EC8712	WD SCC											
81	EC8501	MD SCC											
82	EC-56	SCC											
83	ECa109	MD SCC			no amplification		amplification						
84	CE-48T/VGH	epidermoid carcinoma											
85	CE-69T/VGH	WD SCC											
86	CE-81T/VGH	WD SCC											
87	EC/CUHK 1	WD SCC											
88	EC/CUHK 2	PD SCC											
89	JROECL 21	MD SCC											HLA A, B, C(+), HLA-DR(-)
90	JROECL 24	PD AD (Barrett)											HLA A,B, C(+), HLA-DR(-)
91	JROECL 33	PD AD (Barrett)											HLA A, B, C(+), HLA-DR(-)
92	JROECL 47	PD SCC AD											HLA A, B, C(+), HLA-DR(-)
93	JROECL 50	MD PD AD											HLA A, B, C(+), HLA-DR(-)
94	SK-GT-4	WD AD (Barrett)											MDR1, TGFα, TGFβ, bFGF, PDGF
95	TT	SCC			amplification and overexpression								
(96	TTn	SCC			amplification and overexpression								heterogeneous expression of E-Cadherin
97	TH	PD SCC											mdr(-), p-glycoprotein(-), GST-pi(+) Topoisomerase II(+)
98	TS	MD SCC											

Continued on next page

Table 3a (continued)

No	Cell line	Tumor pathology	erbB	overexpression	CyclinD1	int2	hst1	myc	Ras	MDM2	bFGF	Mage	HLA,MHC
99	SH-1	PD SCC											mdr(-), p-glycoprotein(-),GST-pi
100	KSE-1	PD SCC											
101	KSE-2	PD SCC											
102	EC-Gr	SCC											
103	YES 1	PD SCC											IL6(-)
104	YES 2	MD SCC											IL6(-)
105	YES 3	MD SCC											IL6
106	YES-4	MD SCC											IL6(-)
107	YES-5	MD SCC											IL6(-)
108	YES-6	MD SCC											IL6
109	HPL-EsC-1	SCC											
110	HPL-EsC-2	epidermoid Ca											
111	EC1NU												MRP
112	EC2NU												MRP
113	WSSC												MRP, MDR-1
114	SGF-3	MD SCC											
115	SGF-4	MD SCC											
116	SGF-5	WD SCC											
117	SGF-7	MD SCC											
118	SGF-8a	PD SCC											
119	sGF-8b	PD SCC											
120	SGF-9	WD SCC											
121	SGF11	MD SCC											
122	SGF12	MD SCC											
123	E-oh	MD SCC											
124	E-Um	SCC											
125	E-Th	MD SCC											
126	E1												
127	KE-1	WD SCC											
128	KE-2	PD SCC											MAGE-3/6, HLA class1(A2402/A260
129	KE-3	MD SCC											MAGE-1, MAGE-4/-41, MAGE-3/6,
130	KE-4	PD SCC											
131	KE-5	WD SCC											
132	AEN-1	WD SCC											
133	AEN-2	MD SCC											

Abbreviations used are as follows: NT: not tested; ND: not done. Others see Table 1

Table 3b Tumor suppressor gene, LOH and other genetic changes

No	Cell line	Tumor path	p53	Nucleotide changes**	p21	p16	p15	FHIT	Smad4/DPC4
1	KYSE-30	WD SCC	mutation (Splice acceptor site of intron 6)	agGT to ggGT		mutation (exon 2 codon 119;GAG to TAG)	wild type		
2	KYSE-50	PD SCC	wild type		no expression	hd	hd	wild type	wild type*
3	KYSE-70	PD SCC	mutation (exon 7 codon 251)	ATC to CATC		hd	hd	mutation (exon 8. T to V transition at codon 98	wild type*
4	KYSE-110	PD SCC	mutation (exon 10 codon 337)	ATC to TGC	no expression	wild type	wild type		wild type
5	KYSE-140	MD SCC	mutation (exon 6 codon 193)	CAT to CGT	no expression	hd	hd	wild type*	wild type*
6	KYSE-150	PD SCC	wild type			methylation (Exon 1 alpha)	wild type	mutation (exon.8 T to V transition at codon 98	wild type*
7	KYSE-170	MD SCC	mutation (Exon 8 codon 266)	GGA to CGA		hd	hd	no expression by RT-PCR*	wild type*
8	KYSE-180	WD SCC	mutation (silent mutation)			hd	hd	wild type*	wild type*
9	KYSE-190	WD SCC	mutation (Exon 6 codon 203)	GTG to CTG		wild type	wild type	deletion (exon 5)	
10	KYSE-200	PD SCC	mutation (Exon 10 codon 337)	CGC to TGC	no expression	mutation (exon 2 codon 109; GCC to GC)	wild type		
11	KYSE-220	MD SCC	mutation (exon 7 codon 248)	CGG to TGG		mutation (exon 2 codon 69; GAG to TAG)	wild type	wild type	wild type*
12	KYSE-240	PD SCC	not checked						
13	KYSE-270	WD SCC	mutation (splice donor site of intron 5)	TGgt to TGat	no expression	wild type	wild type	abnormal mRNA	wild type*
14	KYSE-280	MD SCC	not checked						
15	KYSE-330	MD SCC	not checked						
16	KYSE-350	MD SCC	mutation (Exon 7 codon 248)	CGG to CAG	no expression	mutation (exon2; splice acceptor site of intron 1 agGT to tgG	wild type	wild type*	
17	KYSE-360	MD SCC	not checked						
18	KYSE-390	PD SCC	not checked						
19	KYSE-410	PD SCC	wild type		no expression	methylation (Exon 1 alpha)	wild type	mutation (exon 8. T to V transition at codon 98	wild type*
20	KYSE-450	WD SCC	mutation (Exon 5 codon 179)	CAT to CGT		hd	wild type	wild type	wild type*
21	KYSE-510	WD SCC	mutation (Exon 7 codon 241-251)	32bp deletion	no expression	hd	hd	wild type, no expression by RT-PCR	wild type*
22	KYSE-520	MD SCC	mutation (splice acceptor site of intron4)	agTA to tgTA		methylation (Exon 1 alpha)	wild type	mutation (exon 8, T to V transition at codon 98	wild type*
23	KYSE-590	MD SCC	mutation (Exon 8 codon 266)	GGA to CGA		mutation (Exon 2 codon 110; TGG to TGA)	wild type	no expression by RT-PCR*	
24	KYSE-770	MD SCC	mutation (Splice acceptor site of intron 5)	agGT to aaGT	no expression	hd	wild type	wild type*	wild type*
25	KYSE-790	WD SCC	wild type			hd	hd		
26	KYSE-850	MD SCC	mutation (Exon 8 codon 273)	CGT to TGT		hd	hd	wild type*	wild type*
27	KYSE-890	MD SCC	mutation (Exon 6 codon 220)	TAT to TGT		hd	hd	wild type*	wild type*
28	KYSE-960	MD SCC	wild type			hd	wild type	wild type*	no expression by RT PCR
29	KYSE-1040	PD SCC	not checked						
30	KYSE-1140	PD SCC	mutation (Splice donor site of intron 5)	TGgt to TGtt		methylation (Exon 1 alpha)	wild type		wild type*

Continued on next page

Table 3b (continued)

No	Cell line	Tumor path	p53	Nucleotide changes**	p21	p16	p15	FHIT	Smad4/DPC4
31	KYSE-1170	PD SCC	mutation (Exon 6 codon 193, Exon 7 codon 25)	CAT to CTT, ATC to GTC		hd	hd	wild type*	wild type*
32	KYSE-1190	PD SCC	wild type			hd	hd	wild type*	wild type*
33	KYSE-1230	Spindle cell	not checked						
34	KYSE-1240	MD SCC	mutation (Exon 6 codon 194)	CTT to GTT		hd	hd	wild type*	wild type*
35	KYSE-1250	MD SCC	mutation (Exon 6 codon 193)	CAT to CTT		mutation (Exon 1 alpha codon 9-22, 33bp deletion	hd	no expression by RT-PCR*	wild type*
36	KYSE-1260	MD SCC	mutation (Exon 10 codon 342-344)	7 bp deletion		hd	hd		wild type*
37	KYSE-1320	MD SCC							
38	KYSE-1440	PD SCC	wild type			hd	hd		
39	KYSE-1830	WD SCC							
40	KYSE-1860	MD SCC				hd	hd		
41	SUM/c	WD SCC							
42	HSA/c	PD SCC							
43	TE-1	WD SCC	mutation (Exon 8 codon 272)	GTG to ATG		wild type			
44	TE-2	PD SCC				deletion (wild type, Gemma 1996)	wild type (Gemma 1996)		
45	TE-3	WD SCC	wild type			hd	hd		
46	TE-4	WD SCC				deletion			
47	TE-5	PD SCC				deletion			
48	TE-6	WD SCC	mutation (Exon 7 codon 248)	CGG to CAG		deletion			
49	TE-7	AD	wild type			deletion			
50	TE-8	MD SCC				deletion			
51	TE-9	PD SCC	mutation (Exon 8 codon 267)	CGG to CG		deletion			
52	TE-10	WD SCC	mutation (Exon 7 codon 242)	TGC to TAG		deletion			
53	TE-11	MD SCC	mutation (Exon 7 codon 237)	ATG to ATT		deletion			
54	TE-12	MD SCC				deletion			
55	TE-13	PD SCC	wild type			deletion			
56	TE-14	MD SCC							
57	TE-15	WD SCC	mutation (splice acceptor site of intron 5)	tagGTC to taaGTC		deletion			
58	HCE-1	MD WD SCC							
59	HCE-3	MD SCC				hd	hd		
60	HCE-4	MD SCC	mutation (Exon 7, Codon 245)	GGC to GTC		wild type			
61	HCE-5	PD SCC							
62	HCE-6	MD SCC	mutation (Exon 8, Codon 278)	CCT to TCT					
	HCE-7 (NM	MD SCC				point mutation (exon 2), wild type (Liu 1995)	wild type		
63	HCE-8	PD SCC, AD							

Continued on next page

Table 3b (continued)

No	Cell line	Tumor path	p53	Nucleotide changes**	p21	p16	p15	FHIT	Smad4/DPC4
64	HCE-9	SCC							
65	HCU-10	PD SCC							
66	HCU-13	WD SCC							
67	HCU-18	PD SCC							
68	HCU-33	WD SCC							
69	HCU-35	WD SCC							
70	HCU-37	PD SCC							
71	HCU-39	WD SCC							
72	HCU-50	MD SCC							
73	HCU-57	MD SCC							
74	B5	PD SCC							
75	B17	MD SCC							
76	B29	MD SCC							
77	SN	WD SCC							
78	CaEs-17/EC	SCC				wild type	wild type		
79	EC8712	WD SCC							
80	EC8501	MD SCC							
81	EC-56	SCC							
82	ECa109	MD SCC				hd	hd		
83	CE-48T/VG1	epidermoid Ca.							
84	CE-69T/VG1	WD SCC							
85	CE-81T/VG1	WD SCC							
86	EC/CUHK 1	WD SCC							
87	EC/CUHK 2	PD SCC							
88	JROECL 21	MD SCC							
89	JROECL 24	PD AD (Barrett)							
90	JROECL 33	PD AD (Barrett)							
91	JROECL 47	PD SCC, AD							
92	JROECL 50	MD PD AD							
93	SK-GT 4	WD AD	Mutation (Exon 5 codon 175)	CGC to CAC					
94	TT	SCC				wild type	wild type		
95	TTn	SCC				wild type	wild type		
96	TH	PD SCC							
97	TS	MD SCC							
98	SH-1	PD SCC							
99	KSE-1	PD SCC							
100	KSE-2	PD SCC							

Continued on next page

Table 3b (continued)

No	Cell line	Tumor path	p53	Nucleotide changes**	p21	p16	p15	FHIT	Smad4/DPC4
101	EC-Gi	SCC							
102	YES-1	PD SCC							
103	YES-2	MD SCC							
104	YES-3	MD SCC							
105	YES-4	MD SCC							
106	YES-5	MD SCC							
107	YES-6	MD SCC							
108	HPL-Esc-1	SCC							
109	HPL-Esc-1	epidermoid Ca							
110	EC1NU	no information							
111	EC2NU	no information							
112	WSSC	no information							
113	SGF-3	MD SCC							
114	SGF-4	MD SCC							
115	SGF-5	WD SCC							
116	SGF-7	MD SCC							
117	SGF-8a	PD SCC							
118	SGF-8b	PD SCC							
119	SGF-9	WD SCC							
120	SGF-11	MD SCC							
121	SGF-12	MD SCC							
122	E-oh	MD SCC							
123	E-Um	SCC							
124	E-Th	MD SCC							
125	E-1	no information							
126	KE-1	WD SCC							
127	KE-2	PD SCC							
128	KE-3	MD SCC	wild type	exon 5-8					
129	KE-4	PD SCC	mutation	exon 7					
130	KE-5	WD SCC							
131	AEN-1	WD SCC							
132	AEN-2	MD SCC							

Abbreviations used are as follows: NT: not tested; ND: not done; hd: homozygous deletion; *: unpublished data; **: Lower case letters indicate sequences in introns. Others see Table 1

problems in the literature with regard to the expression of the CyclinD1 gene. Nakagawa et al (1995) reported that HCE-4, HCE-7, TT and TTn cell lines demonstrated amplification of the CyclinD1 gene. However, HCE4 and HCE7 were derived from the same patient, and TTn and TT were also derived from the same patient. Jiang et al (1992) reported CyclinD1 amplification and overexpression in the EC 109 cell line. However, another report indicated that EC 109 had no amplification and low expression of this gene (Zhou et al. 1995a). Antisense to CyclinD1 inhibited the growth of HCE-7 cell lines (Zhou et al. 1995a).

In the KYSE series, detailed analysis of *c-myc*, *c-erb*B, *hst*-1 and CyclinD1 genes was reported by Kanda et al (1994). Co-amplification of hst-1, int-2 and CyclinD1 genes is a feature of esophageal cancer, and co-amplification of hst-1 and the cyclinD1 gene was found in 9 KYSE cell lines (9/23, 39%) and amplification of 11q13 was associated with poor prognosis. Amplification of the c-myc gene was observed in 5 KYSE cell lines derived from well-differentiated carcinomas, accompanied by co-amplification of other oncogenes. The c-erbB gene was amplified in 3 KYSE cell lines. Murine double minute 2 (MDM2) gene binds to p53 protein and inhibits its ability to activate transcription. Shibagaki et al (1995) reported amplification of the MDM2 gene in 13 (18.1%) of 72 tumor tissues and in 4 (33.3%) of 12 cell lines of the KYSE series.

Co-amplification of *hst*-1 and *int*-2 gene was also observed in 5 of 13 TE cells examined (Nishihira et al. 1993), and TE 9, 10 and 11 cells demonstrated amplification of the CyclinD1 gene (Nakagawa et al. 1995). Nishihira et al (1993) also reported amplification of the *c-myc* gene in 6 of 11 TE cell lines examined. With regard to ras oncogene, Galiana et al (1993) stated that there were no point mutations of the H-*ras* and K-*ras* genes in 30 fresh surgical specimens including the primary tumors from which the TE cell lines had been derived. However, 3 of the 7 TE cell lines examined contained point mutations and Galiana et al (1993) concluded that *ras* mutation in the TE series was either due to their long term culture or that only a small portion of the original tumors contained such mutations.

Epidermal growth factor (EGF) and transforming growth factor α (TGF α) share an EGF receptor, a product of the c-*erb*B proto-oncogene. Yoshida et al (1990) revealed that expression of mRNA for TGF α was high in all of the 6 TE cell lines examined, while mRNA for EGF was expressed in 3 of 6 lines. Iida et al (1994) reported that all of the 13 TE series expressed FGFβ whereas Okamura et al (1992) showed that TE1, 2, and 5 expressed no mRNA of FGFβ. Shiga et al (1993) described amplification of the *c-erb*B-2 gene in 4 of 60 resected tumors (6.7%) and in 1 out of 12 (8.3%) cell lines in the TE series. In contrast, *c-erb*B-2 overexpression was observed in 7 cell lines in which gene amplification was not found.

Evidence for the involvement of p53 in esophageal SCC and Barrett adenocarcinoma of the esophagus is well known. The TE-1 to 13 cell lines have been reported to contain the wild type p53 gene (exon 5 to 11 of the p53 gene) (Yamada et al 1991). In contrast to this report, Barnas et al (1997) showed that 6 out of 9 of the same TE series had p53 mutations and, surprisingly, with the exception of TE-1, none of the cell lines exhibited p53 binding to DNA. So, Barnas et al concluded that p53 mutation in esophageal cancer is frequent, and that inactivation of p53 may occur through mechanisms other than mutation in the coding sequence of p53. p53 mutation was also observed in 2 out of 4 HCE cell lines (Hollstein et al 1990).

In the KYSE series, Tanaka et al (1996) reported that compared with the incidence in tumors (28/60, 40%) the p53 mutation frequency in cell lines (22/29, 76%) was higher with a different mutation spectrum. Tanaka et al also indicated that there was no correlation between the status of the p53 gene and the ability to establish a cell line.

SK-GT-4 cells (derived from an adenocarcinoma) showed p53 mutation in exon 5 codon 175, but expressed p53 protein in a western blot analysis. SK-GT-4 is more resistant to 5-fluorouracil, mitomycin-C and cisplatin than other cell lines with a wild type p53 gene. Mutant p53 protein levels did not increase in SK-GT-4 cells after exposure to chemotherapeutic drugs. The half-life of p53 in SK-GT-4 cells is reported as 6h (Nabeya 1994).

In esophageal SCC, allelic loss on chromosome 9p was reported to be frequent and subsequent studies have revealed somatic mutations of the p16 gene in 10 to 52% of esophageal carcinomas. In contrast, analysis of the p16 gene revealed that 13 of 14 cell lines in the TE series have homozygous deletion (Igaki 1994, Liu 1995). TE-1 is the only cell line which preserves the wild type p16 gene. Kitahara et al (1996) also reported that out of eight human esophageal carcinoma cell lines (TE series), seven (87.5%) and six (75%) cell lines showed homozygous deletions of the p16 and p15 genes, respectively. Detailed analysis revealed that all the p16-negative cell lines expressed high levels of cyclinD1, CDK4 and $p27^{KIP1}$ protein. In the KYSE series, p16 aberrations were also found in 28 of 30 (93%) cell lines (Tanaka et al 1997). The type of aberrations were as follows: 18 homozygous deletions, 6 point mutations and 4 hypermethylations. Homozygous deletions of the p15 gene were observed in 16 cell lines (53%) and aberrations of the p15 gene were less frequent than those of the p16 gene (Tanaka et al 1997).

Recently, the FHIT (fragile histidine triad) gene was identified and alterations were also reported in 50% (5/10) esophageal carcinomas (Ohta 1996). Zou et al (1997) found that one of 13 primary esophageal tumors expressed no detectable FHIT transcript. In six of 9 KYSE cell lines no FHIT RT-PCR product was detectable. However, genomic PCR and direct sequencing of exons 5-9 revealed a wild type sequence in 8 cell lines. They suggested that the open reading frame of FHIT transcript is not important in esophageal cancer. Further analysis of FHIT in the KYSE series demonstrated that 12 of 20 cell lines have wild type transcription of FHIT gene (Tanaka et al 1998).

DPC4/Smad4 (deleted in pancreatic cancer 4) is a candidate tumor suppressor gene for pancreatic cancer. Lei et al (1996) reported infrequent DPC4 gene mutation in esophageal cancer and one out of 23 of KYSE cell lines showed aberration of this gene (Tanaka et al 1997. unpublished data).

5. CELL LINE APPLICATIONS

In vitro screening tests of anticancer drugs for the TE series was reported by Nishihira et al (1985). MDR and GST-π were not expressed in TE-1, TH and SH-1 cell lines, while a range of levels of topoisomerase II was expressed. There was poor correlation between in vitro and in vivo results. SK-GT-4 cells also expressed MDR-1 mRNA transcript (Nabeya et al 1995).

KSE-1 cell growth was inhibited at a temperature of approximately 42.5°C. Concomitant application of hyperthermia, bleomycin and irradiation had a maximal effect on cell growth, compared to proliferation in cells treated with only one or two of these modalities (Matsuoka et al 1989). The temperature range at which the cells proliferate is from 31° to 39°C for SGF-3 and from 29° to 41°C for SGF-5 cells. No proliferative activity was observed after the cells were exposed to 42°C for 72 h (Saito et al 1990). The effect of hyperthermia on the uptake and subsequent distribution of [^{195m}Pt]cisplatin (CDDP) in KYSE-150 (CDDP sensitive line) and KYSE-170 (CDDP resistant line) were analyzed by Miyahara et al (1993). They reported that thermal enhancement of the cytotoxity of CDDP is caused mainly by acceleration of drug entry into the cell and binding to DNA, probably due to increased permeability.

The growth of estrogen receptor (ER) positive and androgen receptor (AR) positive KSE-1 cells was inhibited by estradiol and dihydrotestosterone (DHT). However, the growth of ER negative and AR negative KSE-2 cells was not influenced by either estradiol or DHT administration (Matsuoka et al 1987, Ueo et al 1990). Some esophageal carcinomas have steroid hormone receptors, but the growth of carcinomas containing estrogen binding components is not always influenced by estradiol.

All-trans-retinoic acid treatment caused a decrease in the c-myc mRNA level and growth rate of the esophageal cancer cell line EC8712 (Feng et al 1992).

Kamata et al (1986) reported that squamous cell carcinoma cell lines (including the TE-series) possess high levels of EGFR and their growth is inhibited by EGF. They also indicated that the sensitivity to the inhibitory effect of EGF correlated well with the elevated level of EGF receptors in SCC cell lines, and higher significance was obtained when data on esophageal

SCCs were excluded. In contrast to this report, Banks-Schlegel et al (1986b) found that certain esophageal carcinoma cell lines (HCE, HCU, TE series) have fewer EGF receptors with increased ligand affinity. They concluded that there was no simple relationship between receptor number, affinity or occupancy and the growth stimulatory effect of EGF. Similar results were obtained in the KYSE series (Shimada et al. 1993a), confirming that the growth response to EGF of esophageal cancer cell lines is complex.

Protein kinase C (PKC) plays a crucial role in signal transduction mediated by hormones and growth factors. Chida et al (1988) reported that TE-1, 2, 8, and 9 had lower PKC activity compared to human epidermal keratinocytes in primary culture. However, Hashimoto et al (1989) found that the PKC activity of eight esophageal cancers was similar to that of the adjacent normal mucosa.

Nakao et al (1995) described the $CD4^-$ $CD8^+$ CTL (KE-4 CTL) cell line. The KE-series (KE-3,4) and TE series (TE-8 to 11) expressed HLA class 1 antigens. The KYSE-series expressed HLA class 1 whereas only two cell lines (KYSE-270 and 450) expressed HLA class 2 (Yamasaki et al 1997). Yamasaki et al (1997) also reported MAGE-1 and MAGE-3 gene expression in 13 out of 31 cell lines (41.9%) and in 15 out of 31 cell lines (48.4%) respectively in the KYSE series. Rockett et al (1997) reported that HLA-A, -B and -C were expressed constitutively, but not HLA-DR, in the JROECL series.

Humoral hypercalcemia associated with solid tumors is caused by a variety of factors, including a parathormone (PTH) like factor. EC-GI cells can produce PTHrp when tumors are implanted in nude mice (Sato et al 1987). KYSE-50 also produces PTHrp and the tumor-bearing mouse became hypercalcemic (Shimada et al 1992).

The CE-series produce CEA (Hu et al 1984). The YES series produce SCC antigen and CEA and YES-2 cells produce SCC antigen (Murakami et al 1991). The YES series also produce several cytokines, including IL-1α, IL-1β, IL-6, IL-7, IL-10, GM-CSF and G-CSF. (Oka et al 1995). In the KYSE-series, 40% of the selected cell lines (12/30) secrete SCC antigen, whereas CEA was secreted from only 3 cell lines. The cytokeratin 19 fragment was secreted from 17/28 (60.7%) cell lines. A recent report by Yamamoto et al (1997) also stated that of the 6 cell lines in the YES series, 5 expressed cytokeratin-19 fragment in their cytoplasm and produced soluble cytokeratin-19 fragment.

There are some cell lines which are derived from the same patient. TTn cell lines are derived from xenografted tumors which were the source of the TT cell lines. There was no morphological difference between these cell lines. HCE-7 was also derived from a xenograft from the same esophageal tumor as that from which the cell line HCE-4 was established. There was no difference in the spectrum of keratin expression between these cell lines, and they shared histological features. SGF8a and SGF8b were derived from same patient, but from different lymph node metastases.

Morikawa et al (1986) isolated 3 sublines from the HPL-EsC-1 cell line. There were major morphological differences amongst the 3 lines: HPL-EsC-1-K showed keratinization in foci and small cell clusters, HPL-EsC-1-S grow in a pavement-like fashion with prominent intercellular bridges and were packed more tightly without piling up on the monolayer, and HPL-EsC-1-M grew in a monolayer and showed little keratinization.

ACKNOWLEDGMENTS

The Author thanks Dr Tetsuro Nishihira (Tohoku University, Japan), Dr. Mitsukazu Saito (Toyama Medical and Pharmaceutical University, Japan), Dr. Hideaki Yamane (Kurume University, Japan), Dr. Hiroyuki Suzuki (Akita University, Japan), Dr. Masayuki Nakamura (Yamaguchi University, Japan), Dr. Kanji Sato (Tokyo Women's Medical College, Japan), Dr Toshifumi Iizuka (Segi Clinic, Japan), Professor Shigeru Morikawa (Shimane Medical University, Japan), Professor EC Chew (The Chinese University of Hong Kong, Hong Kong) and Dr Banks-Schlegel (NIH, USA) for kindly giving information regarding their established cell lines. Unfortunately, the author could not get any information from mainland China or South Africa, as the researchers have relocated.

REFERENCES

Akaishi T. et al. *Nihon Gekagakkai Zasshi*. 85: 1440, 1984
Akaishi T. et al. Esophageal cancer (Kasai M edi). *Excerpta Medica* p35, 1986
Altorki N. et al. *Cancer* 72: 649, 1993
Banks-Schlegel S.P. et al. *Cancer Res* 46: 250, 1986a
Banks-Schlegel S.P. et al. *J Biol Chem* 261: 4359, 1986b
Barnus C. et al. *Int J Cancer* 71: 79, 1997
Bey E. et al. *In vitro* 12: 107, 1976
Botha J H. et al. *JNCI* 76: 1053, 1986
Chew E.C. et al. *Recent Advances in Cancer* 1989: p79, 1989
Chida K. et al. *Biochem Biophys Res Commun* 157: 1, 1988
Doki Y. et al. *Cancer Res* 53: 3421, 1993
Feng L. et al. *Science in China* 35: 445,1992
Galiana C. et al. *Int. J. Cancer* 54: 978, 1993
Gao J. et al. *Clin. Expl. Metastasis* 2: 205, 1984
Gemma A. et al. *Int.J. Cancer* 68: 605, 1996
Haffejee A.A. et al. *South Afr J Surg* 20:259, 1982
Hainaut P. et al. *Recent advances in gastroenterological carcinogenesis* 1. Monduzzi (editor) p77, 1996
Hashimoto Y. et al. *Biochem Biophys Res Commun* 163: 406, 1989
Helden P.D. et al. *Cancer Res* 48: 5660, 1988
Hollstein M.C. et al. *Proc. Natl. Acad. Sci. USA* 87:9958, 1990

Hu C. et al. *JNCI* 72: 577, 1984
Igaki H. et al. *Biocem. Biophys. Res. Commun* 203: 1090, 1994
Iida S. et al. *Biocem. Biophys. Res. Commun* 19: 1113, 1994
Iihara K.et al. *Int. J. Cancer* 55: 364, 1993
Iizuka T. et al. *Jpn. J. Clin. Oncol.* 11: 487, 1981
Jiang W. et al. *Cancer Res* 52: 2980, 1992
Jiang W. et al. *Proc. Natl. Acad. Sci. USA* 90: 9026, 1993
Kamata N. et al. *Cancer Res* 46: 1648, 1986
Kanda Y et al. *Int. J. Cancer* 58: 291, 1994
Kitahara K. et al. *J Exp Ther Oncol* 1: 7, 1996
Kuriya Y. et al. *J. Exp. Med.* 139: 377, 1983
Lei J .et al. *Oncogene* 13: 2459, 1996
Li S. et al. *Proc. CAMS and PUMC* 1: 81, 1986
Liu Q. et al. *Oncogene* 10: 619, 1995
Matsuoka H. et al. *Cancer Res* 47: 4134. 1987
Matsuoka H. et al. *Eur Surg Res* 21: 49, 1989
Matsuoka H. et al. *Jpn J Cancer Res* 84: 336, 1993
Mijahara T. et al. *Jpn J Cancer Res* 84: 336, 1993
Mok C.H. et al. *Anticancer Res* 7: 409, 1987
Morikawa S et al. *Acta Histochem Cytochem* 19: 398, 1986
Murakami T. et al. *Nippon Gekagakkai Zasshi* 92: 1563, 1991
Murase M. et al. *Journal of Surgical Oncoloogy* 61: 223, 1996
Nabeya Y. et al *Int J Cancer* 64: 37, 1995
Naito M. et al. *Jpn J of Surg* 20: 170, 1990
Naitoh H. et al. *Am J Pathology* 146: 1161, 1995
Nakagawa H. et al. *Cancer* 76: 541,1995
Nakamura M. et al. *Arch Jpn Chir* 60: 3, 1991
Nakao M. et al. *Cancer Res* 55: 4248, 1995
Nishihira T. *Jpn J Cancer Res (GANN)* 70: 575, 1979
Nishihira T. et al. *In vitro models of Cancer Research.* M.M.Webber. (ed) 1: 65,1985
Nishihira T. et al. *Diseases of the Esophagus* 125, 1989
Nishihira T. et al. *J Cancer Res Clin Oncol* 119: 441, 1993
Nishihira T. et al. *Atlas of Human Tumor Cell Lines.* Hay et al (ed) p269, 1994
Ohta M. et al. *Cell* 84:587, 1996
Oka M. et al. *Journal of interferon and cytokine research* 15: 1005, 1995
Okamura K. et al. *Biocem. Biophys. Res. Commun* 186: 1471, 1992
Oku K. et al. *Anticancer Res* 11: 1591, 1991
Pan Q. et al. *Proc. CAMS and PUMC* 4: 52, 1989
Rabin H. et al. *Cancer of the Esophagus* 2: 99, 1982
Robinson A.A. et al. *Clinical Oncology* 6: 125, 1980
Robinson K.M. et al. *SA Journal of Medical Sciences* 41: 285,1976
Robinson K.M. et al. *Esophageal cancer. Morio Kasai (ed)* 39,1986
Robinson K.M. et al. *Scanning Electron Microscopy* p231, 1980
Robinson K.M. et al. *South Afr J of Surg* 20: 245, 1982
Robinson K.M. et al. *JNCI* 70: 89, 1983a
Robinson K.M. et al. *J. Pathology* 140: 193, 1983b
Robinson K.M. et al. *Scanning Electron Microscopy* p361, 1983c
Robinson K.M. et al. *South Afr J Surg* 20: 253, 1982
Rockett J.C. et al. *Br J Cancer* 75: 258, 1997
Saito M. et al. *In Vitro Cell. Dev. Biol.* 26: 181, 1990
Saito T. et al. *Cancer* 70: 2402, 1992

Saito T. et al. *Oncology* 51: 440, 1994
Saito T. et al. *Human cell* 7: 55, 1994
Sato K. et al. *Jpn. J. Cancer Res.* 78: 1044, 1987
Saunders R.M. *South Afr Medical Journal* 204, 1978
Shibagaki I. et al. *Clinical Cancer Res* 1: 769, 1995
Shiga K. et al. Anticancer Res 13: 1293, 1993
Shigetomi A. et al. *Human cell* 5: 273, 1992
Shimada Y. et al. *Igaku no Ayumi* 159: 413, 1991a
Shimada Y. et al. *Human Cell* 4: 315, 1991b
Shimada Y. et al. *Cancer* 59: 277, 1992
Shimada Y. et al. *Arch Jpn Chir* 62: 153, 1993a
Shimada Y. et al. *Br. J. Surg.* 80: 605, 1993b
Shimada Y. et al. *Human cell* 7: 193, 1994
Shimada Y. et al. *Diseases of the esophagus.* A.Peracchia (ed) 109, 1995
Shimada Y. et al. *Br J Surg* 83: 1148, 1996
Shinbo T. et al. *Jpn J Gastroenterol Surg* 14: 1646, 1981
Shiozaki H. et al. *Br J Cancer* 71: 250, 1995
Su Y. et al. *Proc. CAMS and PUMC* 3: 84, 1988
Surgical Laboratory for PMC *Chinese Medical Journal* 2: 357, 1976
Takano R. et al. *Am J Path* 137: 393, 1990a
Takano R. et al. *Am J Path* 137: 1007, 1990b
Tanaka H. et al. *Int J Cancer* 65: 372, 1996
Tanaka H. et al. *Int. J Cancer* 70: 437, 1997
Tanaka H. et al. *Cancer Res* 1998 (in press)
Terashima M. et al. *Jpn J Cancer Chemother* 19: 1999, 1992
Ueo H. et al. *Cancer Res* 50: 7212, 1990
Watanabe H. et al. *Biotherapy* 10: 1332, 1996
Whang-Peng J.et al. *Cancer Genet Cytogenet* 45: 101, 1990
Wuu K.D. et al. *Cancer Genet Cytogenet* 20: 279, 1986
Yamada Y. et al. *Cancer Res* 51: 5800, 1991
Yamamoto K. et al. *Cancer* 79: 1647, 1997
Yamamoto T. et al. *Cancer Res* 46: 414, 1986
Yamasaki S. et al. *Proc 88th Annual Meeting* AACR, p631 1997
Ye X. et al. *Science in China* 35: 76, 1992
Yoshida K. et al. *Int. J. Cancer* 45: 131, 1990
Yoshida K. et al. *Cancer Res Clin Oncol* 119: 401, 1993
Zhao X. et al. *Science in China* 38: 580, 1995
Zhou P. et al. *Oncogene* 11: 571, 1995a
Zhou X. et al. *Genes, Chromosom Cancer* 13: 285, 1995b
Zhou X. et al. *Cancer J Sci Am* 2: 221, 1996
Zou T.T. et al. *Oncogene* 15: 101, 1997

Chapter 11

Bladder Cancer

Ruth Knuechel and John RW Masters
Institut fur Pathologie, Universität Regensburg, D-93042 Regensburg, Germany.
Tel: 0049-941-944-6625; Fax: 0049-941-944-6602;
E-mail: ruth.knuechel-clarke@klinik.uni-regensburg.de
and Institute of Urology, University College London, London W1P 7PN, UK.
Tel: 0044-171-504-9366; Fax: 0044-171-504-9366; E-mail: j.masters@ucl.ac.uk

Since the first human bladder cancer cell line (RT4) was described in 1970 (Rigby and Franks, 1970), more than 50 other lines derived from bladder cancers have been established (see Tables 1–3). The bladder cancer cell line most widely used is T24 (Bubenik *et al.*, 1973). T24 has a place in the history of human cancer research, as it was the first human cell line in which an oncogene was identified. The Ha-*ras* proto-oncogene was identified by transfection of DNA from T24 into rodent fibroblasts, resulting in their neoplastic transformation (Reddy *et al.* 1982, Goldfarb *et al.*, 1982). T24 is also well-known for the same reason as HeLa, because it has cross-contaminated many other human cell lines.

1. CULTURE CONDITIONS

Primary cultures of human bladder cancer are easy to produce in conventional serum-containing commercial culture media. However, only a small fraction of these become continuous cell lines. The main causes for the failure of long-term culture are fibroblast overgrowth and tumor cell senescence. The cell lines that have been established are generally easy to maintain as attached monolayers and have relatively rapid population doubling times (24–48 h) (e.g. Fogh, 1978).

J.R.W. Masters and B. Palsson (eds.), Human Cell Culture Vol. 1, 213–230.

Table 1 Clinical origin of cell lines.

Name	Patient age	Patient sex	T	N	M	Grade	Primary site	Specimen site	Culture medium	Culture method	Authentication	Availability	Primary reference	Establishment	First author
5637	68	m	nd	nd	nd	nd	bl	bl	RPMI + 10-20%FBS	D	I,D	1,2,3,4,6	Proc Natl Acad Sci USA 82, 1526-1530, 1985	1974	GB Cannon
1016T	nd	m	nd	nd	nd	1	bl	bl	RPMI + 10%FCS	E	K	O	Natl Cancer Inst. Monogr 9.23-24, 1978	<1977	AY Elliott
192B	nd	m	nd	nd	nd	4	bl	bl	RPMI + 10%FCS	D,E	I	O	Natl Cancer Inst Monogr 9:23-24, 1978	<1977	AY Elliott
253J	53	m	4mf	nd	nd	4	bl, ur, rp.	mln	RPMI + 10%FCS	E,X	I	O	J Natl Cancer Inst 53:1341-1349, 1974	1972	AY Elliott
292W	nd	m	nd	nd	nd	4	rp	rp	RPMI + 10%FCS	E	K	O	Natl Cancer Inst Monogr 9:23-24, 1978	<1977	AY Elliott
486P	61	m	nd,mf	1	nd	4	bl	mln	RPMI + 10%FCS	E	I	O	Cancer Res 37:1279-1289, 1977	<1977	AY Elliott
575A	nd	m	nd	nd	nd	3	bl	bl	RPMI + 10%FCS	E	K	O	Natl Cancer Res Inst Monogr 9:23-24, 1978	<1977	AY Elliott
639V	69	m	nd	nd	nd	3	ur	ur	RPMI + 10%FCS	E	I	O	Cancer Res 37:1279-1289, 1977	1973	AY Elliott
647V	59	m	1	0	0	2	bl	bl	RPMI + 10%FCS	D,E	K	O	Cancer Res 37:1279-1289, 1977	1974	AY Elliott
682B	nd	m	nd	nd	nd	1	bl	bl	RPMI + 10%FCS	E	K	O	Natl Cancer Inst Monogr 9:23-24, 1978	<1977	AY Elliott
743E	nd	m	nd	nd	nd	4	bl	mbr	RPMI + 10%FCS	E	K	O	Natl Cancer Inst Monogr 9:23-24, 1978	<1977	AY Elliott
751G	nd	m	nd	nd	nd	1	bl	bl	RPMI + 10%FCS	E	K	O	Natl Cancer Inst Monogr 9:23-24, 1978	<1977	AY Elliott
BT-1	49	f	nd	nd	nd	3	bl	bl	DMEM + 15%FBS	E	I	O	Urol Res 16:23-29, 1988	1987	W Heckl
BT-B	66	m	a	0	0	3	bl	bl	RPMI + 10%FBS	D	I,D	O	Exp Cell Res 187:185-192, 1990	1988	J van der Bosch
COLO 232	70	m	3a	nd	nd	3	bl	bl	F12 + 10% FBS	D	I		In vitro 40:301-306, 1978	1976	GE Moore
HOK-1	77	m	3	nd	1	3	bl	bl	MCDB151 + DMEM 1:1 + 10%FCS	D,E	I,H	O	Int J Cancer 49:122-128, 1991	1990	FA Offner
HT 1376	58	f	>2	nd	nd	3	bl	blr	EMEM+NEASS+EBSS +10%FBS	D	I	1,2,7	J Natl Cancer Inst 58:881-890, 1977	1973	S. Rasheed
IG	30	m	a	0	0	2	bl	bl	RPMI + 15%FBS	D	nd		Eur Urol 14:65-71, 1985	1985	H Logothetou-Rella
J82 COT	58	m	3	nd	nd	3	bl	bl	EMEM+NEASS+ EBSS+10%FBS	D	D[3]H,I	1	Br J Cancer 38:64-76, 1978	1972	C O'Toole
JTC-29	77	m	a	nd	nd	1	bl	bl	DM-160 + 10%FBS	E	nd		In vitro 19:591-599, 1983	<1983	T Kakuya
JTC-30	46	a	a	nd	nd	1	bl	bl	DM-160 + 10%FBS	E	nd		In vitro 19:591-599, 1983	<1983	T Kakuya
JTC-31	54	m	nd	nd	nd	4	bl	bl	DM-160 + 10%FBS	E[2]	nd		In vitro 19:591-599, 1983	<1983	T Kakuya
JTC-32	73	f	nd	nd	nd	4	bl	bl	DM-160 + 10%FBS	E	nd		In vitro19:591-599, 1983	<1983	T Kakuya
JTC-33	70	m	nd	nd	nd	4	bl	bl	MEM + 10%FBS	E[2]	nd		In vitro 19:591-599, 1983	<1983	T Kakuya
JTC-34	64	m	a	nd	nd	1	bl	bl	DM-160 + 10%FBS	E	nd		In vitro 19:591-599, 1983	<1983	T Kakuya
KK47	50	m	a	nd	nd	1	bl	bl	Ham's F12 + 20%FCS		D,X	I	Jpn J Urol 70:485-494, 1979	1977	H Hisazumi
KW103	44	m	nd	nd	nd	2	bl	bl	Ham's F12 + 20%FCS	D	M		Jpn J Urol 73:1019-1031, 1982	<1982	K Naito
LA-B1	57	m	>2	nd	nd	3	blw	bl	DMEM/Ham's F2 + 20% FCS	D	K	O	Cancer Res 42:2392-2397, 1982	<1982	E Messing
MGH-U3	76	f	a	0	0	1	bl	blr	McCoy 5a+5%FCS	D	I,K		Cancer Res 45:5070-5079, 1985	1977	GR Prout
MGH-U4	57	m	is	0	0	3	bl	blr	McCoy 5a+5%FCS	D	I,K		Cancer Res 45:5070-5079, 1985	1977	GR Prout
NBT-2	64	m	>2	nd	nd	3	bl	bl	Eagle'sMEM+20%FCS	D	K		Jpn J Urol 70:351-357, 1979	<1979	T Yamamoto
NTUB1	70	f	>2	1	0	3	bl	blpw	RPMI1640+15%FCS	D	K		J Formosan Med Assoc 91:608-613, 1991	1989	H-J Yu
PS1	63	m	4	1	0	3	bl	bl	Ham's F12 + 20%FCS	D,X	1		Invest Urol 15:246-252, 1978	1975	EJ Sanford

Continued on next page

Table 1 (continued)

Name	Patient age	Patient sex	T	N	M	Grade	Primary site	Specimen site	Culture medium	Culture method	Authentication	Availability	Primary reference	Establishment	First author
RT112	nd	f	nd	nd	nd	2	bl	bl	nd	D	D[3],I	3	J Natl Cancer Inst 58:1743, 1977	1973	CC Rigby
RT4	63	m	2	nd	0	1+2	bl	bl	McCoy's5a+10%FBS	E	H,I	1,2,5	Br J Cancer 24:746-754, 1970[4]	1967	CC Rigby
SBC-2	82	f	nd	0	0	2	bl	bl	RPMI+10%FBS	D	D,I	4	J Urol 150:1932-1937, 1993	<1993	AJ Ulmer
SCaBER	58	m(B)	3	0	0	2	bl	bld	EMEM+NEASS +EBSS+10%FBS	E	H,I	1	Int J Cancer 17:707-714, 1976	1974	C O'Toole
SW1710	84	f	a	nd	nd	1	bl	bllw	L15 + 15%FBS	E	I		Cancer Res 44:3997-4005, 1984	1977	FA Kyriazis
SW1738	55	m	4mf	nd	1	2+4	bl	bl	L15 + 15%FBS	E	I		Cancer Res 44:3997-4005, 1984	1977	FA Kyriazis
SW780	80	f	1[1]	nd	nd	1	bl	bl	L15 + 15%FBS	E	I		Cancer Res 44:3997-4005, 1984	1974	FA Kyriazis
SW800	54	m	nd	0	0	3	bl	blr	L15 + 15%FBS	E	I		Cancer Res 44:3997-4005, 1984	1974	FA Kyriazis
T24	81	f	nd	nd	nd	3	bl	blr	McCoy 5a+10%FCS	D	D[3],H	1,5,6	Int J Cancer 5:310-319, 1970 Int J Cancer 11:765-773, 1973	1970	J Bubenik
TCCSUP	67	f	>2	nd	lmbo, mbr4	4	bl	bln	EMEM+NEASS++SP +EBSS+0%FBS	D	H,I	1	Br J Cancer 35:142-151, 1977	1974	SK Nayak
TSGH-8301	56	m(C)	a	nd	nd	2	bl	blr	RPMI + 10% FCS	E	I,K	O	J Surg Oncol 37:177-184, 1988	1983	MY Yeh
UCRU-BL-13	62	m	3	nd	1	2	bl	bl	RPMI1640+10% FCS+ insulin +hydroco	X	H,K	O	Int J Cancer 44:276-285, 1989	1988	P Russell
UCRU-BL-17CL	69	f	4b	nd	nd	3	bl	bl	RPMI1640+10%FCS + insulin+hydroco	X	I,H	O	Int J Cancer 41:74-82, 1988	1986	P Russell
UCRU-BL-28	62	m	4a	nd	nd	2	bl	blr	RPMI1640+10%FCS + insulin+hydroco	D,X	K	O	J Urol 150:1038-1044, 1993	1988	P Russell
UM-UC-1	26	m(B)	>2	0	0	2	bl	min	EMEM+NEAA +15%FBS	D	I	O	J Urol 132:834-837, 1984	1982	H Grossman
UM-UC-2	61	m	>1	0	0	3	blr	ur	EMEM+NEAA +15%FBS	D	I	O	J Urol 132:834-837, 1984	1981	H Grossman
UM-UC-3	nd	m	nd	nd	nd	nd	bl	bl	EMEM+SP+10%FBS	D	K	I	J Urol 136:953-959, 1986	1982	H Grossman
UM-UC-4	nd	f	nd	1	nd	2	bl	ura	EMEM+SP+10%FBS	D	K	O	J Urol 136:953-959, 1986	1982	H Grossman
UM-UC-5	nd	f	nd	nd	nd	nd	bl	bl	MEM+20%FCS	D	K	O	J Urol 136:953-959, 1986	1983	H Grossman
UM-UC-6	nd	m	nd	nd	nd	nd	bl	bl	EMEM+SP+10%FBS	D	K	O	J Urol 136:953-959, 1986	1983	H Grossman
UM-UC-7	nd	m	nd						EMEM+SP+10%FBS	D	nd	O	J Urol 136:953-959, 1986	1983	H Grossman

Continued on next page

Table 1 (continued)

Name	Patient age	Patient sex	T	N	M	Grade	Primary site	Specimen site	Culture medium	Culture method	Authentication	Availability	Primary reference	Establishment	First author
VM-CUB-I	nd	m	nd	nd	nd	nd	bl	bl	MEM+15%FBS	E	I		Invest Urol 17:359-363, 1980	1975	RD Williams
VM-CUB-III	nd	m	nd	nd	nd	nd	nbl	bl	MEWM+15%FBS	e	i		Invest Urol 17:359-363, 1980	1975	RD Williams
no name	76	f	1	0	0	3	bl	blt	Ham's F12+10%FCS	E	K	O	Cancer Genet Cytogenet 67:101-107, 1993	1989	B Perissel

Superscript numbers: (1) = a year later T4NXM1G2; (2) = special suspension culture, (3) = Masters *et al.*, 1988, (4) = and further characterization in Int J Cancer 10:77–91, 1972

General abbreviations: nd = not done or not documented;

Special abbreviations in:

headlines: T, N, M = Tumor, Node, Metastasis categories according to the TNM-classification of malignant tumors, International Union Against Cancer (UICC), P. Hermanek *et al.* (Eds.), Springer, Berlin, 1992; grade: categories of cellular atypia, mostly related to: International Classification of tumors No. 10. Histological typing of urinary bladder tumors. Mostofi *et al.* (Eds.) WHO, Geneva, 1973. Some authors use grade 4, trying to indicate complete loss of differentiation, although it is mostly agreed upon now that grade 4 should be included into the category of grade 3. A significant difference between the cell lines is not expected.

patient sex: m= male, f= female, (B)= black race,(C)= chinese race, all others are Caucasians

T: mf = multifocal

M: mbr = metastasis brain, mbo = metastasis bone marrow.

primary site and specimen site: bl = bladder, bld = bladder dome, bllw = bladder lateral wall, blpw = bladder posterior wall, blr = bladder recurrence , blt = bladder trigonum, blw = bladder washing, mln = metastasis lymph node, mbr = metastasis brain, ur = ureter, ura = urachus , rp = renal pelvis.

culture method: E = explant culture, D = culture from dissociated tumor, X = xenograft culture.

availability: all cell banks were checked in 1997; 1: ATCC= American Type Tissue Collection; 2: ECACC= European Collection of Animal Cell Culture; 3: DKFZ= cell bank of the German Society for Cancer Research; 4: DSMZ=Deutsche Sammlung von Mikroorganismen und Zellkulturen GmbH; 5: Russian Academy of Medical Sciences, Moscow, cell bank; 6: RIKEN= Japanese Collection of Research Bioresources cellbank; 7: IZBSS = Instituto Zooprofilafitco Sperimentale, Brescia. O: originator; listed when he/she has been cited by others to be the source for the cell line.

authentication: D = DNA-fingerprinting, H = HLA-typing, I = isoenzyme pattern, K = karyotyping, M = morphology

Table 2 Comparison of cell lines with the organ and tissue of origin, and related xenograft morphology.

Name	Description of tumor pathology	*In vitro* features of cell line in comparison to original tumor	Xenograft pathology
5637	well differentiated TCC	epithelial-like morphology, small malignant cells	NM: moderately well differentiated epidermoid carcinoma G2
1016T	TCC, G 1	epithelial	HCP: like primary tumor
192B	TCC, G 4	epithelial, lack of contact inhibition	HCP: like primary tumor
253J	TCC, multiple tumor, G4	TCC, probably differentiated	G1-2, TCC with glandular metaplasia
292W	TCC	epithelial	HCP: like primary tumor
486P	TCC G 4	epithelial, uniform in size+clusters	HCP: like primary tumor
575A	TCC G 3	epithelial	HCP: like primary tumor
639V	TCC G 3	variety of shape and size, clumping, multilayering	HCP: like primary tumor
647V	TCC G 2, invasive	small and epithelial, multilayering=lack of contact inhibition	HCP: like primary tumor
682B	TCC G 1	epithelial	HCP: like primary tumor
743E	TCC-metastasis, G 4	epithelial	HCP: like metastasis
751G	TCC G 1	epithelial	nd
BT-1	TCC G 3 with squamous carcinoma parts	epithelioid, no contact inhibition	NM: G3 TCC with squamous carcinoma component similar to primary tumor
BT-B	TCC	epithelial	nd
COLO232	TCC G 3	multilayered, compact, epithelial-like	irradiated C3H/HeJ mice: differentiated TCC
HOK-1	G 3 tumor with squamous and glandular differentiation	clusters of closely opposed polygonal cells, no contact inhibition, some cells with multiple nuclei, few spindle cells, partially mucin producing cells	NM: tumors G3 with glandular and squamous differentiation, better differentiated than primary tumor
HT1376	TCC, mostly differentiated	primary: invasive and moderately pleomorphic, grow in soft agar, induce fibrinolytic activity	NM and HCPG3, squamous cell or TCC with extensive squamous metaplasia and high mitotic rate
IG	TCC, papillary, G2	mixed: irregular round and fibroblastic tumor cells, tumor nodules	nd
J82 COT	poorly differentiated, invasive carcinoma, no desmosomes, prominent micro-filaments, varying amounts of RER	mixed epithelial fibroblastic morphology	NM: no growth
JTC-29	papillary TCC, well differentiated	small cells, high mitotic index, partial glandular formation	NM: not like primary: nonpapillary anaplastic
JTC-30	papillary TCC, well differentiated	pavement like, multilayering, multinucleated cells	NM: like primary: papillary differentiated
JTC-31	nonpapillary TCC, anaplastic	intense multilayering, better growth as suspension	NM: like primary: papillary anaplastic
JTC-32	nonpapillary TCC, anaplastic	pavement like, multilayering, multinucleated cells	NM: better differentiated: moderate nonpapillary cystic

Continued on next page

Table 2 (continued)

Name	Description of tumor pathology	*In vitro* features of cell line in comparison to original tumor	Xenograft pathology
JTC-33	nonpapillary TCC, anaplastic	intense multilayering, better growth as suspension	NM: like primary: nonpapillary anaplastic
JTC-34	papillary TCC, well differentiated	small cells, high mitotic index, partial glandular formation	NM: not like primary: papillary anaplastic
KK47	TCC bladder stage A and G 1	pavement pattern and partially overlapping cells, urothelial pattern	NM: pattern like TCC(1)
KW-103	papillary TCC bladder, G 2	cuboidal sheets, urothelial	NM: similar to primary tumor
LA-B1	nonpapillary invasive G3 TCC	malignant cells, consistent with TCC	NM: >2
MGH-U3	recurrent papillary TCC, G 1	uniform polygonal epithelioid, swirling pattern at confluency	NM: similar to primary
MGH-U4	recurrent multifocal severe dysplasia	large round epithelioid cells	NM: no growth
NBT-2	papillary muscle invasive low differentiated urothelial carcinoma	epithelioid, ultrastructurally resembling high grade urothelial carcinoma	HCP: undifferentiated carcinoma like primary tumor
NTBU1	papillary muscle invasive low differentiated carcinoma	epithelioid multilayered cells	NM: 9 out of 10 tumors growing like primary tumor
PS1	large sessile low differentiated muscle and prostate invasive TCC	polygonal sheets, pleomorphic	NM: tumorigenic, with better differentiated than primary tumor, but frankly invasive growth
RT112	nd	differentiated TCC	NM: G1-2 with focal squamous and glandular metaplasia
RT4	recurring papillary tumor	well differentiated transitional cells, possible papilloma	NM and HCP: well differentiated papillary epidermoid carcinoma consistent with urinary bladder primary G1-2
SBC-2	papillary TCC well to moderately differentiated	epithelioid, often triangle shaped	nd
SCaBER	squamous carcinoma	carcinoma consistent with epidermoid carcinoma	NM: epidermoid carcinoma(1)
SW710	papillary TCC, G 1	large polygonal cells with N:C ratio of 1:4-5	NM: no growth
SW1738	TCC, G1-2 with focal areas of G4	moderately elongated cells with N:C ratio of 1:3	NM: G3 polymorphic tumor
SW780	nd	round or ovoid cells with N:C ratio of 2:3	NM: cells arranged in cords with delicate stroma
SW800	TCC, G1	round elongated polygonal cells with N:C ratio of 1:3	NM: cells multilayered around blood vessels and solid cell nests, high mitotic index
T24	nd	epidermoid TCC	NM: nd, however growth in HCP
TCCSUP	TCC, anaplastic, G4	low G TCC, ill defined cells, no contact inhibition	NM: no growth
TSGH-8301	TCC, G2	epithelial monolayer, loss of contact inhibition, occasional giant cells, cellular vacuoles	NM: growth with morphological similarities to *in vitro* culture, desmosomes
UCRU-BL-13	TCC, G2 with focal squamous differentiation	tight clusters of polygonal cells, rare glandular or squamous cells	NM: TCC G2 with focal squamous and glandular differentiation

Continued on next page

Table 2 (continued)

Name	Description of tumor pathology	*In vitro* features of cell line in comparison to original tumor	Xenograft pathology
UCRU-BL-17CL	TCC with glandular differentiation and cuboidal	produce mucin, wide range of morphology, spindle shape G 3 pleomorphic	NM: with glandular and squamous differentiation,
UCRU-BL-28	2nd recurrence as invasive G2 carcinoma with focal glandular differentiation	polygonal sheets, mucin negative	NM: reflects the tumor, is mucin positive
UM-UC-1	lymph node metastases of TCC, G2	attached growth and loss of contact inhibition	NM: similar to metastasis, i.e. G2 TCC
UM-UC-2	G 3 invasive carcinoma in ureter in patient with pTis of the bladder, as a recurrence after papillary tumor	attached growth and loss of contact inhibition	NM: similar to the ureteral tumor, i.e. G3
UM-UC-3	TCC, G3	spindle shaped, undifferentiated	NM: +
UM-UC-4	lymph node metastasis of urachus adenocarcinoma	epithelioid, cuboidal G2	NM: similar to primary, i.e. adenocarcinoma G2
UM-UC-5	squamous carcinoma	epithelioid, G2	NM: morphology like primary, i.e. squamous carcinoma G2 in mice
UM-UC-6	nd, TUR specimen	spindle shaped G3	NM: like primary, i.e. G3 TCC
UM-UC-7	nd, cystectomy specimen	cuboidal, polymorphic, G2	NM: like primary, i.e. G2 TCC
CM-CUB-I	nd	nd	nd
CM-CUB-II	nd	TCC, probably differentiated	NM: G3 squamoid carcinoma
no name	Squamous carcinoma, G3 with keratinization	epithelial-fibroblastic	nd

The text presented in this table is mostly directly transferred from the original literature as cited in table 1, a different origin of data is indicated in the text.

superscript numbers: (1) = Rieger *et al.* 1995

abbreviations: G: nuclear grading ; HCP: hamster cheek pouch of immunosuppressed animals; N:C ratio: nuclear-cytoplasmic ratio, nd = not done or not documented; NM: nude mice; TCC: transitional cell carcinoma; TUR: Transurethral resection

Table 3 Ploidy, chromosome numbers, and important genetic findings

Name	Ploidy	MCNo	Marker	Ch9	p16 (Ch9p)	p15 (Ch9p)	RB (Ch13)	p53 (Ch17)	p21 (Ch6)	ras	p27	Others
5637	hypotriploid	67	14		wt[17]	p15wt (5)	mut[12]	mut[12]				p19wt[16], p18wt[5], structural abnormalities assoc. with leukemia[9]
1016T	aneuploid	60–63	–									
192B	aneuploid	61–63	+									
253J	aneuploid	60–63			HD[13]	HD[13]					wt[13]	
292W	aneuploid	52–54	–									
486P	aneuploid	72–80	–									
575A	aneuploid	106–110	Y									
639V	aneuploid	56–60	+									
647B	aneuploid	66–70	+									
682B	aneuploid	46–48	Y									
743E	aneuploid	63–67	Y									
751G	aneuploid	61–65	Y									
BT-1	aneuploid	76	22									
BT-B	hypertriploid	72	+									
COLO 232	aneuploid	58	10		HD[13]	HD[13]					wt[13]	
HOK-1	near diploid aneuploid	45–48	+		9p–							
HT1376	aneuploid	111	6		w[14]		mut[12]	mut[11,14]	mut[4]			
IG	nd	nd	nd									
J82 COT	triploid	72	5		wt[14]		mut[4]	mut[12]		mut[2]		ECad-mRNA-[12]
JTC-29	near diploid	43	Y									
JTC-30	aneuploid	47	Y									
JTC-31	aneuploid	45	Y									

Continued on next page

Table 3 (continued)

Name	Ploidy	MCNo	Marker	Ch9	p16 (Ch9p)	p15 (Ch9p)	RB (Ch13)	p53 (Ch17)	p21 (Ch6)	ras	p27	Others
JTC-32	aneuploid	95–102	nd									
JTC-33	aneuploid	47	Y									
JTC-34	aneuploid	47	Y?									
KK47	aneuploid	59	2				wt[12]	wt[12]			wt[13]	ECad-mRNA-[12]
KW-103	hyperdiploid	57	+									
LA-B1	aneuploid	nd	nd									
MGH-U3	diploid	49	nd									
MGH-U4	diploid	45	nd	PO			MO					
NBT-2	aneuploid	72, 74	nd									
NTBU1	aneuploid	76	nd									
PS1	triploid	72 (79)	10				wt[12]	PO[12]				
RT112	near diploid	47[13]	nd		mut[15], HD[13]	HD[13]	wt[12]	wt[12]			wt[13]	
RT4	near diploid tetraploid	49/80	4		mut[15] del[14]HD[13]	HD[13]	wt[12]	wt[11,14,5]	wt[4] PO		wt[13]	
SBC-2	near triploid	67	+									
SCaBER	hyperdiploid-hypertetraploid	65/108	11		mut[14]		mut[12]	mut[11,5] wt[14]	wt[4]		wt[13]	
SW1710	near triploid aneuploid	72-83	>5				wt[4]					
SW1738	aneuploid	54–68	>2				wt[4]					
SW780	subtetraploid	85–91	>2				wt[4]					
SW800	near diploid	46–49	5				wt[4]					
T24	hypodiploidy-hypopentaploidy	86	6		9p-, wt[14]		wt[8]	mut[7,5] wt[14]	wt[4]	mut[12]	wt[13]	ECad-mRNA-[12]

Continued on next page

Table 3 (continued)

Name	Ploidy	MCNo	Marker	Ch9	p16 (Ch9p)	p15 (Ch9p)	RB (Ch13)	p53 (Ch17)	p21 (Ch6)	ras	p27	Others
TCCSUP	hypertriploid	72	8		wt[14]		mut[12]	wt[12] mut[7,14]		mut[12]	wt[13]	
TSGH-8301	heterodiploid	50	nd									
UCRU-BL-13	aneuploid (multiple stemlines)	85	+	9–				wt[10]				
UCRU-BL-17C	triploid	70	+					mut[10,6]	PO[6]			
UCRU-BL-28	triploid	73	+					wt[10],	PO[10]			
UM-CU-2	aneuploid	48	3	9-	9p-							
UM-UC-1	aneuploid	68	4	9 UR	9p-							
UM-UC-3	hypertriploid	80	30	no loss	del[14]		wt[12]	mut[11,14] PO[12]				ECad-mRNA-[12]
UM-UC-4	aneuploid	51	8	9-								
UM-UC-5	aneuploid	69	11	9-								
UM-UC-6	aneuploid	51	3	no loss								
UM-UC-7	aneuploid	70	nd									
VM-CUB-I	near tetraploid[13]	75[13]	nd		HD[13]	HD[13]					wt[13]	
VM-CUB-II	diploid[13]	46[13]	nd		HD[13]	HD[13]	wt[12]	mut[12]			wt[13]	
no name	aneuploid	48	2	TRI								TRI 7 and 8

abbreviations: Ch: chromosome, Ecad: E-Cadherin, HD: homozygous deletion, MCNo: mean chromosome numbers, MO: monosomy, mut: mutated, PM: partial monosomy, PO: polymorphism, UR: under-represented, wt: wild type

Superscript numbers are used to indicate source of description of the genetic change or ploidy etc. other than stated by the original supplier or by one of the cell line banks. superscript numbers: 1= Chresta CM *et al.* 1996, 2 = Der CJ *et al.* 1982, 3 = Horowitz JM *et al.* 1989, 4 = Horowitz JM *et al.* 1990, 5 = Kawasaki T *et al.* 1996, 6 = Lukas J *et al.* 1995, 7 = Malkovicz SB *et al.* 1996, 8 = Parada LF *et al.* 1982, 9 = Pfluger KH *et al.* 1986, 10 = Ribeiro JC *et al.* 1996, 11 = Rieger AF *et al.* 1995, 12 = Shih C *et al.* 1982, 13 = Southgate J *et al.* 1995, 14 = Spruck CH *et al.* 1996, 15 = Wu Q. *et al.* 1996, 16 = Zariwala M and Xiong Y., 1996.

Some lines have been adapted to grow in serum-free media, including 647V (Messing *et al.*, 1982), J82 (Bishai *et al.*, 1985) and T24 (Zirvi *et al.*, 1986). One of the factors influencing serum-free culture is cell density, as demonstrated by Ruck *et al.* (1994), who were able to grow 5 out of 9 cell lines (5637, Hu549, SD, TCCSuP, T24) in serum-free medium. Another cell line, LA-B1, was initiated in serum-free medium (Messing *et al.*, 1982). Serum-free conditions have been used to study ectopic hormone production, for example β human chorionic gonadotropin (β-HCG) in various bladder cancer cell lines (Iles and Chard, 1989) and colony stimulating factor production by the cell line 5637 (Pfluger *et al.*, 1986).

2. DO THE CELL LINES AVAILABLE REPRESENT THE CLINICAL DISEASE?

Human bladder cancer is the sixth most frequent type of cancer worldwide (Kantor *et al.*, 1988) and the second most frequent tumor of the genitourinary tract after prostate (Thrasher and Crawford, 1993). Bladder cancer is rare in the first five decades of life, thereafter showing a sharp rise in incidence with increasing age. More men than women are affected (ratio approximately 4:1).

At primary diagnosis 70–80% of lesions are superficial pTa or pT1 tumors with a papillary architecture (Thrasher and Crawford, 1993). These cancers have a 70–80% chance of recurring, but rarely invade. The majority are composed of transitional cells, and are known as transitional cell cancers (TCC). Less than 5% of bladder cancers are squamous, glandular or mixed (Kantor *et al.*, 1988). Carcinoma-in-situ is a rare (less than 5% of bladder cancers) but more aggressive superficial cancer, with an approximately 50% incidence of progression. Approximately 25% of bladder cancers present with advanced disease that has invaded the muscle (pT2 or greater), and for these patients the prognosis is relatively poor.

The characteristics of the bladder cancer cell lines that have been described are shown in Table 1, indicating that cell lines have been established from all stages and grades of disease. Bladder cancer cell lines have been established from primary tumors, as well as recurrent tumors, invasive and non-invasive tumors and metastatic sites (see Table 1). Although urothelium lines the bladder, renal pelvis, ureters, and most of the urethra, only three cell lines are not derived from bladder or bladder cancer metastases: 292W, renal pelvis; UM-UC2, ureter; UM-UC-4, urachus (see Table 1). Three cell lines were established from advanced tumors in relatively young patients (UM-UC-1, HT1197, 253J), although the majority of cell lines were derived from patients within the usual age range. Cell lines from well-differentiated tumors have been established in culture (e.g. RT4 and RT112), and retain features of highly differentiated tumors (see Table 2). However, most of the cell lines are derived from high grade tumors. This is in contrast to the clinical pattern of presentation.

A variety of methods have been used for authentication (proof that the cell line was derived from the proposed tumor of origin) (Table 1). The most reliable methods are DNA fingerprinting and HLA typing. Some cell lines are only identified by limited karyotyping or morphological features, which are inadequate for authentication. Some cell lines, including SD1 (Grups and Frohmueller, 1988), KU-1 and NBT-2 (Ohigashi *et al.*, 1992), NTUB1, BFTCC905 and BFTCC909 (Pu *et al.*, 1996) have not been described in sufficient detail for inclusion in Table 1.

3. PATHOLOGY

The epithelium of the upper urinary tract, bladder and proximal urethra is composed of 3–4 layers of transitional cells called urothelium. The surface layer is composed of large polyploid umbrella cells which have a specialized hinged outer membrane allowing for expansion as the bladder fills. The lumenal membrane is thickened and is consequently known as the asymmetric unit membrane as it contains a layer of protective proteins (uroplakins) preventing the entry of urine. Due to the embryonic derivation of these cells, the potential for squamous and glandular differentiation is maintained, and reflected in mixed differentiation in some cancers as well as benign squamous and glandular differentiation in inflammatory lesions.

The identification of epithelial cells is facilitated by applying lineage-specific antibodies against, for instance, cytokeratins, which are maintained even in spindle-shaped fibroblastoid cancer cell lines (for overview see Fusenig, 1991).

Cytokeratins also serve as markers to indicate the state of differentiation of the urothelium (Moll, 1982), and with increasing tumor grade cytokeratin expression is known to shift to simple epithelial isotypes, including cytokeratin 8, 18, and 19, with the loss of cytokeratin 13 (Moll, 1988). The more recently discovered cytokeratin 20 is limited to superficial cells in normal urothelium, and changes of expression have been linked to dysplasia (Harnden and Parkinson 1996). While cytokeratins are stable cytoskeletal markers, a large number of membrane molecules have been isolated to help characterize the urothelial phenotype (for reviews see Fradet and Cordon-Cardo, 1993, Huland *et al.*, 1991). High specificity is seen with a group of non-blood group related glycoproteins isolated by the group of Fradet, some of which are selectively expressed in superficial tumors (M344, 19A12), and others in invasive tumors (T138 and to a lesser extent T43) (Bergeron *et al.*, 1997, Ravery *et al.*, 1995). The uroplakins are major components of the asymmetric unit membrane, the specialized lumenal surface of the terminally differentiated superficial urothelial cells (Sun *et al.*, 1996, Wu and Sun, 1993).

Squamous carcinoma of the bladder is rare in Europe. It is associated with infection with *Schistosoma haematobium* (bilharziasis or schistosomiasis), and accounts for approximately 80% of bladder cancers in Egypt, where this parasite is common (Ghonheim *et al.*, 1985). The molecular genetics of squamous cancer differ from those of papillary TCC (Tamimi *et al.*, 1996).

Most of the cell lines were established from transitional cell carcinomas. However, some lines are derived from squamous cancers (SCaBER, UM-UC-5) or mixed tumors (HOK-1, UCRU-BL 17 and 28). A rare cancer derived from remnants of the urachus infiltrating the dome of the bladder is represented by the adenocarcinoma cell line UM-UC-4, obtained from a lymph node metastasis.

Most of the cell lines which have been investigated appear to be representative of the primary tumor, both in monolayer culture and as xenografts in nude mice (Table 2). Cell lines derived from well-differentiated tumors, such as RT4, retain the ability to form a differentiated architecture in appropriate 3-dimensional culture systems (Leighton and Matulaitis, 1990). Differentiation is also maintained in three-dimensional heterologous cultures of this cell line with fibroblasts (Schuster *et al.*, 1994). Similar observations have been made with RT4 and RT112 grown on bladder stroma (Booth *et al.*, 1997).

Colony formation in agar does not correlate well with tumorigenicity in mice, although both assays are used to indicate tumorigenic potential (Marshall *et al.*, 1977). The interaction of the tumor cells with the host environment is disrupted by culturing tumor cells, and thus selective paracrine influences of fibroblasts and inflammatory cells are only partially replaced by medium and the cutaneous stroma of immune-deprived mice. It seems that host factors in immune-deprived mice have the ability to induce squamous metaplasia, since squamous changes of TCC cell lines are seen more often in xenografts than in culture or in the primary tumors (see Table 2).

4. MOLECULAR GENETICS

The molecular genetics of bladder cancer have been reviewed (Knowles, 1995; Reznikoff *et al.*, 1996). There is debate whether multifocal tumors are monoclonal in origin or are polyclonal as a result of field cancerization. Monoclonality would indicate that multifocal tumors arise from seeding or migration of tumor cells within the bladder. A list of the genetic alterations observed in bladder cancer cell lines is presented in Table 3.

Changes on chromosome 9 (both arms) are found in the majority of bladder cancers, even in the earliest stages. Some genes, such as the retinoblastoma gene, are modified late in the disease process. The continuous cell lines provide a spectrum of genetic alterations that, in general, are representative of those seen in bladder cancer (Rieger *et al.*, 1995). An attempt is made to pool much of the published data in Table 3, but it is difficult to track down all the published data on the molecular genetics of individual cell lines and there are disparities in the literature. Everyone working in this field is encouraged to check that the particular genetic alteration they wish to study is present and not to make any assumptions.

The tumor suppressor gene p53 is important as it is mutated in the majority of advanced bladder cancers. Stabilized mutant and high levels of wild-type p53 can be detected using immunocytochemistry, but some mutations result in negative immunocytochemistry. Although there are hot spots for mutation, many sites within the gene have been identified. These variants may be responsible for discrepant data on p53 status.

Cell selection occurs during culture. Selective pressure is a possible explanation for the finding that a higher frequency of mutation of the p16 gene locus is seen in bladder cancer cell lines compared to (unmatched) primary tumors (Spruck *et al.*, 1994). However, there are other possible reasons for this finding, such as contamination of the primary tumor with normal tissue, which could result in false negative results in mutation screening (Cairns *et al.*, 1995).

Expression vectors containing the p16 gene have been transfected into RT4 and RT112, under the control of a strong CMV-promoter (Wu *et al.*, 1996). Gene and protein transfer of p16/CDKN2 and cyclin D1 into various tumor cell lines, including the retinoblastoma gene deficient bladder cell line J82, have also been performed (Lukas *et al.*, 1995).

Transfection of normal urothelial cells also contributes to the understanding of early events in bladder cancer carcinogenesis (Reznikoff *et al.*, 1994). These cells can be investigated further by viral transformation of cell lines, and subsequent passages of the transformed cell line through nude mice to obtain subclones with differing degrees of malignancy (Kao *et al.*, 1993).

5. PROBLEMS OF CELL LINE CROSS-CONTAMINATION

The first human cancer cell line to be established, HeLa, cross-contaminated many other cell lines as a result of poor tissue culture technique. Some stocks of bladder cancer cell lines were contaminated with HeLa (Franks and Rigby, 1975). Bladder cancer has an equivalent of HeLa, the cell line T24, which has independently cross-contaminated many stocks of supposedly unique bladder cancer cell lines (O'Toole *et al.*, 1983). The problem became apparent when a number of independent bladder cancer cell lines (including T24, EJ and some stocks of J82) were discovered to carry transforming Ha-*ras* sequences. Isozyme and HLA typing indicated that not only were these stocks of bladder cancer cell lines cross-contaminated (O'Toole *et al.*, 1983), but that other supposedly independently-derived cell lines were also identical to T24 (MGH-U2, HU456, HU961T). These results were subsequently confirmed using locus-specific probes (Masters *et al.*, 1988) and DNA fingerprinting (Christensen *et al.*, 1993). Further confusion has been caused by changing the name of EJ to MGH-U1.

Cell banks that do not verify the identity of cells that they collect have contributed to the problem of cross-contamination. Despite publications in Nature (O'Toole *et al.*, 1983) and other leading cancer journals highlighting cross-contamination of bladder cancer cell lines by T24, the European bank ECACC continued to sell T24 and EJ as independent bladder cancer cell lines.

6. CELL LINES WITH SPECIAL FEATURES

Drug-resistant sublines have been established (see Table 4), some of which retain their resistance in the absence of the drug (Walker *et al.* 1990). Three-dimensional culture is a more representative model than monolayer culture for studying mechanisms which influence drug resistance, such as drug diffusion (Knuechel *et al.* 1989).

Cells have been passaged in immune-deprived mice to determine the biological potential of resultant subclones (Russell *et al.* 1989, Dinney *et al.* 1995, Kovnat *et al.* 1988). For example, 9 clones of the cell line UCRU-BL-17CL were established and six of these grew as xenografts (Brown *et al.*, 1990). Passage through mice can result in faster proliferating, more malignant, and more highly metastatic subclones (Dinney *et al.* 1995).

Another important aspect of cell lines as model systems is the expression of growth factors and their receptors, including epidermal growth factor receptor (EGFR) and the family of fibroblast and insulin growth factor receptors. A broad range of EGFR expression is found amongst bladder cancer cell lines (Brockhoff *et al.*, 96; Messing and Reznikoff, 1987; Messing, 1990). The level of receptors and their functional regulation can be altered by the growth state of the cells, and again three-dimensional culture may be more representative of the tumor in vivo (Mansbridge *et al.*, 1992).

ACKNOWLEDGEMENT

Prof. F. Hofstaedter is gratefully acknowledged for carefully reviewing the manuscript.

Table 4

Resistant cell line	Parent cell line	Resistant to	Cross resistant to	Reference
T24/ADM-1	T24	DXR	ETO	Hasegawa *et al.* 1995
T24/ADM-2	T24	DXR	ETO	Hasegawa *et al.* 1995
KK47/ADM	KK47	DXR	ETO, VCR, CPT11	Hasegawa *et al.* 1995
KK47/ADM	KK47	DXR	EPR, PIR, VBL, VCR, ETO	Kimiya *et al.* 1992
T24/VCR	T24	VCR	DXR	Hasegawa *et al.* 1995
T24/A	T24	DXR	nd	Pu *et al.* 1996
NTUB1/A	NTUB1	DXR	nd	Pu *et al.* 1996
T24/P	T24	CPL	nd	Pu *et al.* 1996
NTUB1/P	NTUB1	CPL	nd	Pu *et al.* 1996
SCaBER/R	SCaBER	BMY	MMC, BM2, BCNU	Singh *et al.* 1995
J82/MMC	J82	MMC	CPL, MEL, MMC analogues	Xu *et al.* 1994
J82NVB	J82	NVB	other VA and TAX	Debal *et al.* 1994
MGH-U1R	MGH-U1	DXR	DAR	McGovern *et al.* 1988
MGH-UIR	MGH-U1	DXR	VBL, ETO	Floyd *et al.* 1990
5637DR5.5	5637	DXR	VBL, ETO, PL	Kim *et al.* 1995
RT112MMC	RT112	MMC	nd	Eickelmann *et al.* 1994
RT112/D21	RT112	DXR	EPR, VBL	Seemann *et al.* 1995
RT112-R	RT112	CPL	CAP, MTX	Walker *et al.* 1990
UM-UC-6-dox	UM-UC-6	DXR	ETO	Shinohara *et al.* 1993

BCNU= bis-chloroethyl-nitrosourea, BM2 = BMY25282, BMY = BMY25067, CAP = carboplatin, CPL = cisplatin, CPT11 = camptothecin derivative, DAR = Daunorubicin, DXR = doxorubicin, EPR = epirubicin, ETO = etoposide, MEL = melphalan, MTX = methotrexate, MMC = mitomycin C, PIR = pirarubicin, TAX = taxoids, VA = vinca alkaloids, VBL= vinblastine, VCR = vincristine, nd = not done

REFERENCES

Bergeron A *et al.*, *Biochem J* 321:889, 1997.
Bishai MB *et al.*, *J Urol* 134:1287, 1985.
Boehle A *et al.*, *J Urol* 150:1932, 1993.
Booth C *et al.*, *Lab Invest* 6:843, 1997.
Bosch van der J *et al.*, *Exp Cell Res* 187:185, 1990.
Brockhoff G *et al.*, *Analyt Cell Pathol* 11:55, 1996.
Brown JL *et al.*, *Br J Cancer* 61:369, 1990.
Bubenik J *et al.*, *Int J Cancer* 5:310, 1970.
Bubenik J., *Int J Cancer* 8:503, 1971.
Bubenik J *et al.*, *Int J Cancer* 11:765, 1973.
Cairns P *et al.*, *Nature Genetics* 11:210, 1995.
Chresta CM *et al.*, *Cancer Res* 56:1834, 1996.
Christensen B *et al.*, *Br J Cancer* 68:879, 1993.
Debal V *et al.*, *Bull Cancer Paris* 81:891, 1994.
Der CJ *et al.*, *Proc Natl Acad Sci USA* 79:3637, 1982.
Dinney CP *et al.*, *J Urol*, 154:1532, 1995.
Eickelmann P *et al.*, *Biol Chem Hoppe Seyler* 375:439, 1994.

Elliott AY, J *Natl Cancer Inst* 53:134, 1974.
Elliott AY *et al.*, *Cancer Res* 37:1289, 1979.
Elliott AY *et al.*, *Natl Cancer Inst Mongr* 9:23, 1978.
Evans DR *et al.*, *J Urol* 117:712, 1977.
Floyd JW *et al.*, *J Urol* 144:169, 1990.
Fogh J, *Natl Cancer Inst Monog* 49:5, 1978.
Fradet Y and Cordon-Cardo C, *Sem Urol* 11:145, 1993.
Franks LM and Rigby C, *Science* 188:168, 1975.
Fusenig NE *et al.*, In Human cancer in primary culture. Ed JRW Masters, pp 55–80. *Kluwer Academic Publishers, London,* 1991.
Ghonheim MA *et al.*, *J Urol* 134:266, 1985.
Goldfarb F *et al.*, *Nature* 296:404, 1982.
Grossman HB *et al.*, *J Urol* 132:834, 1984.
Grossman HB *et al.*, *J Urol* 136:953,1986.
Grups JW and Frohmueller HGW, *Urol Int* 43:265, 1988.
Harnden P and Parkinson C, *Curr Diagn Pathol* 3:109, 1996.
Hasegawa S *et al.*, *Br J Cancer* 71:907, 1995.
Heckl W *et al.*, *Urol Res* 16:23, 1988.
Hisazumi H *et al.*, *Jpn J Urol* 70:485,1979.
Horowitz JM *et al.*, *Science* 243:937, 1989.
Horowitz JM *et al.*, *Proc Natl Acad Sci* 87:2775, 1990.
Huland E *et al.*, *J Urol* 146:1631, 1991.
Iles RK and Chard TJ, *Endocrinol* 123:501, 1989.
Kakuya T *et al.*, *In vitro* 19:591, 1983.
Kao C *et al.*, *Carcinogenesis* 11:2297, 1993.
Kantor AF et al., *Cancer Res* 48:3853, 1988.
Kawasaki T *et al.*, *Int J Cancer* 68:501, 1996.
Kim WJ *et al.*, *Jpn J Cancer Res* 86:969, 1995.
Kimiya K *et al.*, *J Urol* 148:441, 1992.
Knowles MA, *Br J Urol* 75:Suppl 1,57, 1995.
Knuechel R *et al.*, *Cancer Res* 49:1397, 1989.
Kovnat A *et al.*, *Cancer Res* 48:4993, 1988.
Kyriazis KA *et al.*, *Cancer Res* 44:3997, 1984.
Leighton J and Matulaitis R. *Clin Biotech* 2:17, 1990.
Lin C-W *et al.*, *Cancer Res* 45:5070, 1985.
Logothetou Rella H *et al.*, *Eur Urol* 14:65, 1988.
Lukas J. *et al. Cancer Res* 55:4818, 1995.
Malkovicz SB *et al.*, *Oncogene* 13:1831, 1996.
Mansbridge J *et al. J Cell Physiol* 151:433, 1992.
Marshall CJ *et al.*, *J Natl Cancer Inst* 58:1743, 1977.
Masters JRW *et al.*, *Cancer Res* 46:3630, 1986.
Masters JRW *et al.*, *Br, J Cancer* 57:284, 1988.
Masters JRW (Ed). Human cancer in primary culture. *Kluwer Academic Publishers, London,* 1991.
McGovern F *et al.*, *J Urol* 140:410, 1988.
Messing EM *et al.*, *Cancer Res* 42:2392, 1982.
Messing EM and Reznikoff CA *Cancer Res* 47:2230, 1987.
Messing EM, *Cancer Res* 50:2530, 1990.
Moll R *et al.*, *Cell* 31:11, 1982.
Moll R *et al.*, *Am J Pathol* 132:123, 1988.
Moore GE *et al.*, *In vitro* 40:301, 1978.

Naito K *et al., Jpn J Urol* 73:1019, 1982.
Nayak SK *et al., Br J Cancer* 35:142, 1977.
Ohigashi T *et al., J Urol* 147:283, 1992.
O'Toole C *et al., Int J Cancer* 10:77, 1972.
O'Toole CM, *Int J Cancer* 17:707, 1976.
O'Toole C *et al., Br J Cancer* 38:64, 1978.
O'Toole C in Webber MM, Vol.IV, *CRC press*, 1986, pp.103–125.
O'Toole CM *et al., Nature* 301:429, 1983.
Offner FA *et al., Int J Cancer* 49:122, 1991.
Parada LF *et al., Nature* 297:474, 1982.
Perissel B *et al., Cancer Genet Cytogenet* 67:101, 1993.
Pfluger KH *et al., Blut* 53:89, 1986.
Pu Y-S *et al., J Urol* 156:271, 1996.
Rasheed S *et al., J Natl Cancer Inst* 58:881, 1977.
Ravery V *et al., Br J Cancer* 71:196, 1995.
Reddy EP *et al., Nature* 300:149, 1982.
Reznikoff CA *et al., Genes Develop* 8:2227, 1994.
Reznikoff CA *et al., Sem Oncol* 23:571, 1996.
Ribeiro JC *et al., Oncogene* 13:1269,1996.
Rieger AF *et al., Br J Cancer* 72:683, 1995.
Rigby CC and Franks LM, *Br J Cancer* 24:746, 1970.
Ruck A *et al., Anticancer Res* 14:55, 1994.
Russell PJ *et al., Int J Cancer* 41:74, 1988.
Russell PJ *et al., Int J Cancer* 44:276, 1989.
Russell PJ *et al., J Urol* 150:1038,1993.
Sanford EJ *et al., Invest Urol* 16:246,1978.
Schuster U *et al., J Urol* 151:1707, 1994.
Seemann O *et al., J Urol* 22:353, 1995.
Shih C *et al., Cell* 29:161, 1982.
Shinohara N *et al., J Urol* 150:505, 1993.
Singh SV *et al., Cancer Let* 95:49, 1995.
Southgate J *et al., Br J Cancer* 72:1214, 1995.
Spruck CH *et al., Nature* 370:183, 1994.
Sun TT *et al., Mol Biol Rep* 23:3, 1996.
Tamimi Y *et al., Int J Cancer* 68:183, 1996.
Thrasher JB and Crawford ED. *J Urol* 149:957, 1993.
Vilien M *et al., Eur J Cancer Clin Oncol* 19:775, 1983.
Walker MC *et al., J Natl Cancer Inst* 79:213, 1987.
Walker MC *et al., Eur J Cancer* 26:742, 1990.
Williams AR., *Invest Urol* 17:359, 1980.
Wu Q *et al., Int J Cancer* 65:840, 1996.
Wu XR and Sun T-T, *J Cell Sci* 106:31, 1993.
Xu BH *et al., Int J Cancer* 58:686, 1994.
Yamamoto T *et al., Jpn J Urol* 70:351, 1979.
Yeh MY *et al., J Surg Oncol* 37:177–184, 1988.
Yu H-J *et al., J Formosan Med Assoc* 91:608, 1992.
Zariwala M and Xiong Y, *Oncogene* 12:451, 1996.
Zirvi KA *et al., In Vitro Cell Dev Biol* 22:369, 1986.

Chapter 12

Renal Cell Cancer

[1]Thomas Ebert, [1]Aristoteles Anastasiadis, [2]Neil H. Bander
[1]Department of Urology, Düsseldorf Medical School, Heinrich-Heine-University, Moorenstr. 5, D-40225 Düsseldorf, Germany. Present address: Euro-Med-Clinic, Europa Allee 1, 90763 Furth, Germany. Tel: 0049-911-9714-531; Fax: 0049-911-9714-532; E-mail: ebert@euro-med-clinic.de and [2]Bernard and Josephine Chaus Chair, Urologic Oncology, The New York Hospital-Cornell Medical Center, New York, NY 10021, USA. Tel: 001-212-746-5493, Fax: 001-212-746-8941, E-mail: nhbander@mail.med.cornell.edu

About 50% of patients with kidney cancer present with metastases at some time during the course of their disease. The prognosis of these patients is poor, since no treatment modality (hormonal therapy, chemotherapy, radiotherapy, immunotherapy) is effective in metastatic renal cancer.

In 1962, Ishihara and coworkers described the first human renal cancer cell line to be established in continuous culture. Since then, many laboratories have published characterizations of continuous renal cancer cell lines. Today, a variety of lines from primary kidney cancers as well as from various metastatic sites is available.

1. CULTURE CONDITIONS

Most renal cancer cell lines were established from primary tumors (see Table 1). The method we use is as follows. Following removal of the surrounding fat, vessels, connective tissue and renal capsule, the tumor tissue is dissociated mechanically with scissors and/or scalpels in a Petri dish containing 3-5 ml of 0.01 M sodium phosphate, 0.15 M NaCl (PBS), pH 7.2. This mixture is transferred to a sterile 100 ml bottle containing 25 ml of 0.25% trypsin and 0.02% collagenase in PBS. The mixture is stirred at 37°C for 30 min. The cells are then washed twice with fresh complete medium (Eagles minimal essential medium, 1% non-essential amino acids, 2 mM glutamine, 100 µl/ml penicillin, 100 µg/ml streptomycin, 7.5% fetal bovine

J.R.W. Masters and B. Palsson (eds.), Human Cell Culture Vol. 1, 231–249.

serum), with centrifugation at 1000 rpm for 10 min. The cells are then plated in tissue culture flasks (T25 or T75 flasks), using approximately 25 to 50 μl of the cell pellet per 25-cm^2 flask area. In large series, the success rate for obtaining continuous cell lines ranges from 9 to 13% (Ebert et al. 1990, Giard et al. 1973).

Most renal lines grow well in the complete medium described in the previous paragraph. Cultures are usually fed twice weekly and passaged shortly after becoming confluent. Cells are removed from the flask by adding 0.25% trypsin/0.1% EDTA at 37°C (2 to 5 min). When cells start to float off the plastic, complete medium is added and the cells are washed for 10 minutes before resuspension in complete medium. The cells are usually split at a ratio of 1:2 to 1:8.

To study autocrine or paracrine growth regulation of cells, defined serum-free medium is required. Most renal cancer lines can be grown for limited periods using serum-free media. Few cell lines have been tested for long term viability in such defined media.

2. DO THE CELL LINES REPRESENT THE CLINICAL SPECTRUM OF DISEASE?

Epidemiologic studies on carcinoma of the kidney indicate that the disease is more frequent in men. The male-to-female ratio is about 2-3:1 (Bennington and Laubscher, 1968). The largest series of renal cell cancer lines was established by the Laboratory of Human Cancer Immunology at the Memorial Sloan-Kettering Cancer Center in New York (Ebert et al. 1990). In this series of 498 tumors, the male-to-female ratio was 2.69:1. Fifty-one of 363 (14.0%) tumors derived from men and 12 out of 135 (8.9%) tumors derived from women were established as continuous cell lines. In this series, 120 culture attempts were made using kidney cancer metastases. Continuous cell lines were established in 17 (14.2 %) of these. This proportion was similar to 46 successful attempts from 378 primary kidney cancers (12.2%). Medical records of 40 of the 63 patients whose tumors were established in culture were reviewed. Of these 40 patients, 35 had metastatic disease either at the time of surgery or developed it soon thereafter. The remaining 5 patients all had main renal vein involvement at the time of surgery. It was concluded that a tumor must be biologically aggressive to adapt to tissue culture.

In summary, permanent renal cancer cell lines can be established with equal success from primary and metastatic tumors as well as from male and female tissue. The cell lines appear to represent one end of the clinical spectrum, that of aggressive disease. Low stage and low grade tumors are more difficult to grow in culture. For most cell lines a renal origin was confirmed. All the SK-RC lines were typed using monoclonal antibodies against renal

Table 1 Origin of cell lines

Cell line	Patient age	Sex	Stage	Grade	Primary site	Specimen site	Culture method	Primary reference
HN 4	65	m	pT3b NX M0	3	left kidney	left kidney	D	Bergerheim USR et al.
HN 51	54	m	pT3b NX M1	2	left kidney	right parietal lobe	D	Bergerheim USR et al.
KTCTL-26A	77	m	pT2a NX M1	2	kidney	kidney		Högemann I et al.
TK-10	43	m	pT2 NX M0	III-IV	left kidney	left kidney	D	Bear A et al.
TK-164	69	m	pT3b NO M0	3	right kidney	right kidney	D	Bear A et al.
UM-RC-2			Clinical stage IV		kidney		D	Grossman HB et al.
UM-RC-3			Clinical stage IV		kidney		D	Grossman HB et al.
UM-RC-5			Clinical stage III		kidney	kidney	D	Grossman HB et al.
UM-RC-6			Clinical stage I		kidney	kidney	D	Grossman HB et al.
UM-RC-7			Clinical stage III		kidney	kidney	D	Grossman HB et al.
SMKT-R-1	63	f	pT2b	2	kidney	kidney	D	Miyao N et al.
SMKT-R-2	79	m	pT3b	2	kidney	kidney	X	Miyao N et al.
SMKT-R-3	64	m	pT3b	3	kidney	kidney	X	Miyao N et al.
DNT-11	66	m			kidney	adrenal gland	D	Buszello H and Ackermann R
DNT-63	52	m			kidney	adrenal gland	D	Buszello H and Ackermann R
KPK-1	74	f	pT4 NX MX		kidney	kidney	D	Naito S et al.
KPK-13	61	m	pT4 NX M1 (bone)		kidney	kidney	D	Naito S et al.
SRCC-1P	47	m		I-II	kidney	kidney	D	Otani N et al.
SRCC-1M	47	m		I-II	kidney	adrenal gland	D	Otani N et al.
A-498	52	m			kidney	kidney	D	Giard DJ et al.
A-704	78	m			kidney	kidney	D	Giard DJ et al.
CCF-RC1	67	m	pT3a pN0 M1 (bone)		I	left kidney	left kidney	D Hashimura T et al.
CCF-RC2	67	m	pT3a pN0 M1 (bone)			left kidney	left kidney	X Hashimura T et al.
RR cell line					kidney	kidney	D	Leung SW et al.
RS cell line					kidney	malignant ascites	X	Leung SW et al.

Continued on next page

Table 1 (continued)

Cell line	Patient age	Sex	Stage	Grade	Primary site	Specimen site	Culture method	Primary reference
UOK 101					kidney	kidney	D	Anglard P et al.
UOK 102					kidney	kidney	D	Anglard P et al.
UOK 103					kidney	kidney	D	Anglard P et al.
UOK 104					kidney	kidney	D	Anglard P et al.
UOK 105					kidney	kidney	D	Anglard P et al.
UOK 106					kidney	kidney	D	Anglard P et al.
UOK 107					kidney	kidney	D	Anglard P et al.
UOK 108					kidney	kidney	D	Anglard P et al.
UOK 109					kidney	kidney	D	Anglard P et al.
UOK 109LN					kidney	lymph node	D	Anglard P et al.
UOK 110					kidney	kidney	D	Anglard P et al.
UOK 111					kidney	kidney	D	Anglard P et al.
UOK 112					kidney	kidney	D	Anglard P et al.
UOK 113					kidney	kidney	D	Anglard P et al.
UOK 114					kidney	kidney	D	Anglard P et al.
UOK 115					kidney	kidney	D	Anglard P et al.
UOK 116					kidney	kidney	D	Anglard P et al.
UOK 117					kidney	kidney	D	Anglard P et al.
UOK 118					kidney	kidney	D	Anglard P et al.
UOK 119					kidney	kidney	D	Anglard P et al.
UOK 120					kidney	kidney	D	Anglard P et al.
UOK 121					kidney	kidney	D	Anglard P et al.
UOK 121LN					kidney	lymph node	D	Anglard P et al.
UOK 122					kidney	kidney	D	Anglard P et al.
UOK 122LN					kidney	lymph node	D	Anglard P et al.

Continued on next page

Table 1 (continued)

Cell line	Patient age	Sex	Stage	Grade	Primary site	Specimen site	Culture method	Primary reference
UOK 123					kidney	kidney	D	Anglard P et al.
UOK 124					kidney	kidney	D	Anglard P et al.
UOK 124LN					kidney	lymph node	D	Anglard P et al.
UOK 125					kidney	kidney	D	Anglard P et al.
UOK 126					kidney	kidney	D	Anglard P et al.
UOK 127					kidney	kidney	D	Anglard P et al.
UOK 128					kidney	kidney	D	Anglard P et al.
UOK 129					kidney	kidney	D	Anglard P et al.
UOK 130					kidney	kidney	D	Anglard P et al.
UOK 131					kidney	kidney	D	Anglard P et al.
D-RC-1	40	m	pT2 pN0 M0	2	kidney	kidney	D	Schattka S et al.
D-RC-2	57	f	pT3b N0 M0	2	kidney	kidney	D	Schattka S et al.
D-RC-3	47	m	pT2 N0 M0	2	kidney	kidney	D	Schattka S et al.
SK-RC-1		m			kidney	kidney	D	Ebert T et al.
SK-RC-2		m			kidney	kidney	D	Ebert T et al.
SK-RC-4		m			kidney	kidney	D	Ebert T et al.
SK-RC-6		m			kidney	kidney	D	Ebert T et al.
SK-RC-7		m			kidney	kidney	D	Ebert T et al.
SK-RC-8		m			kidney	kidney	D	Ebert T et al.
SK-RC-12		m			kidney	kidney	D	Ebert T et al.
SK-RC-15		m			kidney	kidney	D	Ebert T et al.
SK-RC-21		m			kidney	kidney	D	Ebert T et al.
SK-RC-28		f			kidney	kidney	D	Ebert T et al.
SK-RC-35		m			kidney	kidney	D	Ebert T et al.
SK-RC-37		m			kidney	kidney	D	Ebert T et al.
SK-RC-40		m			kidney	kidney	D	Ebert T et al.

Continued on next page

Table 1 (continued)

Cell line	Patient age	Sex	Stage	Grade	Primary site	Specimen site	Culture method	Primary reference
SK-RC-41		m			kidney	kidney	D	Ebert T et al.
SK-RC-44		m			kidney	kidney	D	Ebert T et al.
SK-RC-47		m			kidney	kidney	D	Ebert T et al.
SK-RC-48		m			kidney	kidney	D	Ebert T et al.
SK-RC-49		m			kidney	kidney	D	Ebert T et al.
SK-RC-51		m			kidney	kidney	D	Ebert T et al.
SK-RC-53		m			kidney	kidney	D	Ebert T et al.
SK-RC-55		m			kidney	kidney	D	Ebert T et al.
SK-RC-56		m			kidney	kidney	D	Ebert T et al.
SK-RC-57		m			kidney	kidney	D	Ebert T et al.
SK-RC-58		f			kidney	kidney	D	Ebert T et al.
SK-RC-59		m			kidney	kidney	D	Ebert T et al.
SK-RC-60		f			kidney	kidney	D	Ebert T et al.
SK-RC-61		m			kidney	kidney	D	Ebert T et al.
SK-RC-62		m			kidney	kidney	D	Ebert T et al.
SK-RC-9		m			kidney	brain	D	Ebert T et al.
SK-RC-13		m			kidney	brain	D	Ebert T et al.
SK-RC-17		m			kidney	soft tissue	D	Ebert T et al.
SK-RC-18		m			kidney	lymph node	D	Ebert T et al.
SK-RC-26a		m			kidney	lung	D	Ebert T et al.
SK-RC-26b		m			kidney	lymph node	D	Ebert T et al.
SK-RC-29		f			kidney	ovary	D	Ebert T et al.
SK-RC-31		m			kidney	lung	D	Ebert T et al.
SK-RC-38		m			kidney	lung	D	Ebert T et al.
SK-RC-39		f			kidney	soft tissue	D	Ebert T et al.
SK-RC-42		m			kidney	bone	D	Ebert T et al.

Continued on next page

Table 1 (continued)

Cell line	Patient age	Sex	Stage	Grade	Primary site	Specimen site	Culture method	Primary reference
SK-RC-45		m			kidney	adrenal gland	D	Ebert T et al.
SK-RC-46		f			kidney	bone	D	Ebert T et al.
SK-RC-52		f			kidney	mediastinal metastasis	D	Ebert T et al.
SK-RC-54		m			kidney	lung	D	Ebert T et al.
OUR-10	32	f		2	kidney	kidney	D	Matsuda M et al.
GKA	62	f			kidney	kidney	D	Sytkowski AJ et al.
NC 65	67	m		II-III	kidney	kidney	D	Höehn W and Schroeder FH
RCC A		f		I	kidney	kidney	D	Gerharz CD et al.
RCC B		f		I	kidney	kidney	D	Gerharz CD et al.
RCC C		f		II	kidney	kidney	D	Gerharz CD et al.
RCC D		m		II	kidney	kidney	D	Gerharz CD et al.
Chromophi-1	64	m	pT3b NX MX	3	kidney	kidney	D	Gerharz CD et al.
Chromophi-2	63	m	pT3a pN1 MX	2	kidney	kidney	D	Gerharz CD et al.
Chromophi-3	64	m	pT3a NX MX	2	kidney	kidney	D	Gerharz CD et al.
Chromophi-4	71	m	pT2 NX MX	1	kidney	kidney	D	Gerharz CD et al.
786-O	58	m	pT3a NX MX		left kidney	left kidney	D	Williams RD et al.
786-P	63	f	pT3b N1 MX		left kidney	left kidney	D	Williams RD et al.
SS78	63	m	TX NX M 1		left kidney	left kidney	D	Schrivastav S et al.
769-P	46	f			left kidney	left kidney	D	Williams RD et al.
#9	48	m			kidney	kidney	D	Ishihara T et al.
Caki-1	49	m			kidney	skin	D	Fogh J and Trempe G
Caki-2	69	m			kidney	kidney	D	Fogh J and Trempe G

Stage: Stage classification of the original paper is given. For TNM Classification several changes were introduced over recent years.
Grade: Grading of the original paper is given. There is no uniform grading system for renal cell carcinoma.
Abbreviations: D = enzymatic dissociation; X = explant culture.

Table 2 Pathology of cell lines

Cell line	Tumor pathology	In vitro features	Growth in nude mice, xenograft pathology
HN 4	Clear cell	DT 63 hr, passage 86, uniform flat and polygonal shape	No
HN 51	Clear cell	DT 75 hr, passage 73, varying size and shape, grow in a dense pattern, no distinct cell borders	Yes, typical clear cell pattern
KTCNL-26A	Clear cell		
TK-10	Papillary	DT 25 hr, mean seeding efficiency 63+/- 9%, epithelial appearance, scant intracytoplasmic vacuoles	No
TK-164	Clear cell	Slow initial growth, doubling time 43 hr, mean seeding efficiency 43+/- 4%, epithelial appearance	No
UM-RC-2		Abundance of coarse granules in the cytoplasm, DT 24 hr, clear cell pattern, P 92, Oil Red O positive	Yes
UM-RC-3		Abundance of coarse granules in the cytoplasm, DT 29 hr, clear cell pattern, P 91, Oil Red O positive	Yes
UM-RC-5		Abundance of coarse granules in the cytoplasm, DT 43 hr, P 27, Oil Red O negative	No
UM-RC-6		Abundance of coarse granules in the cytoplasm, DT 43 hr, P 57, Oil Red O positive	No
UM-RC-7		Abundance of coarse granules in the cytoplasm, DT 53 hr, P 32, Oil Red O positive	No
SMKT-R-1	Clear cell	P 68, fusiform-like cells with round nuclei and nucleoli, lack of contact inhibition, DT 180 hr	Yes, similar to original tumor
SMKT-R-2	Clear cell	P 56, large polygonal cells (round nuclei, abundant cytoplasm) and small compact cells (small nuclei), DT 56 hr	Yes, predominantly granular cells
SMKT-R-3	Papillary	P 41, two distinct types of cells (large and small), lack of contact inhibition, DT 56 hr	Yes, similar to original tumor
DNT-11	Clear cell	Epithelioid polygonal cells, some cells with cytoplasmic vacuoles	No

Continued on next page

Table 2 (continued)

Cell line	Tumor pathology	In vitro features	Growth in nude mice, xenograft pathology
DNT-63		Epithelioid tumor cells with poorly defined borders and some spindle cells	No
KPK 1		Many islands of epithelial cells, monolayer sheet, large round shaped nuclei with a few prominent nucleoli	Yes, similar to original tumor
KPK 13		Epithelial cells	Yes, similar to original tumor
SRCC-1P	Alveolar and clear cell	Polygonal cells and compact cells, lack of contact cell inhibition, cytokeratin and vimentin expression	
SRCC-1M	Alveolar and clear cell	Polygonal cells and compact cells, lack of contact cell inhibition, cytokeratin and vimentin expression	
A-498	Epithelial like	Consistent with papillary epidermoid or adenocarcinoma forms	Yes, undifferentiated carcinomas
A-704	Epithelial like	Adenocarcinoma cells	No
CCF-RC1	Pleomorphic renal cell cancer	Epithelioid polygonal cells in which the nucleus has one or more large nucleoli. Scattered round or dendritic cells are observed	Yes
CCF-RC2	Pleomorphic renal cell cancer	Epithelioid polygonal cells in which the nucleus has one or more large nucleoli. Scattered round or dendritic cells are observed	Yes
RR cell line		DT 20 hr, 49% G1-Phase, 38% S-Phase and 13% G2- and M-Phases	Yes, poorly differentiated tubular structures and small solid nests
RS cell line		DT 37 hr, 52% G1-Phase, 24% S-Phase and 24% G2- and M-Phases	Yes, poorly differentiated neoplastic epithelial cells
UOK 101	Clear cell		No
UOK 102	Clear cell		No
UOK 103	Clear cell/granular cell	Typical spindle cell morphology	No
UOK 104	Clear cell		No
UOK 105	Clear cell/granular cell/ sarcomatoid cells		Yes, clear cell/granular cell
UOK 106	Clear cell		No
UOK 107	Clear cell		No
UOK 108	Clear cell		No

Continued on next page

Table 2 (continued)

Cell line	Tumor pathology	In vitro features	Growth in nude mice, xenograft pathology
UOK 109	Clear cell/granular cell	Epithelial cells	No
UOK 109LN	Clear cell/granular cell		No
UOK 110	Clear cell	Epithelial cells	No
UOK 111	Clear cell		No
UOK 112	Papillary cell	Cells growing in swirls which are very similar to normal kidney epithelial cells	No
UOK 113	Clear cell/granular cell		Yes, clear cell
UOK 114	Clear cell	Unusual shape	No
UOK 115	Clear cell		No
UOK 116	Clear cell		No
UOK 117	Clear cell/granular cell/ sarcomatoid cell		Yes, clear cell/granular cell
UOK 118	Clear cell		No
UOK 119	Clear cell/granular cell/ sarcomatoid cell		No
UOK 120	Clear cell/granular cell/ sarcomatoid cell		Yes, clear cell/granular cell/papillary cell
UOK 121	Granular cell		No
UOK 121LN	Granular cell		No
UOK 122	Clear cell/granular cell		No
UOK 122LN	Granular cell		No
UOK 123	Clear cell/granular cell/ sarcomatoid cell		No
UOK 124	Clear cell/granular cell	Very poorly defined borders	Yes, clear cell/granular cell
UOK 124LN	Clear cell/granular cell		No
UOK 125	Clear cell	Very poorly defined borders	No
UOK 126	Clear cell		No

Continued on next page

Table 2 (continued)

Cell line	Tumor pathology	In vitro features	Growth in nude mice, xenograft pathology
UOK 127	Clear cell/granular cell/ sarcomatoid cell		Yes, clear cell/granular cell/sarcomatoid cell
UOK 128	Clear cell	Very poorly defined borders	No
UOK 129	Clear cell		No
UOK 130	Clear cell/granular cell		No
UOK 131	Clear cell/granular cell		Yes, clear cell/granular cell
D-RC-1	Clear cell	Expression of renal tubule antigen(s), Polygonal cells and spindle cells, heterogeneous shape, lack of contact inhibition, DT 24 hr (23-26)	Yes, clear cell
D-RC-2	Clear cell	Expression of renal tubule antigen(s), Epithelial, polygonal cells, excentric singular nucleus, lack of contact inhibition, DT 42 hr (40-45)	Yes, clear cell
D-RC-3	Clear cell	Expression of renal tubule antigen(s), Epithelial, polygonal cells, excentric singular nucleus, lack of contact inhibition, DT 56 hr (52-61)	Yes, clear cell
SK-RC-1		Expression of renal tubule antigen(s), DT 42 hr	Yes
SK-RC-2		Expression of renal tubule antigen(s), DT 60 hr	Yes
SK-RC-4		Expression of renal tubule antigen(s)	No
SK-RC-6		Expression of renal tubule antigen(s), DT 60 hr, SA	Yes
SK-RC-7		Expression of renal tubule antigen(s)	
SK-RC-8		Expression of renal tubule antigen(s)	
SK-RC-12		Expression of renal tubule antigen(s)	
SK-RC-15		Expression of renal tubule antigen(s)	Yes
SK-RC-21		Expression of renal tubule antigen(s), SA, DT 38 hr	No
SK-RC-28		Expression of renal tubule antigen(s), SA	
SK-RC-35		Expression of renal tubule antigen(s) DT 30 hr	Yes
SK-RC-37		Expression of renal tubule antigen(s)	Yes
SK-RC-41		Expression of renal tubule antigen(s), SA	
SK-RC-44		Expression of renal tubule antigen(s)	Yes

Continued on next page

Table 2 (continued)

Cell line	Tumor pathology	In vitro features	Growth in nude mice, xenograft pathology
SK-RC-47		Expression of renal tubule antigen(s), SA	
SK-RC-48		Expression of renal tubule antigen(s)	Yes
SK-RC-49		Expression of renal tubule antigen(s), DT 42 hr	Yes
SK-RC-51		Expression of renal tubule antigen(s)	Yes
SK-RC-53		Expression of renal tubule antigen(s)	Yes
SK-RC-55		Expression of renal tubule antigen(s)	
SK-RC-56		Expression of renal tubule antigen(s), SA, DT 48 hr	Yes
SK-RC-57		Expression of renal tubule antigen(s), SA, DT 48 hr	Yes
SK-RC-58		Expression of renal tubule antigen(s), SA, DT 56 hr	Yes
SK-RC-59		Expression of renal tubule antigen(s), SA, DT 56 hr	Yes
SK-RC-60		Expression of renal tubule antigen(s), SA	Yes
SK-RC-61		Expression of renal tubule antigen(s), SA, DT 48 hr	Yes
SK-RC-62		Expression of renal tubule antigen(s), DT 96 hr	No
SK-RC-9		Expression of renal tubule antigen(s), SA	No
SK-RC-13		Expression of renal tubule antigen(s), DT 42 hr	Yes
SK-RC-17		Expression of renal tubule antigen(s), SA	
SK-RC-18		Expression of renal tubule antigen(s), SA, DT 42 hr	Yes
SK-RC-26a		Expression of renal tubule antigen(s)	
SK-RC-26b		Expression of renal tubule antigen(s)	
SK-RC-29		Expression of renal tubule antigen(s), SA, DT 36 hr	Yes
SK-RC-31		Expression of renal tubule antigen(s), SA	Yes
SK-RC-38		Expression of renal tubule antigen(s), SA	Yes
SK-RC-39		Expression of renal tubule antigen(s), SA, DT 42 hr	Yes
SK-RC-42		Expression of renal tubule antigen(s), SA, DT 48 hr	Yes
SK-RC-45		Expression of renal tubule antigen(s), SA, DT 48 hr	Yes
SK-RC-46		Expression of renal tubule antigen(s)	
SK-RC-52		Expression of renal tubule antigen(s), SA, DT 36 hr	Yes

Continued on next page

Table 2 (continued)

Cell line	Tumor pathology	In vitro features	Growth in nude mice, xenograft pathology
SK-RC-54		Expression of renal tubule antigen(s)	
OUR-10	Clear cell/granular cell	Large polygonal cells in which the nucleus has one or more nucleoli; Scattered round or dendritic cells	No
GKA	Clear cell	Epithelioid appearance with large nuclei containing one to 2 nucleoli; occasionally, multinucleate cells; cytoplasm with small granules	
NC 65	Clear cell/papillary cell	Round and spindle shaped cells, large nuclear size, multiple nucleoli	Yes, similar to original tumor
RCC-A	Clear cell	DT 27 hr, spindle-shaped, fibroblast like cells, no glycogen deposits	No
RCC-B	Clear cell	DT 64 hr, polygonal, epithelial-like cells, more voluminous than RCC-D, extensive cytoplasmic glycogen deposits	Yes, clear cell carcinoma grade III
RCC-C	Clear cell	DT 51 hr, spindle-shaped fibroblast like cells	Yes, clear cell carcinoma grade III
RCC-D	Clear cell	DT 104 hr, polygonal, epithelial-like cells	No
Chromophi-1	Chromophilic	DT 24 hr, cells with polygonal shape, tightly apposed	No
Chromophi-2	Chromophilic	DT 40 hr, polygonal shape, separated by irregular shapes bridged by cytoplasmic microspikes	No
Chromophi-3	Chromophilic	DT 40 hr, polygonal shape, separated by irregular shapes bridged by cytoplasmic microspikes	Yes, tubulopapillary growth pattern
Chromophi-4	Chromophilic	DT 51 hr, polygonal shape, separated by irregular shapes bridged by cytoplasmic microspikes	No
786-O	Clear cell	Uniform, epithelioid morphology, globular cells with indistinct border and high nuclear-to-cytoplasmic ratio, concentric growth	Tumors in immunosuppressed hamsters, histology compatible to RCC
786-P	Clear cell	Uniform, epithelioid morphology, globular cells with indistinct border and high nuclear-to-cytoplasmic ratio, concentric growth	Tumors in immunosuppressed hamsters, histology compatible to RCC
SS78	Small cell	Very large and vacuolated cells, epithelioid with polygonal to round nuclei and abundant cytoplasm, lack of contact inhibition	Yes, similar to original tumor
769-P	Clear cell	Globular cells with indistinct cell borders and a high nuclear-to-cytoplasm ratio	Tumors in immunosuppressed hamsters, tumors in athymic mice
Caki-1		Epithelial like	Yes, similar to original tumor
Caki-2		Epithelial like	Yes, similar to original tumor

DT: Doubling Time; hr: hour; SA: growth in soft agar; P: passage number

Table 3 Molecular genetics of cell lines

Cell line	Type of change
UOK 102	LOH on 3p
UOK 104	LOH on 3p
UOK 105	LOH on 3p
UOK 106	LOH on 3p
UOK 107	LOH on 3p
UOK 108	LOH on 3p
UOK 109LN	LOH on 3p
UOK 110	LOH on 3p
UOK 111	LOH on 3p
UOK 113	LOH on 3p
UOK 114	LOH on 3p
UOK 115	LOH on 3p
UOK 116	LOH on 3p
UOK 117	LOH on 3p
UOK 118	LOH on 3p
UOK 119	LOH on 3p
UOK 121	LOH on 3p
UOK 121LN	LOH on 3p
UOK 122	LOH on 3p
UOK 122LN	LOH on 3p
UOK 123	LOH on 3p
UOK 124LN	LOH on 3p
UOK 125	LOH on 3p
UOK 126	LOH on 3p
UOK 127	LOH on 3p
UOK 128	LOH on 3p
UOK 129	LOH on 3p

Continued on next page

Table 3 (continued)

Cell line	Type of change
UOK 130	LOH on 3p
D-RC-1	del (3)(p13p25), Tetraploid karyotype, (7p+), del(11)(p11.2), 31 metaphases
D-RC-2	del (3)(p12/13), Tetraploid karyotype, 31 metaphases
D-RC-3	t(3;8)(p11;q11:2/13)
KTCTL-26A	Deletion on 3p14, gain of copies of chromosomes 5 and 7
HN 4	One clonal translocation, der(10)t(3;10)(q13;p12) and del(2)(p13-14)
HN 51	Triploid metaphases, no consistent clonal aberrations
TK-10	Modal chromosome number 51, Y chromosome present in all karyotypes
TK-164	Modal chromosome number 8.2, Y chromosome present in all karyotypes
UM-RC-2	Median number of chromosomes 57
UM-RC-3	Median number of chromosomes 55
UM-RC-5	Median number of chromosomes 48
UM-RC-6	Median number of chromosomes 71
UM-RC-7	Median number of chromosomes 84
SMKT-R-1	Median number of chromosomes 86
SMKT-R-2	Median number of chromosomes 81, Y chromosome
SMKT-R-3	Median number of chromosomes 77, Y chromosome
DNT-11	Hyperdiploid DNA content with a Di=3.0
DNT-63	Population A hyperdiploid with a Di=3.0, Population B hyperdiploid Di=3.0
KPK-1	100% aneuploid, number of chromosomes 48-62
KPK-13	Triploid, number of chromosomes 68-69, distinctive Y chromosomes
A-498	Hypertriploid cell line. Modal chromosome number 74
A-704	(P59) Diploid to hyperdiploid, hypertriploid to hypertetraploid with abnormalities
CCF-RC1	Modal chromosome number 83, two marker chromosomes: M1:I(12q) and M2, Y chromosome absent
CCF-RC2	Modal chromosome number 73 (P12) and 77-78 (P35), 4 marker chromosomes
RR cell line	Aneuploid chromosome counts in the triploid range, modal number 72

Continued on next page

Table 3 (continued)

Cell line	Type of change
RS cell line	Aneuploid chromosome counts in the tetraploid range, modal number 94, nine marker chromosomes were found
OUR-10	Chromosome number 31-90, hypodiploid modal number of 39 and 40; nonrandom loss of 3 chromosomes
GKA	Modal chromosome number: 45 with some degree of aneuploidy and some tendency toward tetraploidy, modal karyotype 45, XX; absent chromosome 17
RCC-A	multiple changes
RCC-B	multiple changes
RCC-C	multiple changes
RCC-D	multiple changes
Chromophi-1	multiple changes
Chromophi-2	multiple changes
Chromophi-3	multiple changes
Chromophi-4	multiple changes
786-O	Hypertriploid karyotype with modal number 76-88. 4 isochromosomes of the size of large C-group chromosomes
786-P	Hypodiploid karyotype with a modal number of 45. Marker: a large, submetacentric chromosome M
SS78	Aneuploid cells, modal number of 80, 5-8 different abnormal chromosomes and two marker chromosomes
769-P	Hypodiploid karyotype at passage 12 with a modal chromosome number of 45
#9	Modal number of chromosomes: 71
Caki-1	Hypertriploid, large meta- or polycentric markers
Caki-2	Hypopenta- to hypohexaploid, large subtelocentric markers

antigens. Rigorous proof, however, that a particular cell line has originated from a given tissue (for example by DNA fingerprinting) is available in few cell lines.

3. PATHOLOGY

Most renal cell carcinomas are histopathologically of the clear cell type. This is also true for most of the primary tumors which have been established in culture (see Table 2). Gerharz et al (1996) described 4 lines established from chromophilic renal cell carcinoma. Renal cancer cell lines take several morphological forms. Some are epithelioid with distinct cell borders, others have irregular or poorly defined borders and others consist of spindle cells. Some cell lines contain cytoplasmic vacuoles or granules. None of the renal cancer lines exhibits contact inhibition. All these features are also typical of the histological appearance of renal cell cancers. In the MSKCC series, the antigenic pattern of the cell lines remained stable for as many as 80 or more passages. Many of the cell lines form xenografts whose histology resembles that of the tumor tissue from which the cell lines were derived.

Some cell lines form colonies in soft agar. Anchorage-independent growth is a feature of transformed cells (Macpherson and Montagnier, 1964), but does not always correlate with growth in immuno-compromised animals.

4. MOLECULAR GENETICS

Defects on the short arm of chromosome 3 are associated with the development of kidney cancer. Although first described in hereditary renal cancer (Cohen et al. 1979), it was shown that sporadic renal cancers frequently demonstrate deletions on chromosome 3p (Carroll et al. 1987). A tumor suppressor gene, the von-Hippel-Lindau (VHL) gene, was mapped to chromosome 3p25-26 (Seizinger et al. 1991). LOH of this area is observed in the majority of clear cell kidney cancers. These changes are also observed in continuous renal cancer cell lines. Apart from major chromosomal alterations a variety of defects on many genes have been described in renal cancers and cell lines (Eickelmann et al. 1994, Sel et al. 1996, van der Leede et al. 1995). Their role in renal cancer has not yet been defined.

To conduct functional assays for tumor suppressor genes, the gene can be inserted in tumor cells. Cell line SK-RC-47 was transfected with various constructs (Clairmont et al. 1994, van der Leede et al. 1995). The yield of stable transfected cells is higher with this cell line than the other SK-RC lines tested.

5. OTHER FEATURES

No cross-contaminations are known. To exclude mycoplasma infection, regular testing of cell cultures is mandatory.

Many renal cell lines have been described with special features. Hypercalcemia in nude mice has been observed when cell lines SK-RC-1, 15, 30, 38, 39, 40, 43, 45 and 46 form xenografts. Hypercalcemia is due to the production of parathyroid hormone-like protein (PTH-like protein) by the tumor (Weir et al. 1988). Interleukin-6 (Il-6) is produced by some renal cancer cell lines (Gastl et al. 1993). Sytkowski et al. (1983) established a renal carcinoma cell line from a patient with erythrocytosis and showed that it can produce erythropoeitin (EPO).

There are some cell lines which were established from different tumor tissues from the same patient. SK-RC-26A and SK-RC-26B are kidney cancer cell lines obtained from a metastasis to the lung and from a metastasis to a lymph node, respectively. SK-RC-44 is a cell line derived from a primary kidney cancer, whereas SK-RC-45 was cultured from an adrenal metastasis of the same patient. Another four pairs of lines have been described by Anglard et al. (1992). The cell lines UOK 109 and UOK 109 LN, UOK 121 and UOK 121LN, UOK 122 and UOK 122 LN, and UOK 124 and UOK 124LN were derived from 4 primary kidney cancers and their corresponding lymph node metastases.

REFERENCES

Anglard P et al, *Cancer Res* 52: 348, 1992
Bander NH, *Cancer* 53: 1235, 1984
Bander NH, *Uremia Invest* 8: 263, 1985
Bear A et al, *Cancer Res* 47: 3856, 1987
Bennington JL and Laubscher FA, *Cancer* 21: 1069, 1968
Bergerheim USR et al, *Cancer Genet Cytogenet* 86: 95, 1996
Buszello H and Ackermann R, *Anticancer Research* 14: 173, 1994
Carroll PR et al, *Cancer Genet Cytogenet* 26: 253, 1987
Clairmont A et al, *Cancer Res* 54: 1319, 1994
Cohen AJ et al, *N Engl J Med* 301: 592, 1979
Ebert T et al, *Cancer Res* 50: 5531, 1990
Ebert T, Thieme Copythek, Stuttgart, *New York*, 1993
Eickelmann P et al, *Carcinogenesis* 15: 219, 1994
Finstad CL et al, *Proc Natl Acad Sci, USA* 82: 2955, 1985
Fogh J and Trempe G, *Human Tumor Cells* In Vitro. NY, NY, Plenum 1975
Gastl G et al, *Int J Cancer* 55: 96, 1993
Gerharz CD et al, *Br J Cancer* 74: 1605, 1996
Gerharz CD et al, *Virchows Arch* 424: 403, 1994
Giard DJ et al, *J Natl Cancer Inst* 51: 1417. 1973

Grossman HB et al, *J Surg Oncol* 28: 237, 1985
Hashimura T et al, *Cancer Res* 49: 7064, 1989
Hernandez A et al, *Cancer Gene Therapy* 4: 59, 1997
Hoehn W and Schroeder FH, *Invest Urol* 16: 106, 1978
Högemann I et al, *Cancer Genet Cytogenet* 78: 175, 1994
Holt et al, *J Biol Chem* 268: 20639, 1993
Ishihara T et al, *Cancer Res* 22: 375, 1962
Leung SW et al, *Cancer* 71: 2276, 1993
Macpherson I and Montagnier L, *Virology* 23: 291, 1964
Matsuda M et al, *Cancer Res* 39: 4694, 1979
Miyao N et al, *Urol Res* 17: 317, 1989
Naito S et al, *J Urol* 128: 1117, 1982
Otani N et al, *J Urol* 149: 1182, 1993
Schattka S et al, *Investigative Urology* 5: 35, 1993
Seizinger BR et al, *PNAS* 88: 2864, 1991
Sel S et al, *Cancer Letters* 101: 205, 1996
Shrivastav S et al, *In Vitro* 17: 1117, 1981
Sytkowski AJ et al, *Cancer Res* 43, 1415, 1983
Van der Leede BJ et al, *Int J Oncol* 6: 391, 1995
Weir et al, *J Clin Invest* 81: 818, 1988
Whaley et al, *Am J Hum Genet* 55: 1092, 1994
Williams RD et al, *In Vitro* 12: 623, 1976

Chapter 13

Skin Cancer (Non-Melanoma)

Petra Boukamp
German Cancer Research Center, Im Neuenheimer Feld 280, D-69120, Heidelberg, Germany.
Tel: 0049-6221-424516, Fax: 0049-6221-424551, E-mail: p.boukamp@dkfz-heidelberg.de

Skin cancer is currently of interest because of the increasing incidence of tumors diagnosed by dermatologists. Because these cancers are not generally reported in tumor registries, the numbers are likely to be underestimated. About 600,000 new cases, representing one third of all new cancers, are diagnosed annually in the USA (reviewed in Corona, 1996), making non-melanoma skin cancer the most frequent type of cancer. The increased incidence is probably due to the depletion of the ozone layer, the shield protecting us against UV-C and most UV-B, in addition to an increase in outdoor activities. Based on data from a health plan in the Portland area (Oregon, USA), the incidence of skin carcinomas increased 2.6 fold in men and 3.1 fold in women from 1960 to 1980 (for review see Mortimer, 1991).

The increasing incidence applies to both types of skin cancer, squamous cell carcinomas (SCC) and basal cell carcinomas (BCC). However, the number of BCC far exceeds that of SCC. Mortality from skin cancers is low and mostly due to SCC. The number of metastasizing SCC is estimated at 3 to 10%. Metastatic BCC are hardly ever seen (less than 1 per 4000 cases). The recurrence rate, on the other hand, is high, approximately 35% at three years and approximately 50% at five years following first diagnosis (for review see Corona, 1996). Furthermore, the occasionally widespread local damage, particularly with BCC, remains a serious problem.

The incidence of skin cancer in the white population is strongly associated with UV exposure, with the highest incidence rate reported for Australia, particularly for British-born migrants who went to Australia before the age of 18. An increase in both the incidence and mortality rates, particularly of SCC, is being seen in renal transplant recipients (for review see Mortimer, 1991) and a Dutch study reported an overall incidence of SCC 250 times higher in

J.R.W. Masters and B. Palsson (eds.), Human Cell Culture Vol. 1, 251–257.

renal transplant patients than that in the general population (Harteelt et al., 1990). Both BCC and SCC develop predominantly in sun-exposed parts of the body such as head and neck, hands and legs. However, an increasing number of trunk lesions have also been reported (see Corona, 1996). In addition to the epidemiological evidence, molecular evidence is accumulating that UV radiation is the cause of skin cancer. Brash and coworkers demonstrated that a high proportion of BCC and SCC carry mutations in the p53 tumor suppressor gene which are C to T transitions in CC sites. Of these mutations, 10% are CC to TT double base changes, indicative of UV-B-induced mutations (Brash et al., 1991; Ziegler et al, 1994).

SCCs usually arise from the interfollicular epidermis as infiltrating sheets of islands of squamous epithelium with variable degrees of differentiation, dysplasia and proliferative activity. Classification was originally based on the degree of differentiation as shown by the degree of keratinization (horn pearl formation and single dyskeratotic cells). However, to predict the malignant potential of the tumor other features are important, including the degree of pleomorphism, the number of mitotic figures and the growth pattern (depth of penetration, way of spreading). In well-differentiated tumors the tissue architecture is quite regular with mitotic cells mainly located basally with horn pearls in the center of the epithelial islands. BCCs, on the other hand, are believed to be derived from hair follicles and typically consist of lobules of basal-like epithelial cells with limited peripheral palisading, surrounded by a characteristic fibrous stroma. Five subtypes are described with a classification depending on the clinical behavior and general growth pattern. Approximately 75% of all BCCs are of the nodular type in which large lobules form a circumscribed tumor mass (for review see McKee, 1996). The lobules can be of various shapes and sizes and are generally interconnected and do not usually differentiate.

1. CULTURE CONDITIONS

Few skin cancer cell lines exist. Most lines were established in the early 1980s and all these were derived from SCC (see Table 1). A limiting factor is the low success rate of establishment of skin cancer cell lines, and this applies to both SCC and BCC (Rheinwald, personal communication and own unpublished results). For skin cancers there is no magic formula, either for a successful primary culture (explant culture versus enzymatically dispersed cells) or for continuous cultivation (medium, trypsinization protocol or feeder support) (Rheinwald and Beckett, 1981; Wu et al., 1982; Tilgen et al., 1983). This low success rate is strange as normal keratinocytes from human skin easily adapt to culture conditions.

Until recently, attempts to establish cell lines from BCC failed completely, even with more sophisticated culture techniques (Rees, personal communication and own unpublished results). BCC cells rapidly adhere and can be maintained as short-term cultures, but eventually terminally differentiate, similar to normal skin keratinocytes (reviewed in Grando et al., 1996, and own unpublished observations). It was suggested that interaction with specific stroma was required to keep the cells in an undifferentiated state. However, attempts to culture BCC also failed using fibroblasts isolated from BCC explants as feeder cells (own unpublished observations). Recently, a BCC cell line was reported for the first time (Chiang et al., 1994), and in a second report Grando and coworkers described an easy and reproducible way to establish long-term cultures from BCC by using lack of differentiation as a selection marker (Grando et al., 1996). However, the number of cell lines and their history have yet to be described. In this report it was suggested that BCC cells may be growth-inhibited by normal keratinocytes and that these cells come to dominate the cultures (Grando et al., 1996). The aneuploidy rate in BCC is low compared to SCC, indicating that BCC have a more normal genotype. Grando et al ignore the fact that BCC are heterogeneous in their tumor phenotypes. Therefore, it cannot be excluded that, by selecting for an abnormal phenoype and growth in soft agar, only a subfraction of possibly differentiation-deficient BCC adapt to the culture conditions and survive. Thus, it remains to be elucidated whether the differentiating phenotype of BCC in culture is real or a culture artefact due to overgrowth by normal skin keratinocytes which eventually senesce.

Table 1 Clinical features

	Patient age/sex	Prior treatment	Primary site	Biopsy	Reference	Availability *
SCL-I	74/female	none	face	primary tumor	Boukamp et al. 1982	+
SCL-II	91/male	radiation	face	primary tumor	Tilgen et al. 1983	+
SCC-12	60/male	immuno-suppressed	face	primary tumor	Rheinwald and Beckett, 1981	+
SCC-13	56/female	radiation	face	primary tumor	Rheinwald and Beckett, 1981	+
SCC-1CB	62/male	none	?	primary tumor	Watanabe et al. 1989	?
HSC-1	75/male	none	hand/back	primary tumor	Kondo and Aso, 1981	+
HSC-5	75/male	none	ear	primary tumor	Hozumi et al. 1990	?
BCC-1/KMC	65/female	?	face/thermal traumatic scar	primary tumor	Chiang et al. 1994	?

*cell lines available from primary investigators

2. PATHOLOGY

Although all cell lines retain the phenotype of the tumor from which they were derived, the few lines available cannot represent the entire spectrum of tumor phenotypes seen in patients. Furthermore and most interestingly, the differentiated phenotype predominates, suggesting that these cells, similar to normal keratinocytes, adapt more rapidly to culture conditions than the abnormal undifferentiated skin carcinoma cells.

Cells were either established using explant cultures (all SCC lines) with (SCC-12, SCC-13, HSC-1, HSC-5) or without feeder support (SCL-I and SCL-II) (Kondo and Aso, 1981; Rheinwald and Beckett, 1981, Boukamp et al., 1982; Tilgen et al., 1983; Hozumi et al., 1990). In the case of SCL-II cells, the initial explant cultures and early passages were maintained in the presence of conditioned medium from SCL-I cells (Tilgen et al., 1983). To establish the SCC-1CB cell line, part of the tumor specimen was transplanted first subcutaneously into nude mice. After 4 passages *in vivo*, part of the xenograft was used for explant culture (Watanabe et al., 1989). For all cell lines high calcium-containing media were used. The BCC cell line, BCC-1/KMC, was initiated on collagen type IV-coated dishes in KGM medium containing 0.1 mM calcium (Chiang et al., 1994)

Three cell lines were described in the mid 1970s by Sinkovics. Except for this report no further details are known. According to the paper, line 2214 was derived from a tumor on the hand of a woman, but it is unclear whether the cell line is derived from the primary tumor or an axillary metastasis. It has an epithelial morphology and at passage 30 cell stocks were frozen. Line 2253 was derived from a lymph node metastasis from a SCC. This culture has 2 phenotypes: densely packed epithelial cells and stratifying, large multinucleated cells. Stocks were frozen at passage 30. Line 3860 was derived from a locoregional lymph node metastasis from a well differentiated SCC arising as an invasive acanthosis in a gasoline burn scar. The cells grew as a densely packed monolayer and stocks were frozen at passage 30.

Table 2 Pathology

	Tumor differentiation	Xenograft pathology	Tumorigenicity	*In vitro* differentiation
SCL-I	moderate	moderate	100%	no morphological evidence
SCL-II	low	low	64% (early passages) 0% (later passages)	no morphological evidence
SCC-12	high	high	100%	expression of diff. markers
SCC-13	high	high, cyst	100%	expression of diff. markers
SCC-1CB	moderate	moderate	tumorigenic	not described
HSC-1	low	–	non-tumorigenic	no morphological evidence
HSC-5	high	–	non-tumorigenic	no morphological evidence
BCC-1/KMC	low		100%	not characterized

Most studies on cell differentiation have used the SCC-13 cell line, which shows a high degree of epidermal differentiation, although it is partly defective in its regulation of terminal differentiation. When deprived of anchorage, keratinocytes are committed to differentiate, forming cornified envelopes. In the SCC lines the induction rate of cornified envelopes is significantly reduced (Rheinwald and Becket, 1981). Furthermore, the coordinated expression of markers following the induction of differentiation by various means is slightly abnormal (Boukamp et al., 1985; Rubin et al., 1989). According to their degree of epidermal differentiation, the cell lines can be ranked in the order SCC-13 > SCC-12 > SCL-I > SCL-II, with SCL-II being completely undifferentiated. The phenotypes of the cells are highly stable as subcutaneous xenografts, and in surface transplants where the cells are transplanted onto the muscle fascia and are exposed to air like skin (Boukamp et al., 1990), and also in raft or organotypic cultures *in vitro* where the cells are similarly allowed to form a 3-dimensional architecture by growing on a collagen gel at the air to medium interface (Boukamp et al., 1985; Kopan and Fuchs, 1989). The differentiation states of the HSC-1 and BCC-1/KMC cell lines have not been reported. For the SCC-12 cell line two clones were described: F.2a and B.2, of which only clone B.2 is tumorigenic (Spellman et al., 1995).

3. MOLECULAR AND CYTOGENETICS

Although the authentication of the skin cancer cell lines is incomplete, no cross-contamination is known for any of these lines. HLA typing was described for the BCC-1/KMC cells with HLA-A2, A24, B27, B35, Cw3, DR2, and DR12 (Chiang et al., 1994). SCL-I and SCL-II were identified by DNA fingerprinting and karyotyping (Boukamp et al., 1982; 1988; Tilgen et al., 1983). For the SCC lines none of these profiles are known. However, as an additional marker for identification, the mutation spectrum of the p53 tumor suppressor gene could be used. The SCL and SCC lines have been analyzed. SCL-I contains a C to T transition in both alleles of codon 196 causing a stop codon and in the SCL-II cells one allele carries a T to G transversion in codon 132 causing an amino acid change from asparagine to lysine (Boukamp et al., in preparation). Both SCC lines have one mutated p53 allele. SCC-12 carries a T to G transversion at codon 216 causing valine to be replaced by glycine. In SCC-13 cells, a typical UV mutation is seen with a C to T transition at codon 258 and glutamine is replaced by lysine. In SCC-12 cells the normal allele is still expressed, but wild type p53 was not detected in SCC-13 cells (Brash et al., 1991; Burns et al., 1993). For HSC-1 and the BCC cell line the p53 status is unknown.

Another common genetic event in nonmelanoma skin cancer is loss of heterozygosity (LOH) from chromosome 9p (Quinn et al., 1994). This LOH is reflected in the cell lines. Additionally, we identified 3p LOH in skin cancers, and this too is reflected in the SCC and SCL lines (Boukamp et al., in preparation). Mutational activation of the ras gene, on the other hand, is a rare event in skin cancers (Ananthaswamy and Pierceall, 1992). A codon 12 Ha-*ras* mutation has so far only been detected in the SCC-12 cell line (Boukamp et al., in preparation).

The HSC-1 cells show amplification and overexpression of the EGF receptor (Kamata et al., 1986; Yamamoto et al., 1986). These cells have therefore been used for a number of studies, including the effects on EGF on cell-matrix interaction (Fujii, 1996), the production of pro-inflammatory cytokines (Maruyama et al., 1995), expression and function of E-cadherin (Fujii et al., 1996) and the identification of EGF-sensitive tyrosine phosphorylation sites of phospholipase C-gamma (Wahl et al., 1990).

Little is known about the skin cancer cell lines as recipients for gene transfer. This might in part be due to their being difficult to handle. For example, the SCC lines require feeder support for long-term growth. It may also be because of the availability of a number of *in vitro* transformed skin keratinocyte lines which are used on a larger scale. My limited experience suggests that SCC-12 cells are difficult to transfect. SCL-I cells, on the other hand, were used as recipients for single chromosomes and individual genes and found to be reasonably good recipients (Bleuel et al, submitted).

Table 3 Molecular and cytogenetics

	HLA-typing	DNA fingerprinting	Ploidy/ modal chr. no	CGH*	p53	EGF-R
SCL-I	nd	+ (Boukamp et al. 1988)	hypotetraploid/ 86-88	9p-,	mut	nd
SCL-II	nd	+ (Boukamp et al. 1988)	hypodiploid/ 42-45	3p-, 9q+	mut	+ (Kemeney et al. 1993)
SCC-12	nd	nd	hypotetraploid/ 79-91	3p-, 9p-, 9q-	mut	no expression EGF, ?EGF-R
SCC-13	nd	nd	hyperdiploid/ 44-50	3p-, 9pter-	mut	nd
SCC-1B	nd	nd	hyperdiploid 56	nd	nd	nd
HSC-1	nd	nd	hypotetraploid/ 80	nd	nd	overexpression of EGF-R (see below)
HSC-5	nd	nd	hypotetraploid 67-80 (76)	nd	nd	nd
BCC-1/ KMC	+ (Chiang et al. 1994)	nd	triploid/ 57-74 (69)	nd	nd	nd

* Comparative genomic hybridization data; nd = not described

REFERENCES

Ananthaswamy HN and Pierceall WE, Prog Clin Biol Res 376: 61-84, 1992
Bleuel K et al. submitted
Boukamp P et al., J Natl Cancer Inst 68: 415-427, 1982
Boukamp P et al., Cancer Res. 45: 5582-5592, 1985
Boukamp P et al., J Cell Biol 106: 761-771, 1988
Boukamp P et al., Differentiation 44: 150, 1990
Brash DE et al., Proc Natl Acad Sci USA 88: 10124-10128, 1991
Burns JE et al., Br J Cancer 67: 1274-1248, 1993
Chiang L-C et al., Kaohsiung J Med Sci 10: 170-176, 1994
Corona R, Ann 1st Super Sanita 32: 37-42, 1996
Fujii K, J Invest Dermatol 107: 195-202, 1996
Fujii K et al. Exp Cell Res 223: 50-62, 1996
Grando SA et al., Arch Dermatol 132: 1185-1193, 1996
Harteelt MM et al., Transplantation 49: 506-509, 1990
Hozumi Y et al, J Dermatol 17: 143-148, 1990
Kamata N et al., Cancer Res 46: 1648-1653, 1986
Kemeny L et al., Acta Derm Venerol 73: 37-40, 1993
Kondo S and Aso K, Br J Dermatol 105: 125-132, 1981
Kopan R and Fuchs E, J Cell Biol 109: 295-307, 1989
Maruyama K et al. J Dermatol 22: 901-906, 1995
McKee, Pathology of the Skin, 2nd ed, 1996
Mortimer P, Curr Opin Oncol 3: 349-354, 1991
Quinn AG et al., Cancer Res 54: 4756-4759, 1994
Rheinwald JG and Beckett MA, Cancer Res 41: 1657-1663, 1981
Rubin AL et al., J Cell Physiol 138: 208-214. 1989
Sinkovics et al., In Methods in Cancer Research 14: 257-272. Acad. Press NY, 1978
Spellman JE et al., J Surgical Res 58: 165-174, 1995
Tilgen W et al., Cancer Res 43: 5995-6011, 1983
Wahl MI. J Biol Chem 265: 3944-3948, 1990
Watanabe F et al, Nippon Hinyokika Gekagakkai Zasshi 99: 793-800, 1989
Wu Y-J et al., Cell 31: 693–702, 1982
Yamamoto T et al., Cancer Res 46: 414-416, 1986
Ziegler A et al., Nature 372: 773-776, 1994

Chapter 14

Melanoma: The Wistar Melanoma (WM) Cell Lines

Mei-Yu Hsu, David E. Elder*, and Meenhard Herlyn
*The Wistar Institute, 3601 Spruce Street, Philadelphia, PA 19104. Tel: 001-215-898-3950; Fax: 001-215-898-0980, E-mail: herlynm@wista.wistar.upenn.edu and Department of Pathology and Laboratory Medicine, University of Pennsylvania**

The easy accessibility and long-term clinical and histopathological observation of cutaneous melanoma has led to the definition of 5 major steps of tumor progression in the human melanocytic system. The lesions representing each step are: (i) common acquired nevus, (ii) dysplastic nevus, (iii) radial growth phase (RGP) primary melanoma, (iv) vertical growth phase (VGP) primary melanoma and (v) metastatic melanoma (MM) (1–3). The development of culture techniques that allowed the establishment and propagation of cells derived from different stages of tumor progression was a major advance in melanoma research. The relatively high successful culture rate in VGP and metastatic lesions has made available multiple cell lines from the same individual, presenting both types of lesions either at the same time or as the disease progresses. Detailed comparison and characterization of these melanocytic cells reflects a stepwise development of a malignant phenotype in accordance with clinical progression. This chapter summarizes some of the commonly used and better characterized melanoma cell lines, from a total of more than 350 established in this laboratory since 1978.

1. ESTABLISHMENT OF CONTINUOUS CELL LINES

Tissue source and patient background (Table 1): Specimens are collected from the Department of Surgical Pathology at the Hospital of the University of Pennsylvania. Samples are stored and transported at 4°C in transport medium containing antibiotics. Isolation is performed within a few hours

J.R.W. Masters and B. Palsson (eds.), Human Cell Culture Vol. 1, 259–274.

Table 1 Clinical information, culture method and availability of cell lines

Cell line	DOB sex/age	AJCC stage/ pathology[a]	Level/ thickness[b] (mm)	Mitotic rate	Site[c]	Status[d]	Culture method[e]	Authenti-cation[f]	Availa-bility[g]	Primary reference	Isolation date
A. Primary melanoma cell lines with RGP-like phenotype (early primary melanoma):											
WM35	10/22/54 F 24	Stage 1 SSM RGP/VGP	II/0.69	0.0	scalp/neck	NED 06/30/92	D	D, H	C, F	(4, 5)	11/17/78
WM1552C	02/03/16 M 72	Stage 3 SSM RGP/VGP	IV/5.92	n.a.	buttock	DOD 08/1/88	D	n.a.	C, F	(6)	02/08/88
WM1575	05/31/43 M 45	STAGE X SSM RGP/VGP	n.a./1.15	n.a.	thigh	DOD 08/6/93	D	n.a.	F	(6)	03/14/88
WM1650	12/28/57 M 30	Stage 4 SSM RGP/VGP	IV/1.14	n.a.	scalp/neck	DOD 12/27/90	D	n.a.	C, F	(7)	08/04/88
WM1789	11/08/31 M 58	Stage 1 SSM RGP/VGP	III/0.82	n.a.	back	NED 05/22/95	D	D	C, F	(6)	08/17/89
WM3208V	09/25/03 F 91	Stage x n.a. RGP/VGP	n.a./n.a.	n.a.	buttock	DOD 07/2/96	D	n.a.	C, F	(6)	08/10/94
WM3211	08/4/20 M 74	Stage 1 RGP/VGP	n.a./n.a.	n.a.	ankle	n.a.	D	D	C, F	(6)	08/10/94
B. Primary melanoma cells with VGP-like phenotype:											
WM39	n.a. M	n.a.	n.a.	n.a.	n.a.	n.a.	D	n.a.	C, F	(3, 8)	12/15/78

Continued on next page

Table 1 (continued)

Cell line	DOB sex/age	AJCC stage/ pathology	Level/ thickness (mm)	Mitotic rate	Site	Status	Culture method	Authenti- cation	Availa- bility	Primary reference	Isolation date
WM75	09/17/30 M 48	Stage 3 SSM RGP/VGP	IV/6.25	3.8	L back/ chest	DOD 06/24/82	D	D, H	C, F	(8–11)	07/17/79
WM98-1	12/18/29 M 50	Stage 2 SSM RGP/VGP	IV/5.40	6.5	chest	DOD 07/19/86	D	D, H	C, F	(3, 10–12)	11/30/79
WM115	03/1/25 F 55	Stage n.a. SSM RGP/VGP	III/2.24	10.3	groin R leg	DOD 02/2/82	D	D, H	C, F	(6, 8–11)	04/30/80
WM278	04/14/19 F 62	Stage 2 NM VGP only	IV/3.7	n.a.	n.a.	DOD 07/26/88	D	D, H	F	(3, 5, 6, 11)	11/11/81
WM740V	04/22/48 F 34	Stage 2 SSM RGP/VGP	III/3.24	n.a.	chest	DOD 01/23/84	D	D	F	(11)	n.a.
WM793B	05/30/46 M 37	Stage 1 SSM RGP/VGP	II/0.55	n.a.	sternum	NED 06/23/97	D	D, H	C, F	(3, 6, 8, 10–12)	01/07/83
WM853-2	12/03/12 M 70	Stage 2 SSM RGP/VGP	IV/3.75	n.a.	back	DOD 07/5/86	D	n.a.	F	(11)	05/25/83
WM902B	n.a.	Stage n.a. SSM VGP	n.a.	n.a.	n.a.	n.a.	D	D, H	C, F	(3, 8, 10)	11/14/83

Continued on next page

Table 1 (continued)

Cell line	DOB sex/age	AJCC stage/ pathology	Level/ thickness (mm)	Mitotic rate	Site	Status	Culture method	Authenti-cation	Availa-bility	Primary reference	Isolation date
WM983A	04/6/30 M 54	Stage n.a. URGP RGP/VGP	IV/25.0	n.a.	abdomen	DOC 02/7/85	D	D	F	(10,11)	08/13/84
WM1341D	10/18/00 M 85	Stage n.a. LMM RGP/VGP	IV/3.70	n.a.	n.a.	DOC 12/11/86	D	n.a.	C, F	n.a.	07/28/86
WM1361A	06/16/36 M 50	Stage x SSM RGP/VGP	IV/1.22	2.2	forearm	DOD 04/24/87	D	D	F	(10, 11)	08/27/86
WM3268V	05/26/25 F 70	Stage 3	n.a.	n.a.	leg	NED 05/5/97	D	n.a.	C, F	n.a.	n.a.

C. Metastatic melanoma cell lines for which cells isolated from the primary sites are available:

Cell line from metastasis/primary	Metastatic site	Culture method	Authentication	Availability	Primary reference	Isolation date
WM373/WM75	skin	D	D, H	F	(8–11)	11/05/82
WM165-1,2/WM115	skin	D	D, H	F	(6, 9–11)	01/16/81
WM239A, B /WM115	R groin LN or R leg skin	D	D, H	C, F	(6, 8–11)	08/07/81
WM266-1,2,3,4 /WM115	266-4 R thigh skin	D	D, H	F	(6, 8, 9, 11)	10/19/81
WM1617/WM278	LN R axilla	D	D	C, F	(6, 11)	06/13/88

Continued on next page

Table 1 (continued)

Cell line from metastasis/primary	Metastatic site	Culture method	Authentication	Availability	Primary reference	Isolation date
WM858/WM740V	LN R axilla	D	D	F	(9, 11)	06/07/83
WM873-1,2,3/WM740V	LN R axilla	D	D	F	(9)	08/03/83
WM983B, C/WM983A	LN	D	D	F	(6, 10, 11)	08/13/84
WM1361B, C/WM1361A	skin	D	D	F	(6, 10, 11)	n.a.
WM3268Met2/WM3268V	n.a.	D	n.a.	C, F	n.a.	n.a.

D. Metastatic melanoma cell lines without cells from primary lesions available:

Cell line	DOB sex/age	AJCC pathology stage/phase	Level/ thickness (mm)	Mitotic rate	Primary site	Specimen site	Culture method	Authenti-cation	Availa-bility	Primary reference	Isolation date
WM9	n.a. M	n.a.	n.a.	n.a.	n.a.	LN axilla	D	H	C, F	(3, 8, 13)	08/03/78
WM46	08/18/36 F 41	Stage 1 SSM RGP/VGP	IV/1.36	3.2	back	skin	D	H	F	(4)	02/02/79
WM164	02/27/59 M 21	Stage 3 NM VGP only	III/2.70	1.0	arm	LN	D	D	C, F	(4, 6, 8, 12, 14)	01/10/81
WM209	09/24/29 M 50	Stage 2 NM VGP only	V/5.0	0.0	face/ear	LN	D	n.a.	F	n.a.	05/13/81
WM852	08/5/20 M 60	Stage 3 NM VGP only	III/1.20	3.0	abdomen	skin	D	H	C, F	(3, 8, 12)	05/02/83

Continued on next page

Table 1 (continued)

Cell line	DOB age/sex	AJCC pathology stage/phase	Level/ thickness (mm)	Mitotic rate	Primary site	Specimen site	Culture method	Authenti- cation	Availa- bility	Primary reference	Isolation date
WM1158	01/27/52 M 32	Stage 2 SSM RGP/VGP	III/2.75	n.a.	scalp/neck	LN	D	H	C, F	(15–17)	10/11/85
WM1852	n.a.	n.a.	n.a.	n.a.	n.a.	n.a.	D	n.a.	F	n.a.	12/29/89
WM2067	12/15/46 M 41	Stage 2 SSM RGP/VGP	III/3.06	n.a.	back	LN axilla	D	n.a.	C, F	n.a.	08/29/91
WM3214	01/18/47 F	n.a.	n.a.	n.a.	n.a.	Ax?	D	n.a.	F	n.a.	09/30/94
WM3258	05/15/24 F 71	n.a.	n.a.	n.a.	n.a.	n.a.	D	n.a.	F	n.a.	10/19/95
451Lu	02/27/59 M 21				WM164	lung (selection in mice)	D	D	C, F	(18)	n.a.
1205Lu	05/30/46 M 37				WM793	lung (selection in mice)	D	D, H	C, F	n.a.	n.a.

[a] Disease staging based on the manual for staging of cancer by the American Joint Committee on Cancer (AJCC). SSM: superficial spreading melanoma. NM: nodular melanoma. LMM: lentigo malignant melanoma. URGP: unclassified radial growth phases.

[b] Anatomic levels of invasion is based on Clark's microstaging criteria.

[c] L: left; R: right; LN: lymph node.

[d] Disease status as of August, 1997: NED: no evident disease; DOD: dead of disease; n.a.: information not available.

[e] D: tissue dissociation.

[f] D: DNA finger printing; H: HLA typing.

[g] C: cells in culture available; F: frozen vials available.

following surgery. In general, primary and early metastatic lesions are from untreated patients prior to surgery. However, patients with MM may have undergone chemotherapy or radiation therapy. Thus, cell lines derived from metastases may have been exposed to DNA-damaging agents (11).

2. PRIMARY CULTURE AND CELL LINE MAINTENANCE

Radial growth phase primary melanomas: Due to technical problems resulting from the small specimen size available for isolation, RGP melanoma cell lines are rarely established (3, 19). Epidermal melanoma cells are isolated from affected skin in a procedure similar to that described for normal melanocytes (20, 21). Briefly, skin samples are cut into small pieces and incubated in 0.25% trypsin-HBSS (Hanks buffered salt solution) at 4°C overnight. The epidermis is then peeled apart from the dermis with forceps and subject to treatment with cell dissociation solution containing 1.25 U/ml dispase, 0.1% (w/v) hyaluronidase, and 10% fetal calf serum (FCS) in MCDB153 medium supplemented with 2 mM $CaCl_2$ and mixed with L-15 medium at a 4:1 (v/v) ratio for 2–4 hr at 37°C, depending on the degree of cell disaggregation. After washing with HBSS, the resulting pellet is resuspended in W489 medium consisting of 4 parts MCDB153 supplemented with 2 mM Ca^{++} and 1 part L-15, 5 μg/ml insulin, 5 ng/ml EGF, 40 μg/ml bovine pituitary extract, and 2% fetal bovine serum with or without 10 ng/ml 12-O-tetradecanoyl phorbol-13-acetate (TPA) and then seeded in 24 well plates precoated with 1% gelatin in physiological saline. Melanoma cells at the dermal/epidermal junction and in the dermis are also collected by mechanically scraping the junctional area and by enzymatic digestion of the dermis with collagenase, hyaluronidase and dispase (22, 23).

Vertical growth phase primary melanomas: The culture success rate ranges from 30–70% for VGP melanomas, depending on the size of the lesion (9, 11). Two isolation techniques are employed for the VGP lesions. Thin, small specimens are subject to enzymatic dissociation as described for RGP lesions. Cells in thick or large samples are released mechanically by fine mincing with crossed scalpels in glass petri dishes. The resulting single cells and cell clumps are seeded at a range of densities into 24 well plates precoated with 1% gelatin in saline.

Metastatic melanomas: The most common sources of metastatic samples are lymph node metastases and, less frequently, cutaneous nodules. These samples are treated in the way described for thick or large VGP lesions with a high success rate of 75–80%. Both VGP and metastatic cells are initiated in the same medium as RGP cells but in the absence of TPA (11). The medium is changed twice a week and cells are split upon reaching confluence. Bovine pituitary extract and EGF in the medium can then be gradually removed once the cell lines are established.

3. PATHOLOGY (Table 2)

Morphology: No significant morphological differences exist between cultures of primary and metastatic melanoma cells (3, 9, 24). Morphological alteration may occur in a cell line over 10 or more passages due to the emergence of subpopulations. Such changes may be observed in the original mass cultures or in the subsequently derived clones, indicating morphological instability in melanoma (11). Moreover, the morphology may vary depending on growth conditions. For example, in the presence of TPA multiple dendricity is typical of melanoma cells of all stages (25).

Pigmentation: RGP cells are often pigmented, especially when grown in TPA-containing medium. Over time in culture, cells may lose their pigmentation and become pigmented only when not proliferating (11). VGP primary and MM cells are less frequently pigmented (20–30%) (25). However, cells of all stages contain premelanosomes and melanosomes as identified by electron microscopy and, in approximately 20% of the cultures, pigmentation can be induced experimentally by differentiation inducers such as dimethyl sulfoxide (DMSO) (25).

Growth characteristics: RGP primary cells have either an extended or an indefinite life span (26). Often they have properties in common with benign nevus cells, including growth dependency on exogenous mitogens such as insulin (or IGF-1) and bFGF, and low colony forming efficiency (CFE) in soft agar (ranging from 0.001–8%) (25, 26). Their response to phorbol esters is heterogeneous and, with few exceptions, most cultures are not stimulated.

Both VGP and MM cells grow permanently in culture (27). Metastatic cells grow more rapidly than VGP primary cells and have shorter doubling times. They reach higher densities, more often detach spontaneously from the substrate, and display growth autonomy regardless of exogenous stimulants. Metastatic cells have high CFE's of up to 50%, whereas cells derived from VGP primary lesions have intermediate CFE's of 4–20%.

VGP cells can be adapted gradually to grow in serum-free medium (28), but insulin is still essential for continuous proliferation (8). Once adapted, these cells have population doubling times of 24–72 hr, which are indistinguishable from those in serum-containing medium. Interestingly, metastatic cells exhibit growth autonomy independent of exogenous growth factors. We

Table 2 Tumor pathology and cellular characteristics in vitro and in vivo

Cell line	In vitro characteristics					In vivo characteristics	
	Morphology[a]	Pigmentation	Population doubling[b]	Growth factor and phorbol ester dependency[c]	Soft agar growth (CFE)	Tumorigenicity[d]	Invasion and metastasis[e]
RGP							
WM35	E	–	n.t.	I	~1.0%		–
WM1552C	S	–	n.t.	I	8.32%	S	n.t.
WM1575	S	–	n.t.	n.t.	n.t.	N	n.t.
WM1650	D	+	n.t.	I, F, P	<0.001%	SN	n.t.
WM1789	S	+	n.t.	I	<0.1%	S	n.t.
WM3208V	S	+	n.t.	I, F	1.28%	N	n.t.
WM3211	E	–	n.t.	I, F	0.24%	N	n.t.
VGP							
WM39	S	–	n.t.	I	19.1%	+	n.t.
WM75	S	+	0.36 (55 h)	I	10.8%	S	–
WM98-1	S	–	n.t.	I	n.t.	n.t.	n.t.
WM115	S	–	0.22 (94 h)	I	6.0%	S	n.t.
WM278	E	–	n.t.	I	4.2%	+	–
WM740V	S	–	n.t.	I	n.t.	n.t.	n.t.
WM793B	S	+	0.29	I	3.71–7.2%	S	–
WM853-2	S	–	n.t.	I	13.23%	+	n.t.
WM902B	E	–	0.31	I	6.5–14.17%	S	n.t.
WM983A	E	–	0.51	I	75%	RR	n.t.
WM1341D	S	–	n.t.	I	n.t.	n.t.	n.t.
WM1361A	E	–	0.35	I	0.8%	S	n.t.
WM3268V	S	+	n.t.	I	n.t.	n.t.	n.t.
Metastatic							
WM9	E, S	–	n.t.	–	31.82%	+	–

Continued on next page

Table 2 (continued)

Cell line	In vitro characteristics					In vivo characteristics	
	Morphology[a]	Pigmentation	Population doubling[b]	Growth factor and phorbol ester dependency[c]	Soft agar growth (CFE)	Tumorigenicity[d]	Invasion and metastasis[e]
WM46	D, S	+	n.t.	–	n.t.	+	n.t.
WM164[f]	S	–	(71–222 h)	–	2.0%	n.t.	–
WM165-1	E	–	(120 h)	–	n.t.	n.t.	n.t.
WM209	D	–	n.t.	n.t.	n.t.	n.t.	n.t.
WM239A	E, S	–	0.38 (96 h)	–	9.0%	R	n.t.
WM266-4	E	–	n.t.	–	n.t.	n.t.	n.t.
WM373	S	+	0.3 (24 h)	–	29.8%	R	n.t.
WM852[g]	E, S	–	n.t.	–	22.7%	+	–
WM858	E	–	n.t.	–	n.t.	n.t.	n.t.
WM873-1	E	–	n.t.	–	n.t.	n.t.	n.t.
WM983B	E	–	0.65	–	85%	RR	n.t.
WM983C	E	–	n.t.	–	72%	n.t.	n.t.
WM1158[h]	S	+	n.t.	–	n.t.	n.t.	n.t.
WM1361C	E	–	0.29	–	1%	S	n.t.
WM1617	E	–	n.t.	n.t.	n.t.	n.t.	n.t.
WM1852	E	–	n.t.	n.t.	n.t.	n.t.	n.t.
WM2067	S	–	n.t.	n.t.	n.t.	n.t.	n.t.
WM3214	CS	–	n.t.	n.t.	n.t.	n.t.	n.t.
WM3259	S	–	n.t.	n.t.	n.t.	n.t.	n.t.
WM3268 met2	S	+	n.t.	n.t.	n.t.	n.t.	n.t.
451Lu	S	–	(93–94 h)	–	17%	+	+
1205Lu	S	–	n.t.	–	25%	+	+

[a] E: epithelioid; S: spindle; D: dendritic; CS: clumps in suspension; [b] Figures shown are population doubling per day and population doubling time in hours at log phase; [c] I: insulin; F: bFGF; P: phorbol ester; n.t.: not tested; [d] Total of 5×10^6 cells/mouse were injected. Tumor growth was monitored after 10 weeks: S = <0.5 g; R = 0.5 to 1.0 g; RR = >1.0 g; + = presence of tumors but sizes not available; - = no tumor growth; [e] Invasiveness is assessed by a modified Boyden chamber assay (10); metastatic potential was determined by lung or liver metastasis following subcutaneous injection of tumor cells; [f] Phenotypic instability; [g] Several additional cell lines from skin metastases available, ras mutations; [h] Diploid with triple translocation

have successfully maintained cells derived from metastases for more than 2 years in culture by supplying only amino acids, vitamins, salts, and sugars (29). This remarkable independence is probably due to the production of endogenous factors that are self- or mutual-stimulatory through an autocrine or paracrine pathway. Aberrant secretion of growth factors in melanoma includes upregulation of TGF-α (30), TGF-β (31), PDGF-A and -B (32), IL-1 (33), bFGF (34), and MGSA (35). There is evidence that production of some of these factors is beneficial to the tumor (36-38).

Unlike some RGP cells where phorbol ester is growth-stimulatory, VGP and metastatic melanoma cells are inhibited by TPA. Although the mechanism of this reversal in response to phorbol ester during melanoma progression is not clear, there is evidence of differences in protein kinase C activation upon TPA stimulation in melanocytic cells (39).

In vivo: tumorigenicity and metastasis: VGP and MM cells form tumors in immunodeficient mice when injected subcutaneously (25-27). Most cell lines grow locally without spontaneous metastasis. However, due to their genetic instability, highly metastatic variants can be selected by continuous passage in vitro (28) and in animals (18). RGP melanoma cells either do not grow in vivo, or form small tumor nodules weighing less than 100 μg by 100 days (6).

451Lu and 1205Lu were derived from lung metastases of WM164 (MM) and WM793 (VGP) respectively, after subcutaneous injection into immunodeficient mice. These cells are highly invasive and exhibit spontaneous metastasis to lung and liver (18).

WM3214 grows in suspension as cell clumps with less than 2% of cells attaching to the substratum. When injected intravenously into immunodeficient mice, these cells formed lung metastases but were not tumorigenic after subcutaneous injection (unpublished data).

4. CYTOGENETICS AND HLA TYPING

Cytogenetic studies of human melanomas (including multiple specimens of different disease stages collected over time from the same patient) have provided direct evidence on the relation between somatic genetic changes and the biologic and clinical evolution of the disease.

Frequent genetic alterations, which involve but are not limited to chromosomes 1, 6, 7, and 10, have been reported in human melanoma (Table 3) (5, 40–44). These non-random abnormalities are deletions, translocations, or amplifications (5, 43). Other abnormalities detected in the listed cell lines include: del2p (21) in WM793 (44); del2p (21), 3p+ (43), +5, 9p+ (21), del9(p13), del12(p13), -13, 15p+ (43), +18 (43), +22 in WM983A, B, and C;

del2p (13), del4(q21) in WM35; t(2;?)(q37;?); -8, t(11;?)(p15;?), t(12;?)(q24;?), -18, t(22;?)(p13;?) in WM1341D; 4p+ in WM46; t(5;?)(q34;?), t(11;?)(p11;?), t(12;?)(q11;?) in WM1552C; 9p+ in WM373 (43); 9p+, 19p+ in WM75 (43); 11p+, 16q+ in WM115, 165–1, 239A and B, 266–1 and -3 (43); 3p-; t(3p;?); 12p+ (13), t(13;22)(p11;p11) in WM9; t(12;?)(p12;?), 14p+ in WM1361A; 12p+ in WM740V; 16q+ in WM39 (44).

Comparison of primary and metastatic cells from the same patient often reveals identical abnormalities, but metastatic cells frequently exhibit additional abnormalities. The progressive accumulation of genetic changes in addition to preexisting abnormalities in cultured metastatic cells suggests a clonal evolution of human melanoma.

Cell lines may have the entire chromosome 6 missing (WM46), or portions of either the short arm (WM115 and the multiple metastases from the same patient [WM165, 239 and 266]) or the long arm (WM9). Many melanoma cell lines also carry an extra chromosome 7 (WM9, 39, 115, 873 and 1361). Deletions on chromosome 9 are found in WM983 cells and WM1158 has a three-way translocation involving chromosome 10.

The HLA type of several cell lines has been studied (Table 4). These cell lines have been extensively used for immunological analyses.

5. PHENOTYPIC INSTABILITY

In general, cultured melanoma cells maintain their properties in a stable manner (9), indicated by chromosomal analyses of these cells at the time of isolation, after 12 passages in culture and after prolonged (more than 6 months) culture. However, spontaneous transformation towards a more malignant phenotype may occur in RGP primary melanoma cells, as seen in WM35 (4). These cells have an indefinite life span and their xenografts grow slowly in immunodeficient mice.

6. MULTIPLE CELL LINES FROM THE SAME PATIENT

WM373 (MM) was isolated from a cutaneous metastasis of the same patient as WM75 (VGP). WM165-1 and -2, WM239A and B, and WM266-1, 2, 3, 4 (MM) were established from lymph node and cutaneous metastases of an individual who developed metastatic disease from the lesion used to derive WM115 (VGP). WM1617 (MM) was from the same patient as WM278. Further lesions from the patient from whom WM740V (VGP) originated gave rise to WM858, and 873-1,2 and 3 (MM). The patient with the lesion used to derive WM983A (VGP) had lymph node metastases

Table 3 Common chromosomal abnormalities in human melanoma cell lines

Chromosome	Type of abnormalities	Involved cell lines
1	+1	WM39, 793
	+1t(1;19)(q12;p13)	WM39, 793
	1p+	WM983A, B, C
	1p-	873-1,2,3, [del1(p11)], 902B [del1(p11)], 853-2 [del1(p11)], 46 [del1(p12)], 9 [del1(p22)], 39[del1(p22)], 858
	1q-	WM983A, B, C [del1(q11)], 873-2 [del1(q11)], 278 [del1(q24)], 858 [del1(q42)], 373 [random]
	t(1;13)(q25;q34)	WM39
	t(1;14)(p11;q11)	WM1361A, C
	t(1;?)(p22;?)	WM1552C
	t(1;?)(p35;?)	WM46
6	-6	WM46
	6q+	WM983A, B, C [6q+(27)]
	6p-	WM115 [del6(p12)], 165-1 [del6(p12)], 239A, B [del6(p12)], 266-1, -3 [del6(p12)], 873-1 [del6(p21)]
	6q-	WM9 [del6(q12)], 740V [del6(q15)], 1361A, C [del6(q15)], 35 [del6(q21)], 858 [del6(q21)], 873-1, -2, -3 [del6(q21)], 902B [del6(q21)], 983A, B, C [del6(q25)], 278 [inconstant], 853-2, 39
	t(5;6;10)(q31;q14;q23q25)	WM1158
7	+7	WM9, 39, 115, 164, 165-1, 239A, B, 266-1, -3, -4, 793, 853-2, 858, 873-1, -2, -3, 1361A, C
	+7p+	WM39 [+7p+(p15)]
	+7q-	WM983A, B, C [+7q-(q11)]
	iso7p	WM793
	iso7q	WM35
	+t(7;?)(p11;?)	WM1341D
	7q+	WM902B [7q+(q36)]
	7q-	WM75 [del7(q22)], 373 [del7(q22)]
9	9p+	WM983A, B, C [9p+(p21)], 75, 373
	9p-	WM983A, B, C [del9(p13)]

Continued on next page

Table 3 (continued)

Chromosome	Type of abnormalities	Involved cell lines
10	10q+	WM75, 373
	-10	WM115, 165-1, 239A, 266-1, 740V, 858, 873-1
	t(10;13)	WM39, 873-1
	t(5;6;10)(q31;q15;q23q25)	WM1158

Table 4 HLA types of melanoma cell lines[a]

Cell line	HLA type
WM793	A1, A29, B57(17), B35, DR5.5
WM75	A2, A29, B12w44, DR4, DR7
WM98	A1, A3, B8, DR3
WM164	A24, B7, C7, DR13, DQ1, DQ6, DRw52
WM1158	A11, A24, B16, B60(40), C3, DR13, DR4, DQ3, DQ6, DRw52, DRw53
WM35	A2, B18, B51, C2, C1203, DRβ107, DR1602, DQβ, DQ0303.05, DRβ4*, DR0101.5*02
WM3248	A1, A3, B35, B56, C1, C4, DR1, DR11, DQ3, DQ5, DRb52
WM902B	A1, A29, B08, B07, C0701, C15, DRβ1, 0801, DRβ1 1104, DQβ1 0301, DQβ.04, DRβ3*0202 (DRw52)

[a] We thank F. Marincola, NCI, Bethesda, MD., and D. Guerry, University of Pennsylvania, PA, for information on HLA type

removed at the time the primary lesion was excised: WM983B and C (MM) were isolated from these metastatic lesions. Metastases were detected in the patients from whom WM1361A (VGP) and WM3268V (VGP) were derived. The cells from the metastatic lesions were designated WM1361B and C, and WM3268met2, respectively.

These cell pairs/groups were routinely monitored by DNA fingerprinting and HLA typing. Identical results were obtained from within these pairs/ groups which further confirmed the origin of these cell lines (6).

REFERENCES

1. Clark, W. H. J. Brit. J. Cancer 64: 631, 1991.
2. Clark, W. H. J. et al. Hum. Pathol. 15: 1147, 1984.
3. Herlyn, M. et al. Cancer Res. 45: 5670, 1985.
4. Herlyn, M. et al. Cancer Res. 40: 3602, 1980.
5. Balaban, G. et al. Cancer Genet. Cytogenet. 11: 428, 1984.
6. Satyamoorthy, K. et al. Melanoma Res. 7: S35, 1997.
7. Easty, D. J., Herlyn, M., and Bennett, D. C. Int. J. Cancer 60: 129, 1995.
8. Rodeck, U. et al. Int. J. Cancer 40: 687, 1987.
9. Herlyn, M. et al. J. Natl. Cancer Inst. 74: 283, 1985.
10. Kath, R. et al. Cancer Ther. Control 1: 179, 1990.
11. Györfi, T. and Herlyn, M. In: R. J. Hay et al. (eds.), Atlas of Human Tumor Cell Lines. p. 413, Acad. Press, 1994.
12. Valyi-Nagy, I. et al. Int. J. Cancer 54: 159, 1993.
13. Zehngebot, L. M. et al. Cancer Immunol. Immunother. 16: 30, 1983.
14. Herlyn, D. et al. Cancer Res. 50: 2296, 1990.
15. Becker, D., Beebe, S. J., and Herlyn, M. Oncogene 5: 1133, 1990.
16. Menrad, A. et al. Cancer Res. 53: 1450, 1993.
17. Parmiter, A. H. and Nowell, P. C. Cancer Treatment & Res. 43: 47, 1988.

18. Herlyn, D. et al. In: L. Nathanson (ed.) Melanoma Research: Genetics, Growth Factors, Metastases, and Antigens. p. 105. Boston, MA: Kluwer Academic Publishers, 1991.
19. Herlyn, M. et al. Adv. Cancer Res. 54: 213, 1990.
20. Hsu, M.-Y. and Herlyn, M. In: G. E. Jones (ed.) Methods in Molecular Medicine: Human Cell Culture Protocols. p. 9. Totowa, NJ: Humana Press Inc., 1996.
21. Herlyn, M et al. Cancer Res. 47: 3057, 1987.
22. Mancianti, M. L. et al. J. Invest. Dermatol. 90: 134, 1988.
23. Mancianti, M. L. et al. Am. J. Pathol. 136: 817, 1990.
24. Shih, I.-M. and Herlyn, M. J. Invest. Dermatol. 100: 196s, 1992.
25. Herlyn, M et al. Pigment Cell 8: 166, 1987.
26. Valyi-Nagy, I. et al. In: O. Sudilovsky, et al. (ed.) Boundaries between promotion and progression during carcinogenesis. p. 315. New York, NY: Plenum Press, 1991.
27. Mancianti, M. L. and Herlyn, M. In: C. J. Conti, et al. (ed.) Skin Tumors: Experimental and Clinical Aspects. p. 369. New York, NY: Raven Press, 1989.
28. Kath, R. et al. Cancer Res. 51: 2205, 1991.
29. Herlyn, M et al. Lab. Invest. 56: 461, 1987.
30. Marquard, H. and Todaro, D. J. Biol. Chem. 257: 5220, 1982.
31. DeLarco, J. E., Pigott, D. A., and Lazarus, J. A. Proc. Natl. Acad. Sci. USA 82: 5015, 1985.
32. Westermark, B. et al. Proc. Natl. Acad. Sci. USA 83: 7197, 1986.
33. Bennicelli, J. L. et al. Cancer Res. 49: 930, 1989.
34. Halaban, R., et al. Oncogene Res. 3: 177, 1988.
35. Richmond, A. et al. EMBO J. 7: 2025, 1988.
36. Shih, I.-M. and Herlyn, M. In Vivo. 8: 113, 1994.
37. Rodeck, U. and Herlyn, M. Cancer Met. Rev. 10: 89, 1991.
38. Rodeck, U. et al J. Invest. Dermatol. 97: 20, 1991.
39. Brooks, G. et al. Cancer Res. 51: 3281, 1991.
40. Becher, R. et al. Cancer Res. 43: 5010, 1983.
41. Pathak, S., Drwinga, H. L., and Hsu, T. C. Cytogenet. Cell Genet. 36: 573, 1983.
42. Trent, J. M., Rosenfeld, S. B., and Meyskens, F. L. Cancer Genet. Cytogenet. 8: 177, 1983.
43. Balaban, G. et al. Cancer Genet. Cytogenet. 19: 113, 1986.
44. Parmiter, A. H. et al. Cancer Res. 46: 1526, 1986.

Chapter 15

Melanoma: Brussels Melanoma Cell Lines

Francis Brasseur
Ludwig Institute for Cancer Research, Brussels Branch, 74 avenue Hippocrate, B-1200, Brussels, Belgium, Tel: 0032-2-764-7456; Fax: 0032-2-762-9405; E-mail: brasseur@licr.ucl.ac.be

Human melanoma cell lines have been widely used for investigations in tumor immunology. Metastatic melanoma cells are relatively easy to adapt to culture, providing tumor lines that are convenient stimulator and target cells for T lymphocytes. Here we present several metastatic melanoma cell lines that have been established in our laboratory to be used for the identification of tumor antigens recognized by autologous cytolytic T lymphocytes (CTL). These antigens consist of a complex between a peptide, produced through degradation of an intracellular protein, and a human leukocyte antigen (HLA) class I molecule (HLA-A, -B or -C) which presents the peptide to CTL.

Autologous anti-tumor CTL can be obtained by stimulating *in vitro* blood-derived lymphocytes, or tumor-infiltrating lymphocytes, with irradiated tumor cells of the same patient in the presence of T cell growth factors. The responder T cell populations, which consist mostly of $CD8^+$ T cells, lyse specifically the tumor cells (reviewed in [1]). From these responder populations, it is possible to isolate tumor-specific CTL clones that can be maintained *in vitro* for long periods of time. By using an approach based on gene transfection and recognition of the transfectants by the relevant CTL clones [2], we have cloned several genes encoding tumor antigens expressed on melanoma lines. This accomplishment has led to the molecular characterization of these antigens. The first characterized human tumor antigen was found to be a nine amino acid peptide derived from the protein encoded by a new gene called *MAGE-1* ('M' for melanoma and 'AGE' for antigen) and bound to HLA-A1 molecules [3–4].

Since then, the number of identified antigens has increased steadily. On the basis of their pattern of expression, the antigens that we and others have so

J.R.W. Masters and B. Palsson (eds.), Human Cell Culture Vol. 1, 275–282.

far characterized on melanomas fall into three groups: (1) highly specific tumor antigens encoded by genes, such as *MAGE-1*, *MAGE-3*, *BAGE* and *GAGE*, that are expressed in a significant proportion of tumors of various histological types but not in normal adult tissues except for testis; (2) differentiation antigens encoded by genes that are expressed only in melanomas, but also in normal melanocytes, such as tyrosinase, Melan-$A^{MART\text{-}1}$, $gp100^{Pmel17}$, and $gp75^{TRP1}$; and (3) antigens unique to individual tumors, or restricted to very few tumors, that appear through tumor-specific mutations in genes that are expressed ubiquitously (reviewed in [5–6]).

Since the identification of human tumor antigens and their encoding genes, the search for an effective specific immunotherapy as an adjuvant treatment for cancer has entered a new phase. First, it is now possible to select patients whose tumors express a given antigen. To be eligible, the tumor should express the relevant peptide-encoding gene along with the appropriate HLA class I specificity, and this can be tested readily by reverse transcription-PCR on RNA extracted from a tumor biopsy. Second, the knowledge of the molecular nature of tumor antigens allows the rational design of highly specific vaccine preparations. These may consist of engineered cells expressing the antigens, or of antigenic peptides combined with appropriate adjuvants. As the genes encoding the antigens are available, recombinant proteins or recombinant defective viruses can also be prepared and used for immunization. Inoculation of DNA encoding the antigen is yet another possibility. It is too early to predict the clinical outcome of such strategies based on tumor antigens targeted by CTL. Preliminary results with a peptide encoded by the *MAGE-3* gene are, however, promising [7]. It is our hope that anti-tumor responses will be obtained in some patients in the early clinical trials, and that the careful study of the lymphocytes and the tumor cells of these patients will produce a rich harvest of additional tumor-associated antigens, and a better understanding of what constitutes an effective anti-tumor response. With this in view, it is important to derive cell lines from the tumors of patients who enter clinical trials of immunotherapy. These are mostly melanoma patients at the present time, and we are making a systematic effort to adapt their tumor cells to culture by using the procedures described below.

1. CULTURE CONDITIONS

Medium – Melanoma cells are cultured in Iscove's Modified Dulbecco's Medium containing penicillin (200 U/ml) and streptomycin (100 μg/ml), supplemented with fetal calf serum (10% v/v) from a selected batch, L-arginine (116 μg/ml), L-asparagine (36 μg/ml), L-glutamine (219 μg/ml), and with the additives present in HITES medium [8], which are hydrocortisone (10 nM), insulin (5 μg/ml), transferrin (100 μg/ml), 17β-estradiol (10 nM), and sodium selenite (30 nM). Cultures are kept at 37°C in a humidified atmosphere of 6% CO_2 in air.

Primary culture – Melanoma specimens are collected at surgery into sterile physiological saline and stored on ice until they reach the laboratory. They are processed immediately. Obvious non-tumor tissue and areas of gross necrosis are removed. Where there is sufficient material, a tumor piece is snap frozen in liquid nitrogen and stored at –80°C. The remainder is transferred to a Petri dish with culture medium and disaggregated mechanically, either by sieving the tissue through a stainless steel mesh (Cellector, mesh 60, Bellco Biotechnology, Vineland, NJ) or, for cutaneous biopsies in particular, by mincing the tissue into pieces of approximately 1 mm^3 using crossed scalpels. After centrifugation of the resulting suspension at 100 g for 4 min, the pellet, which usually contains many tumor cell aggregates and small tissue fragments in addition to single cells, is plated at relatively high density onto freshly irradiated (100 Gray from a cesium source) 50%-confluent mouse fibroblast feeders (BALB/c 3T3 cells; [9]), in 2–4 25 cm^2 culture flasks containing 5 ml of medium. Where there is sufficient material, a portion of the cell suspension is mixed with dimethyl sulfoxide (10% v/v) and preserved at –80°C. Cultures are given a 1:1 medium change twice a week. The old medium is spun briefly (1 min at 40 g), the supernatant containing tissue debris and dead cells is discarded, and the pellet, consisting usually of cell aggregates with a bright appearance under the inverted microscope, is replated onto the original flask or onto fresh 3T3 feeders in a new flask. If no sign of melanoma cell growth is observed after 1 month of culture the flasks are discarded.

Subculture – Cultures are only passaged when they are nearly confluent. Incubation at 37°C in 0.05% trypsin – 0.02% EDTA solution (2 ml/25 cm^2) for 2–5 min is an easy and effective method for removing cells from the culture surface. This method does not appear to be harmful to melanoma cells. After addition of an equal volume of fresh culture medium to inactivate the trypsin, the cell suspension is spun at 200 g for 5 min and the pellet is replated onto 25 cm^2 flasks in the presence or absence of 3T3 feeders, at split ratios not exceeding 1:4. On subsequent passages of the cultures, progressively increased split ratios are tested, up to 1:10.

The use of 3T3 feeders is an effective way of obtaining some enhancement of attachment and spreading of melanoma cells, and of reducing growth of contaminating fibroblasts. Other procedures which aim to eliminate fibroblasts from tumor cell cultures, including mechanical scraping, differential trypsinization, the use of reduced serum concentration, or the isolation of tumor cells by limiting dilution cloning, have given poor results with melanoma in our experience. Mechanical scraping using a rubber policeman has been useful to recover well-isolated melanoma cell colonies from cultures with low contents of fibroblasts. In situations where melanomas grow as aggregates of cells loosely attached to the culture surface, selectively passaging the tumor cells is easy since gentle pipetting of the cultures will release them into the medium without removing fibroblasts from the culture surface. However, when using this method, one should be aware of the risk of Epstein-Barr virus-transformed B lymphoblastoid lines emerging, especially with cultures derived from lymph-nodes, because such lines grow as floating cells.

In our experience, 5–10 passages are usually required before melanoma cells can be deprived of 3T3 feeders and thereafter give rise to a fibroblast-free culture which will eventually establish a permanent cell line. Using these procedures we achieve a success rate of around 35% for establishment of melanoma lines from metastatic lesions.

Xenografts – Cell lines can be derived *in vitro* from melanoma specimens grown as xenografts in athymic nude mice for 1–2 passages. We found that of 21 metastatic melanoma lesions that could not be established directly in culture, 16 (76%) were established as xenografts in mice. Some of these tumors (3 out of 7) were then successfully cultured from xenografted material to give rise to melanoma cell lines.

Isolation of TIL - Tumor-infiltrating lymphocytes (TIL) that may be present in primary cultures are isolated by subjecting the supernatant to density-gradient centrifugation using Lymphoprep (600 g for 20 min), 1 to 3 days after the initiation of culture. TIL and the few melanoma cells that lie at the medium/Lymphoprep interface are transferred to culture medium containing human serum (10% v/v) instead of fetal calf serum. After washing by centrifugation (200 g for 5 min), the cell pellet is cultured in medium containing human serum (10% v/v) and interleukin-2 (25 U/ml) to allow TIL to proliferate (HITES additives are not necessary). After a few days of culture, the cells are harvested, counted, washed by centrifugation, resuspended in ice-cold freezing medium (45% Iscove's medium plus antibiotics, 45% human serum, and 10% dimethyl sulfoxide) and preserved at –80°C in vials containing $0.5–10 \times 10^6$ cells.

2. CONTINUOUS CELL LINES

We have established cell lines from metastatic melanomas of stage III-IV patients, and transformed blood-derived lymphocytes of some of these patients with Epstein-Barr virus to obtain lymphoblastoid cell lines (Table 1). Melanoma lines LB33-MEL.A, LB33-MEL.B and LB33-MEL.C originate from metastases that were removed from patient LB33 in 1988, 1993 and 1994, respectively. MZ2-MEL.3.0 and MZ2-MEL.3.1 are clonal sublines that we derived from the original MZ2-MEL line established by Dr A. Knuth (Mainz, Germany).

Table 1 Cell lines of metastatic melanoma[1]

Cell line designation	Patient sex and age		Primary site	Specimen site	Pathological staging[2]	Culture method[3]	Doubling time (h)[4]	Patient HLA class I phenotype	EBV-B line[5]	References
BB74-MEL	F	49	Head and neck	Adrenal gland	N1M1b IV	D	48	A1 A29 B35 B44 Cw*1601		
LB24-MEL	F	61	Lower limb	In-transit metastasis	N2bM0 III	D	55	A2 A25 B18 B*4402 Cw5		[10]
LB30-MEL	M	61	Trunk	Axillary lymph-node	N+M0 III	D	45	A3 A11 B27 B70 Cw2 Cw7		
LB33-MEL	F		Lower limb					A*2402 A*68012 B*1302 B*4402 Cw*0601 Cw*0704	yes	[11–13]
LB33-MEL.A		42		Subcutaneous metastasis, trunk	M1a IV	X	40			
LB33-MEL.B		47		Small bowel	M1b IV	D	52			
LB33-MEL.C		48		Muscular metastasis, trunk	M1b IV	D	65			
LB34-MEL	F	65	Lower limb	In-transit metastasis	N2bM0 III	D	36	A1 A24 B7 B35 Cw4 Cw7		
LB39-MEL	F	41	Upper limb	Subcutaneous metastasis trunk	M1b IV	X	72	A2 A24 B7 B51 Cw1 Cw7	yes	[14]
LB41-MEL	M	41	Head and neck	Cervical lymph-node	N2aM0 III	D	48	A2 A32 B13 B60 Cw3 Cw6	yes	
LB45-MEL	M	56	Trunk	Axillary lymph-node	N+M0 III	X	38	A1 B8 Cw7	yes	
LB265-MEL	F	71	Trunk	Cervical lymph-node	M1a IV	D	>100	A2 A3 B7 B62 Cw7 Cw9		[15]
LB278-MEL	F	59	Unknown	Axillary lymph-node	M1a IV	D	100	A1 B8 B49		
LB373-MEL	F	32	Lower limb	In-transit metastasis	N2cM0 III	D	50	A2 A11 B22 B*4402 Cw5 Cw9	yes	
LB929-MEL	M	77	Lower limb	Subcutaneous metastasis, trunk	M1b IV	D	64	A3 A30 B*4403 B57 Cw6 Cw*1601		
LB951-MEL	F	41	Trunk	In-transit metastasis	M1b IV	D	64	A2 A24 B37 B56 Cw1		
LB1319-MEL	M	72	Lower limb	Inguinal lymph-node	N+M0 III	D	52	A1 A2 B39 B44 Cw5	yes	
LB1622-MEL	F	49	Lower limb	In-transit metastasis	N2cM0 III	D	68	A1 A2 B35 B51 Cw4	yes	
MZ2-MEL	F		Unknown	Adrenal gland	M1b IV	D		A1 A29 B37 B*4403 Cw6 Cw*1601	yes	[3–4,16–21]
MZ2-MEL.3.0						C	30			
MZ2-MEL.3.1						C	30			
TOU2-MEL	M	46	Trunk	Submuscular metastasis, trunk	M1b IV	X	86	A1 A26 B38 B39		

[1] The HLA class I phenotypes of some lines are known to be altered (for details, see text). [2] AJCC classification, 1992 [22]. [3] D, direct culture from clinical material; X, culture from nude mouse xenograft; C, clonal subline of MZ2-MEL. [4] Doubling time calculated according to the formula 7.225 x J / Log_{10} (N/N_0), where N_0 is the number of cells seeded in the culture flask and N is the number of cells in the flask after J days of culture. [5] A lymphoblastoid cell line obtained by transformation of blood lymphocytes with Epstein-Barr virus is available.

Most of the melanoma lines that we have established express several genes of the *MAGE, BAGE* and *GAGE* families, and the genes encoding tyrosinase, Melan-A and gp100. Remarkably, the patterns of gene expression found in the melanoma lines and the corresponding original melanoma lesions appear similar (Table 2). The antigens encoded by these genes are presented by various HLA class I molecules (reviewed in [6]). Gene *MAGE-1* encodes two different antigenic peptides presented by HLA-A1 and HLA-Cw16. *MAGE-3* encodes three distinct peptides that bind to HLA-A1, A2 and B44. The *BAGE* gene encodes a peptide presented by HLA-Cw16. Genes *GAGE-1* and 2 code

Table 2. Expression of genes *MAGE, BAGE, GAGE, Tyrosinase, Melan-A* and *gp100* in cell lines and fresh biopsies of melanomas[1]

	MAGE						*BAGE*	*GAGE*		*Tyrosi-nase*	*Melan-A*	*gp100*
	1	*2*	*3*	*4*	*6*	*12*		*1–2*	*3–6*			
Cell lines												
BB74-MEL	+++	+++	+++	+++	+++	+	+	–	+	+++	+++	+++
LB24-MEL	+++	+++	+++	+	++	+++	+++	+++	+++	+++	+++	++
LB30-MEL	+++	+++	+++	+++	+++	+++	+	++	+	–	–	–
LB33-MEL.A	–	+++	+++	–	+++	+	–	–	–	–	–	–
LB33-MEL.B	–	+++	+++	–	+++	+	+	–	–	+++	++	++
LB33-MEL.C	+	+++	+++	–	+++	+++	+	–	–	+++	+++	+
LB34-MEL	+	+++	+++	+++	+++	+	+	–	–	++	+	–
LB39-MEL	+	+	+++	–	+++	++	–	+++	++	+++	+++	+
LB41-MEL	–	–	–	–	+	+	–	–	–	–	–	–
LB45-MEL	–	–	–	–	–	–	+	–	–	++	+	++
LB265-MEL	–	–	–	–	–	–	–	–	–	+++	+++	+++
LB278-MEL	–	++	+++	+	+++	++	+	+++	+++	+	–	–
LB373-MEL	+++	+++	+++	+++	+++	+++	+++	–	–	+	+	+
LB929-MEL	+	+++	+++	–	+++	++	+++	+++	+++	+++	+++	+
LB951-MEL	+	+	+	–	+++	+	–	–	–	+++	+++	+++
LB1319-MEL	+	+++	+++	–	+++	+	–	–	–	+++	+++	+
LB1622-MEL	+++	+++	+++	+++	+++	++	+++	+++	+++	+	+	+
MZ2-MEL.3.0	+++	+++	+++	–	+++	+	+++	+++	+++	+	–	+
TOU2-MEL	++	–	+	–	–	–	–	–	–	+++	+++	+++
Biopsies												
BB74-MEL	++	+++	+++	+++	+	++	+	+	+	+++	+++	–
LB33-MEL.B	–	+++	+++	–	+++	++	++	–	–	+++	++	+
LB33-MEL.C	+	+++	+++	–	+++	+	–	–	–	+++	+++	–
LB265-MEL	–	–	–	–	–	–	–	–	–	+++	+	+++
LB929-MEL	+	+++	+++	–	+++	+++	+++	+++	+++	+++	++	++
LB951-MEL	+	+	++	–	+++	+	–	–	–	+++	+++	+++
LB1319-MEL	–	++	+++	–	+++	+	–	–	–	+++	+++	+
LB1622-MEL	+++	+++	+++	+++	+++	++	+++	+++	+++	+++	++	+
TOU2-MEL	+	–	+	–	–	+	–	–	–	+++	+++	+++

[1] Expression was determined by reverse transcription-PCR with specific primers and scored according to band intensity of PCR products fractionated in agarose gels. The symbols represent expression levels that were approximately the following, in per cent of those found in the reference cell lines (melanoma MZ2-MEL.3.0 for *MAGE-1, 2, 3, 6, BAGE* and *GAGE*, sarcoma LB23-SAR for *MAGE-4*, melanoma LB373-MEL for *MAGE-12*, and melanoma SK-MEL-23 for *Tyrosinase, Melan-A* and *gp100*): –, <1% (no gene expression detected using our conditions); +, 1–10%; ++, 10–50%; +++, 50–200%.

for the same antigenic peptide presented by HLA-Cw6. Most of the melanocytic differentiation antigens encoded by *Tyrosinase, Melan-A* and *gp100* are presented by HLA-A2, but other HLA-peptide combinations have been found.

Loss of HLA expression is a frequent phenomenon in tumors. We have not systematically assessed whether the HLA class I phenotypes of our melanoma lines are altered. Loss of an HLA haplotype (A29, B44, Cw16) has been found in subline MZ2-MEL.3.1, but not in subline MZ2-MEL.3.0. LB33-MEL.A cells express the products of all six HLA class I alleles, whereas LB33-MEL.B cells have retained expression of a single HLA class I allelic product (A24). FACS analysis with monoclonal antibodies W6/32 (anti-HLA-A, -B, -C) and BB7.2 (anti-HLA-A2) has shown that HLA class I expression in BB74-MEL and LB1622-MEL cells is lost completely. This loss of expression was also observed in immunostained cryostat sections of the lesions from which these two lines originate (F. Garrido, personal communication). Absence of HLA-A2 expression, but not of HLA class I expression, in LB41-MEL cells has also been shown using the monoclonal antibodies W6/32 and BB7.2. Another consideration is that tumor antigens presented by HLA-A2 are recognized by the relevant CTL clones on LB24-MEL, LB39-MEL, LB265-MEL and LB373-MEL, indicating that these cells express functional HLA-A2 molecules. Similarly, CTL recognize specific antigens presented by HLA-B44 on LB33-MEL.A, LB33-MEL.C and LB373-MEL, and by HLA-A1 on LB34-MEL.

DNA fingerprinting has shown that each melanoma line reported in Table 1 is of the same genetic origin as blood mononuclear cells of the patient. Samples of Mycoplasma screened cells (MycoTect kit, Gibco BRL) of all these lines but one (the original MZ2-MEL) are available from our frozen stocks.

ACKNOWLEDGMENTS

I am particularly grateful to Madeleine Swinarska who established several melanoma cell lines described in this chapter. I would also like to thank Dr Marie Marchand for providing clinical data, and Dr Pierre van der Bruggen for his helpful comments.

REFERENCES

1. Boon T et al, *Annu Rev Immunol*, 12:337, 1994.
2. De Plaen E et al, in: I Lefkovits (ed), *The Immunology Methods Manual*, pp 692-718, Academic Press Ltd, 1997.

3. van der Bruggen P et al, *Science*, 254:1643, 1991.
4. Traversari C et al, *J Exp Med*, 176:1453, 1992.
5. Van Pel A et al, *Immunological reviews*, 145:229, 1995.
6. Boon T and van der Bruggen P, *J Exp Med*, 183:725, 1996.
7. Marchand M et al, *Int J Cancer*, 63:883, 1995.
8. Carney D et al., *Proc Natl Acad Sci USA*, 78:3185, 1981.
9. Vessière-Louveaux F et al, *Int J Cancer*, 35:231, 1985.
10. Brichard V et al, *J Exp Med*, 178:489, 1993.
11. Lehmann F et al, *Eur J Immunol*, 25:340, 1995.
12. Coulie PG et al, *Proc Natl Acad Sci USA*, 92:7976, 1995.
13. Ikeda H et al, *Immunity*, 6:199, 1997.
14. Coulie P G et al, *J Exp Med*, 180:35, 1994.
15. Zarour H et al, *J Invest Dermatol*, 107:63, 1996.
16. Van den Eynde B et al., *Int J Cancer*, 44:634, 1989.
17. van der Bruggen P et al, *Eur J Immunol*, 24:2134, 1994.
18. Gaugler B et al, *J Exp Med*, 179:921, 1994.
19. Boël P et al, *Immunity*, 2:167, 1995.
20. Van den Eynde B et al., *J Exp Med*, 182:689, 1995.
21. Brichard V et al, *Eur J Immunol*, 26:224, 1996.
22. AJCC (American Joint Committee on Cancer), *Manual for Staging of Cancer*, 4th ed, pp 143-148, J B Lippincott, Philadelphia, 1992.

Chapter 16

Melanoma: The Milan Melanoma Cell Lines

Andrea Anichini, Roberta Mortarini, Claudia Vegetti, Alessandra Molla, Alessandra Borri, Giorgio Parmiani

Division of Experimental Oncology D, Istituto Nazionale per lo Studio e la Cura dei Tumori, Via Venezian I, Milan 20133, Italy. Tel: 0039-2-239-0630; Fax: 0039-2-236-2692; E-mail: anichini@istitutotumori.mi.it

Establishment of melanoma cell lines from surgical specimens has been carried out in our laboratory for over 15 years. These cell lines have been used to study T cell-mediated response to autologous tumor and for investigating several aspects of the biology of human melanoma. Most of the lines have been characterized for expression of HLA class I and II antigens, adhesion molecules, integrins and melanoma-associated antigens. Some of these lines are being evaluated by RT-PCR for expression of genes coding for known tumor antigens recognized by cytotoxic T lymphocytes, such as the MAGE family (see Chapter 15) and differentiation antigens (Melan-A/Mart-1, Tyrosinase, TRP-1, gp100). Selected lines have been characterized for expression of genes coding for cytokines and growth factors. Some lines have been evaluated for susceptibility to anti-proliferative or antigen-modulating effects of various cytokines. Interaction with extracellular matrix and with endothelial cells has also been investigated, as well as metastatic activity in nude mice.

1. CULTURE CONDITIONS

All the cell lines established in our laboratory and described in this chapter were isolated from surgical specimens of patients admitted to our Institute. The current success rate in deriving stable cell lines from fresh tumor samples is approximately 35%. The number of live tumor cells is critical, and every effort should be made to minimize the processing time of the specimen. Well-differentiated, highly melanotic tumors are generally hard to establish in culture.

J.R.W. Masters and B. Palsson (eds.), Human Cell Culture Vol. 1, 283–292.

Melanoma cells from skin, lymph nodes or subcutaneous nodules are isolated by mechanical disaggregation as described (1). Using a scalpel, the specimen is cut into small fragments (approximately 3 x 3 mm) in a Petri dish containing RPMI 1640 medium supplemented with antibiotics. Each fragment is then gently squeezed in the barrel of a syringe. The resulting cell suspension is filtered through sterile gauze to remove large fragments, and then gently centrifuged. When there is abundant connective tissue, enzymatic digestion is preferred (1). The tissue fragments are resuspended in collagenase on a magnetic stirrer at room temperature for up to 2 h. Red blood cells are lysed with ammonium chloride. Separation of live cells is achieved by centrifuging up to 5 mL of the cell suspension layered on 2.5 mL of a Ficoll gradient.

Primary cultures are routinely set up in either 25 cm^2 flasks or 24 well plates in RPMI 1640 supplemented with 10% fetal calf serum (FCS). The presence of many adherent cells with spindle to dendritic morphology as early as 12 h after initial seeding is a good indication of success. Fibroblasts, often present as contaminants in primary cultures, can be removed by adding cholera toxin to the culture medium.

The primary culture is not split until the cells approach confluence, which can take between 7 and 30 days. Once the primary culture has reached confluence, the line is likely to become established as, in our experience, well proliferating primary cultures will consistently progress to a continuous cell line. Established cell lines are easy to maintain in culture. RPMI 1640 with 10% FCS is the medium of choice and a seeding concentration of 1–2 x 10^4/mL in 10 mL (for 25 cm^2 flasks) or in 30 mL (for 75 cm^2 flasks) is recommended. Cultures should be split before confluence, since most melanoma cell lines will show signs of degeneration soon after confluence is reached, and subsequently it is difficult for these cells to recover.

In our hands serum-free culture of melanoma cell lines is not easy, in contrast to the Wistar experience (see chapter 14). In most instances growth is markedly reduced or even totally abolished. Serum-free media can be used for a few days if necessary. Some cell lines can be adapted to continuous culture in serum-free medium, but these are exceptional.

2. PATHOLOGY

Table 1 describes 70 melanoma cell lines established in our laboratory. Most of the cell lines were isolated from lymph node metastases (48 lines), some from primary tumors (8 lines, indicated as "Skin", in the specimen site column). Other sites include subcutaneous metastatic nodules (12 lines) and metastases in visceral organs (2 lines). Regional lymph nodes are the most frequent site of metastasis, accounting for the fact that 68 % of the cell lines

Table 1 Melanoma cell lines established at the Division of Experimental Oncology D, Istituto Nazionale per lo Studio e la Cura dei Tumori, Milan, Italy

Cell line (1)	Cell Line (2)	Patient Age/Sex (3)	TNM Category (4)	Path Stage and Grade (5)	Primary Site	Specimen Site	Culture Method (6)	Authentication	Availability	Primary References (8)
INT-MEL-1	5810	84/F	pT4, N0, M0	IIB	Left leg	Skin, left leg	D	HLA-Typing	(7)	①
INT-MEL-2	1684	29/F	pT4, N1, M0	IIIA	Left thigh	Lung	D	HLA-Typing	(7)	①
INT-MEL-3	1684/1	43/F	pT4, N1b, M0	IIIA	Adominal skin	Lymph node	D	HLA-Typing	(7)	①
INT-MEL-4	9874	68/M	pT?, N1b, M0	IIIA	Right leg	Lymph node	D	HLA-Typing	(7)	①
INT-MEL-5/1*		60/F	pT2, N1b, M0	IIIA	Left ankle	Lymph node	D	HLA-Typing	(7)	②
INT-MEL-5/2*		61/F	pT2, N1b, M1	IV	Left ankle	S.c. nodule	D	HLA-Typing	(7)	②
INT-MEL-5/3*		61/F	pT2, N1b, M1	IV	Left ankle	S.c. nodule	D	HLA-Typing	(7)	②
INT-MEL-6		42/F	pT2, N2b, M0	IIIB	Right thigh	Lymph node	D	HLA-Typing	(7)	②
INT-MEL-7		53/M	pT?, N1b, M1	IV	Unknown	Lymph node	D	HLA-Typing	(7)	②
INT-MEL-8	3046	62/M	pT3, N2c, M1	IV	Left arm	S.c. nodule	D	HLA-Typing	(7)	①
INT-MEL-9	4840	27/F	pT3, N2, M1	IV	Right shoulder	Omentum	D	HLA-Typing	(7)	①
INT-MEL-10	1858/8	63/F	pT?, N2c, M0	IIIB	Back	Lymph node	D	HLA-Typing	(7)	①
INT-MEL-11	1329/4	19/F	pT3, N1, M0	IIIA	Lumbar region	Lymph node	D	HLA-Typing	(7)	①
INT-MEL-12	1053/8	56/F	pT4, N1, M0	IIIA	Left leg	Skin, Left leg	D	HLA-Typing	(7)	①
INT-MEL-13	8959	51/M	pT3, N1, M0	IIIA	Back	Lymph node	D	HLA-Typing	(7)	①
INT-MEL-14		51/M	pT3, N2c, M0	IIIB	Left knee	Lymph node	D	HLA-Typing	(7)	②
INT-MEL-15/1*	1402/1	22/M	pT4, N0, M0	IIB	Lumbar region	Skin, Lumbar r.	D	HLA-Typing	(7)	①
INT-MEL-15/2*	1402/R	22/M	pT4, N1c, M1	IV	Lumbar region	Skin, Lumbar r.	D	HLA-Typing	(7)	①
INT-MEL-16/1*	9923/P	65/F	pT4, N1c, M0	IIIA	Right malleolus	Skin, malleolus	D	HLA-Typing	(7)	①
INT-MEL-16/2*	9923/M	65/F	pT4, N1c, M0	IIIA	Right malleolus	Lymph node	D	HLA-Typing	(7)	①
INT-MEL-17	4405	83/F	pT4, N0, M0	IIB	Left cheek	Skin, cheek	D	HLA-Typing	(7)	①
INT-MEL-18		54/M	pT4, N1c, M0	IIIA	Left cheek	Lymph node	D	HLA-Typing	(7)	②
INT-MEL-19	1493/2	65/M	pT4, N2a, M0	IIIB	Left thigh	Lymph node	D	HLA-Typing	(7)	①
INT-MEL-20	1007	70/M	pT4, N0, M0	IIB	Left leg	Skin, left leg	D	HLA-Typing	(7)	①

Continued on next page

Table 1 (continued)

Cell line (1)	Cell Line (2)	Patient Age/Sex (3)	TNM Category (4)	Path Stage and Grade (5)	Primary Site	Specimen Site	Culture Method (6)	Authentication	Availability	Primary References (8)
INT-MEL-20	1007	70/M	pT4, N0, M0	IIB	Left leg	Skin, left leg	D	HLA-Typing	(7)	①
INT-MEL-21		41/M	pT?, N1c, M0	IIIA	Right knee	Lymph node	D	HLA-Typing	(7)	②
INT-MEL-22		39/F	pT3, N2a, M0	IIIB	Upper back	Lymph node	D	HLA-Typing	(7)	②
INT-MEL-23	1344/3	32/M	pT3, N2c, M1	IV	Left ear skin	Lymph node	D	HLA-Typing	(7)	①
INT-MEL-24		68/M	pT4, N2a, M0	IIIB	Back	Lymph node	D	HLA-Typing	(7)	②
INT-MEL-25		37/F	pT3, N1a, M0	IIIA	Left arm	Lymph node	D	HLA-Typing	(7)	②
INT-MEL-26		44/M	pT?, N1b, M0	IIIA	Unknown	Lymph node	D	HLA-Typing	(7)	②
INT-MEL-27		51/M	pT4, N2a, M0	IIIB	Back	Lymph node	D	HLA-Typing	(7)	②
INT-MEL-28		70/M	pT4, N1, M0	IIIA	Right thigh	Lymph node	D	HLA-Typing	(7)	②
INT-MEL-29		45/M	pT3, N1c M0	IIIA	Back	Lymph node	D	HLA-Typing	(7)	②
INT-MEL-30		45/M	pT4, N2, M0	IIIB	Right sole	Lymph node	D	HLA-Typing	(7)	②
INT-MEL-31		66/F	pT3, N1b, M0	IIIA	Right foot	Lymph node	D	HLA-Typing	(7)	②
INT-MEL-32		24/F	pT2, N1, M0	IIIA	Right leg	Lymph node	D	HLA-Typing	(7)	②
INT-MEL-33		62/M	pT2, N1c, M0	IIIA	Neck	Lymph node	D	HLA-Typing	(7)	②
INT-MEL-34		52/M	pT?, N1b, M0	IIIA	Unknown	Lymph node	D	HLA-Typing	(7)	②
INT-MEL-35		34/M	pT4, N2a, M0	IIIB	Back	Lymph node	D	HLA-Typing	(7)	②
INT-MEL-36		52/M	pT3, N1b, M0	IIIA	Sternal region	Lymph node	D	HLA-Typing	(7)	②
INT-MEL-37		41/M	PT3, N1b, M0	IIIA	Right leg	Lymph node	D	HLA-Typing	(7)	②
INT-MEL-38/1*		45/M	pT2, N1b, M0	IIIA	Trunk	Lymph node	D	HLA-Typing	(7)	②
INT-MEL-38/2*		45/M	pT2, N1b, M0	IIIA	Trunk	Lymph node	D	HLA-Typing	(7)	②
INT-MEL-39	1693/8	40/M	pT4, N2a, M0	IIIB	Right sole	Lymph node	D	HLA-Typing	(7)	①
INT-MEL-40		53/F	pT?, N1a, M0	IIIA	Right thigh	Lymph node	D	HLA-Typing	(7)	②
INT-MEL-41		51/F	pT4, N1a, M0	IIIA	Right sole	Lymph node	D	HLA-Typing	(7)	②

Continued on next page

Table 1 (continued)

Cell line (1)	Cell Line (2)	Patient Age/Sex (3)	TNM Category (4)	Path Stage and Grade (5)	Primary Site	Specimen Site	Culture Method (6)	Authentication	Availability	Primary References (8)
INT-MEL-42		48/F	pT4, N1a, M0	IIIA	Left thigh	Lymph node	D	HLA-Typing	(7)	②
INT-MEL-43	1509	51/M	pT3, N1a, M1	IV	Right thigh	S.c. nodule	D	HLA-Typing	(7)	①
INT-MEL-44		34/M	pT3, N1b, M0	IIIA	Back	Lymph node	D	HLA-Typing	(7)	②
INT-MEL-45		63/F	pT3, N1b, M0	IIIA	Breast skin	Lymph node	D	HLA-Typing	(7)	②
INT-MEL-46		578/M	pT3, N2b, M0	IIIB	Heel	Lymph node	D	HLA-Typing	(7)	②
INT-MEL-47/1*		44/F	pT3, N1a, M0	IIIA	Left leg	Lymph node	D	HLA-Typing	(7)	②
INT-MEL-47/2*		45/F	pT3, N1a, M1	IV	Left leg	S.c. nodule	D	HLA-Typing	(7)	②
INT-MEL-47/3*		45/F	pT3, N1a, M1	IV	Left leg	S.c. nodule	D	HLA-Typing	(7)	②
INT-MEL-48/1*		51/M	pT4, N1c, M0	IIIA	Left thigh	Lymph node	D	HLA-Typing	(7)	②
INT-MEL-48/2*		51/M	pT4, N1c, M1	IV	Left thigh	S.c. nodule	D	HLA-Typing	(7)	②
INT-MEL-49		51/F	pT?, N1b, M0	IIIA	Left gluteus	Lymph node	D	HLA-Typing	(7)	②
INT-MEL-50	1022/1	72/F	pT?, N1b, M1	IV	Right arm	Lymph node	D	HLA-Typing	(7)	①
INT-MEL-51		70/F	pT4, N1b, M0	IIIA	Heel	Lymph node	D	HLA-Typing	(7)	②
INT-MEL-52	4024	53/F	pT2, N2b, M0	IIIB	Left leg	S.c. nodule	D	HLA-Typing	(7)	①
INT-MEL-53		58/F	pT3, N0, M0	IIA	Scalp	Skin, Scalp	D	HLA-Typing	(7)	②
INT-MEL-54		69/F	pT3, N1b, M0	IIIA	Right heel	Lymph node	D	HLA-Typing	(7)	②
INT-MEL-55		52/F	pT3, N1b, M1	IV	Back	Lymph node	D	HLA-Typing	(7)	②
INT-MEL-56		35/M	pT3, N2a, M0	IIIB	Right arm	Lymph node	D	HLA-Typing	(7)	②
INT-MEL-57		41/M	pT?, N2b, M0	IIIB	Right sole	Lymph node	D	HLA-Typing	(7)	②
INT-MEL-58	9229	43/M	pT3, N1b, M0	IIIA	Right shoulder	Lymph node	D	HLA-Typing	(7)	①
INT-MEL-59/1*	9742/1	60/F	pT3, N2, M0	IIIB	Right shoulder	Lymph node	D	HLA-Typing	(7)	①
INT-MEL-59/2*	9742/1	60F	pT3, N2, M1	IV	Right sole	S.c. nodule	D	HLA-Typing	(7)	①

Continued on next page

Table 1 (continued)

Cell line (1)	Cell Line (2)	Patient Age/Sex (3)	TNM Category (4)	Path Stage and Grade (5)	Primary Site	Specimen Site	Culture Method (6)	Authentication	Availability	Primary References (8)
INT-MEL-60/1*	665/1	59/F	pT3, N2b, M1	IV	Left leg	S.c nodule	D	HLA-Typing	(7)	①
INT-MEL-60/2*	665/2	59/F	pT3, N2b, M1	IV	Left leg	S.c. nodule	D	HLA-Typing	(7)	①
INT-MEL-60/3*	665/3	59/F	pT3, N2b, M1	IV	Left leg	S.c. nodule	D	HLA-Typing	(7)	①

(1) The Code "INT-MEL-##" is a recent nomenclature adopted to clearly identify the source of these lines. *Cell lines indicated as "INT-MEL-## / #" were derived from independent lesions of the same patient.
(2) Name of the cell lines as found in papers listed in references 1–21.
(3) Age of the patient at the time of surgery for removal of lesion that originated the cell line.
(4) "pT" code refers to the histological classification of the primary lesion that in many instances was removed from each patient months or years before the lesions that gave origin to the cell line. "N" and "M" codes refer to the nodal and distant metastasis situation of the patient at the time one of the removed lesions was established as a cell line. In some instances thickness of the primary tumor is unknown ("pT?") due to the fact that the lesion had not been removed at our Institute.
(5) The AJCC/UICC classification is used here, and refers to the clinical stage at time the cell line was isolated from the surgical specimen.
(6) Cells were isolated by mechanical dissociation followed by culture in RPMI-1640+10%FCS.
(7) Contact Dr. Andrea Anichini (e-mail: Anichini@istitutotumori.mi.it) at Division of Experimental Oncology D, Istituto Nazionale Tumori, Milan.
(8) Cell lines described in published papers of our Division (see references 1–21) are identified as "①" , while cell lines never described in the literature are identified as ②

were derived from nodal specimens. The relative paucity of cell lines from primary lesions is due to a bias in tissue availability, as primary lesions are often small and the whole surgical specimen is needed for evaluation by the pathologist. No cell lines were derived from patients with a pT1 primary tumor at initial diagnosis, and most were derived from patients that, at the time of initial diagnosis, already had advanced primary cancers (pT3 or pT4). This bias towards thicker primaries reflects the well known relationship between primary tumor thickness and the risk of progression in melanoma.

The clinical spectrum of the disease is not fully represented in our panel of cell lines since no lines from radial growth phase primary melanomas were obtained. In addition, few cell lines were derived from metastases in visceral organs, because usually these are found in terminally ill patients and are rarely removed surgically.

When comparing melanoma cell lines from different patients or from distinct lesions in the same patient, heterogeneity for almost any biological parameter is the rule. This applies to cell morphology, degree of pigmentation, proportion of adherent cells, doubling time, karyotype, cell surface antigen and gene expression (see 1–21). This heterogeneity can also extend to clones derived from the same lesions. This is true mainly when clones are isolated from tumor lines kept in culture for only a few passages. Once established, melanoma clones are stable for all the biological parameters that we have investigated (2, 4, 7–8, 10, 12–21), both in vitro and in vivo.

Selected melanoma cell lines have been grown as xenografts in nude mice (Table 2) (see references 3, 11, 21). Interestingly, neither primary or metastatic lesions produced metastases after subcutaneous injection, even though tumor take approached 100%. In contrast, in some animals injected intraperitoneally, local growth and metastasis to distant organs were observed (see Table 2).

3. MOLECULAR AND CYTOGENETICS

Table 3 shows the genetic changes in the cell lines that have been studied. The multiple lesions and clones from the patient originating the cell lines INT-MEL-60/1–3 (patient 665 in the published references) have been studied in detail. The tumors harbor a point mutation at position 61 of the N-*ras* gene (14–15). Distinct lesions established as cell lines or analyzed as fresh tumor samples are heterogeneous for the mutation, and this is reflected in a well characterized set of clones isolated from the lesion INT-MEL-60/2 (14). Some clones have the mutation in the N-*ras* gene, while others have the wild-type sequence (14).

Point mutations of N- or K-*ras* genes have been found in two additional lines (INT-MEL-16/1 and INT-MEL-1). Interestingly, one of these lines (INT-MEL-16/1) bears two distinct mutations in the N- and K-*ras* genes. N-and

Table 2

Cell line	Tumor pathology	In vitro features	Xenograft pathology
INT-MEL-12	Primary Melanoma	Mostly adherent	No growth after s.c. implantation. Growth in all injected animals after i.p. injection (as metastases to the diaphragm).
INT-MEL-20	Primary Melanoma	Mostly adherent	No growth after s.c. implantation.
INT-MEL-15/1	Primary Melanoma	Mostly adherent	No growth after s.c. implantation. Metastases to liver, pancreas and diaphragm after i.p. injection. Metastases to liver and pancreas after i.s. injection.
INT-MEL-15/2	s.c. Metastatic melanoma	Mostly adherent	No growth after s.c. implantation. Metastases to liver, pancreas and diaphragm after i.p. injection. Metastases to pancreas and ovaries after i.s. injection.
INT-MEL-60/1	s.c. Metastatic melanoma	Mostly adherent	No growth after s.c. implantation. Metastases to lymph nodes and intestine after i.p. injection. Metastases to lymph nodes and lungs after i.s. injection.
INT-MEL-60/2	s.c. Metastatic melanoma	Mostly adherent	No growth after s.c. implantation. No growth after i.p. implantation. Metastases to liver after i.s. injection.

Table 3

Main genetic changes	Cell lines
Point Mutation of N-*ras* gene (Gln to Arg at position 61)	INT-MEL-60/1
Point Mutation of N-*ras* gene (Gln to Arg at position 61)	INT-MEL-60/2
Point Mutation of N-*ras* gene (Gln to Arg at position 61)	INT-MEL-60/3
Point Mutation of N-*ras* gene (Gly to Ser at position 12)	INT-MEL-16/1
Point Mutation of K-*ras* gene (Gln to Leu at position 61)	INT-MEL-16/1
Point Mutation of N-*ras* gene (Gln to Arg at position 61)	INT-MEL-1
Loss of expression of HLA class I antigens	INT-MEL-16/1
Loss of expression of HLA class I antigens	INT-MEL-16/2
Deletions and translocations involving various regions of chromosomes #1, #6 and #7	INT-MEL-12
Deletions and translocations involving various regions of chromosomes #1, #6 and #8	INT-MEL-15/1
Deletions involving regions of chromosome #6	INT-MEL-20
Duplications of regions of chromosome #1, deletions involving regions of chromosome #6	INT-MEL-60/1
Deletions involving regions of chromosome #6	INT-MEL-58

K-*ras* gene mutations were never found in the EBV-transformed autologous B cells. In clones from lesion INT-MEL-60/2, the presence of the N-*ras* mutation correlated with high susceptibility to lysis by LAK cells and with increased expression of adhesion molecules and integrins (10, 12, 14). Furthermore, INT-MEL-60/2 clones bearing the N-*ras* mutations constitutively express the genes coding for IL-1a, IL-6 and TNF-α, and transfection of the mutated genes into clones bearing the wild-type form of N-*ras* converted these cells to IL-1a, IL-6 and TNF-α expressors (15). Other cell lines (INT-MEL-16/1, INT-MEL-16/2) have lost HLA-class I expression.

All of the cell lines described in this chapter are mycoplasma-free as determined, in most cases, by commercially available ELISA assays for the detection of the most common *Mycoplasma* types. Cross-contamination was excluded by HLA typing of patient PBL and tumor cells.

A number of cell lines have been characterized for susceptibility to the anti-proliferative activity of cytokines (9). A specific set of clones isolated from the cell line INT-MEL-58 has been analyzed for susceptibility to the modulation of HLA antigens by IFN-gamma (8). The results indicated that induction of HLA-DR and HLA-DQ antigens was independently regulated, as one clone could be induced to express both antigens while the other only expressed HLA-DR after IFN-gamma treatment (8). Expression of genes coding for several cytokine and growth factors has been determined by RT-PCR on at least 20 different lines (22). Selected cell lines and clones from lesion INT-MEL-60/2 were characterized for adhesion and/or proliferation to extracellular matrix components, including laminin and fibronectin (see references 12, 13, 16, 17 for details). These cell lines have also been characterized for the expression of adhesion receptors of the integrin family (12, 13, 16, 17).

REFERENCES

1. Anichini A., et al., *Int. J. Cancer*, 35:683, 1985.
2. Anichini A., et al., *J. Exp. Medicine*, 163:215, 1986.
3. Rodolfo M., et al., *Invasion Metastasis*, 8:317, 1988.
4. Anichini A., et al., *J. Immunology*, 142:3692, 1989.
5. Anichini A., et al., *J.Exp. Med.*, 177:989, 1993.
6. Anichini A., et al., *J. Immunol.*, 156:208, 1996.
7. Anichini A., et al., *Int. J. Cancer*, 38:505, 1986.
8. Anichini A., et al., *J. Immunology*, 140:183, 1988.
9. Mortarini R., et al., *Int. J. Cancer*, 45:334, 1990.
10. Anichini A., et al., *Int. J. Cancer.*, 46:508, 1990.
11. Sozzi G., et al., *Cancer Genet. Cytogenet.*, 44:61, 1990.
12. Mortarini R., et al., *Int. J. Cancer*, 47:551, 1991.
13. Mortarini R., et al., *Cancer Res.*, 52:4499, 1992.
14. Lupetti R., et al., *Melanoma Res.*, 4: , 1994.
15. Castelli C., et al., *Cancer Res.*, 54:4785, 1994.
16. Mortarini R., et al., *Cancer Res.*, 55:4702, 1995.
17. Lupetti R., et al., *Int. J. Cancer*, 65:1, 1996.
18. Martin-Padura I., et al., *Cancer Res.*, 51: 2239, 1991.
19. Jonjic N., et al., *Eur.J. Immunol.*, 22:2255, 1992.
20. Jonjic N., et al., *Am. J. Pathol.*, 141:1323, 1992.
21. Garofalo A, et al., *Cancer Res.*, 55: 414, 1995.
22. Mattei, S., et al., *Int. J. Cancer*, 56: 853, 1994.

Human Cell Culture

1. J. R. W. Masters and B. Palsson (eds.): *Human Cell Culture, Vol. I.* 1998
ISBN 7923–5143–6

KLUWER ACADEMIC PUBLISHERS – DORDRECHT / BOSTON / LONDON

The manufacturer's authorised representative in the EU is Springer Nature Customer Service Centre GmbH, Europaplatz 3, 69115 Heidelberg, Germany. If you have any concerns regarding our products, please contact ProductSafety@springernature.com

Printed and bound by CPI Group (UK) Ltd, Croydon, CR0 4YY
22/09/2026
02220945-0010